B.V. Gnedenko I.N. Kovalenko

Introduction to Queueing Theory

Second Edition
Revised and Supplemented

Translated by Samuel Kotz

Birkhäuser
Boston · Basel · Berlin

B.V. Gnedenko
Department of Probability and
Mathematical Statistics
Moscow University
Moscow, USSR

I.N. Kovalenko
Department of Probability and
Mathematical Statistics
Moscow University
Moscow, USSR

Samuel Kotz
Department of Management Science
and Statistics
University of Maryland
College Park, MD 20742, USA

Library of Congress Cataloging-in-Publication Data
Gnedenko, Boris Vladimirovich, 1912–
[Vvedenie v teoriiu massovogo obsluzhivaniia. English]
Introduction to queueing theory / by B.V. Gnedenko and I.N. Kovalenko.—2nd ed., rev. and supplemented.
p. cm.—(Mathematical modeling; 5)
Translation of: Vvedenie v teoriiu massovogo obsluzhivaniia.
Includes index.
ISBN 0-8176-3423-1.—ISBN 3-7643-3423-1
1. Queueing theory. I. Kovalenko, Igor' Nikolaevich. II. Title.
III. Series: Mathematical modeling (Boston, Mass.); no. 5.
T57.9.G5813 1987
519.8′2—dc20 89-7306

Printed on acid-free paper.

ISBN 0-8176-3423-1
ISBN 3-7643-3423-1

Typeset by Asco Trade Typesetting Ltd., Hong Kong.
Printed and bound by R.R. Donnelley & Sons, Harrisonburg, Virginia.
9 8 7 6 5 4 3 2 1

Contents

Introduction to the Second Edition 1
Introduction to the First Edition 2

Chapter 1. Problems of Queueing Theory under the Simplest Assumptions 8
1.1. Simple Streams 8
1.1.1. Historical Remarks 8
1.1.2. The Notion of a Stream of Homogeneous Events 9
1.1.3. Qualitative Assumptions and Their Analysis 9
1.1.4. Derivation of Equations for Simple Streams 11
1.1.5. Solution of the Equations 13
1.1.6. Derivation of the Additional Assumption from the Other Three Assumptions 14
1.1.7. Distribution of Times of Events of a Stream 16
1.1.8. The Intensity and Parameter of a Stream 17
1.2. Service with Waiting 18
1.2.1. Statement of the Problem 18
1.2.2. The Servicing Process as a Markov Process 19
1.2.3. Construction of Equations 20
1.2.4. Determination of the Stationary Solution 21
1.2.5. Some Preliminary Results 23
1.2.6. The Distribution Function of the Waiting Time 24
1.2.7. The Mean Waiting Time 26
1.2.8. Example 27
1.3. Birth and Death Processes 28
1.3.1. Definition 28
1.3.2. Differential Equations for the Process 29
1.3.3. Proof of Feller's Theorem 29
1.3.4. Passive Redundancy without Renewal 32
1.3.5. Active Redundancy without Renewal 33
1.3.6. Existence of Solutions for Birth and Death Equations 35
1.3.7. Backward Equations 36

1.4. Applications of Birth and Death Processes in Queueing Theory 37
- 1.4.1. Systems with Losses ... 37
- 1.4.2. Systems with Limited Waiting Facilities ... 40
- 1.4.3. Distribution of the Waiting Time until the Commencement of Service ... 41
- 1.4.4. Team Servicing of Machines ... 42
- 1.4.5. A Numerical Example ... 42
- 1.4.6. Duplicated Systems with Renewal (Passive Redundancy) . 44
- 1.4.7. Duplicated Systems with Renewal (Active Redundancy) .. 46
- 1.4.8. Duplicated Systems with Renewal (Partially Active Redundancy) ... 46

1.5. Priority Service ... 47
- 1.5.1. Statement of the Problem ... 47
- 1.5.2. Problems with Losses ... 48
- 1.5.3. Equations for $p_{ij}(t)$ ... 49
- 1.5.4. A Particular Case ... 51
- 1.5.5. The Possibility of Failure of the Servers ... 53

1.6. General Principles of Constructing Markov Models of Systems . 53
- 1.6.1. Homogeneous Markov Processes ... 53
- 1.6.2. Characteristics of Functionals ... 55
- 1.6.3. A General Scheme for Constructing Markov Models of Service Systems ... 56
- 1.6.4. The HyperErlang Approximation ... 57

1.7. Systems with Limited Waiting Time ... 57
- 1.7.1. Statement of the Problem ... 57
- 1.7.2. The Stochastic Process Describing the State of a System for $\tau = \text{const}$. ... 58
- 1.7.3. System of Integro-differential Equations for the Problem . 59
- 1.7.4. Various Characteristics of Service ... 62
- 1.7.5. Distribution of the Queue Length ... 63
- 1.7.6. Waiting Time Bounded by a Random Variable ... 64

1.8. Systems with Bounded Holding Times ... 65
- 1.8.1. Statement of the Problem and Assumptions ... 65
- 1.8.2. A Stochastic Process Describing the Service ... 66
- 1.8.3. Stationary Distributions ... 67
- 1.8.4. Holding Time in a System Bounded by a Random Variable ... 67

Chapter 2. The Study of the Incoming Customer Stream ... 69

2.1. Some Examples ... 69
- 2.1.1. The Notion of the Incoming Stream ... 69
- 2.1.2. Feed of Components from a Hopper ... 71
- 2.1.3. A Regular Stream of Customers ... 72
- 2.1.4. Streams of Customers Served by Successively Positioned Servers ... 72

2.1.5. A Wider Approach to the Notion of the Incoming Stream 74
2.1.6. Marked Streams 75
2.2. A Simple Nonstationary Stream 76
2.2.1. Definition of a Simple Nonstationary Stream 76
2.2.2. Equations for the Probabilities $p_k(t_0, t)$ 77
2.2.3. Solution of the System (7) 78
2.2.4. Instantaneous Intensity of a Stream 79
2.2.5. Examples 80
2.2.6. The General Form of Poisson Streams without Aftereffects 81
2.2.7. A System with Infinitely Many Servers 82
2.3. A Property of Stationary Streams 83
2.3.1. Existence of the Parameter 83
2.3.2. A Lemma 84
2.3.3. Proof of Khinchin's Theorem 85
2.3.4. An Example of a Stationary Stream with Aftereffects 86
2.4. General Form of Stationary Streams without Aftereffects 88
2.4.1. Statement of the Problem 88
2.4.2. The Existence of the Limits $\lim_{t \to 0} \frac{\pi_k(t)}{t}$ 89
2.4.3. Equations for the General Stationary Stream without Aftereffects 91
2.4.4. Solution of Systems (3) and (4) 92
2.4.5. A Special Case 93
2.4.6. The Generating Function of the Stream 94
2.4.7. Concluding Remarks 96
2.5. The Palm-Khinchin Functions 97
2.5.1. Definition of the Palm-Khinchin Functions 97
2.5.2. Proof of the Existence of the Palm-Khinchin Functions 98
2.5.3. The Palm-Khinchin Formulas 100
2.5.4. Intensity of a Stationary Stream 102
2.5.5. Korolyuk's Theorem 103
2.5.6. The Case of Nonorderly Streams 104
2.6. Characteristics of Stationary Streams and the Lebesgue Integral 105
2.6.1. A General Definition of Mathematical Expectation 105
2.6.2. A Refinement of the Notion of Orderliness 106
2.6.3. Existence of the Parameter of a Stream 106
2.6.4. Dobrushin's Theorem 107
2.6.5. The Existence of the Palm-Khinchin Function 107
2.6.6. The k-Intensity of a Stream 108
2.7. Basic Renewal Theory 108
2.7.1. Definition of Renewal Processes (Renewal Streams) 108
2.7.2. A Property of Renewal Streams 109
2.7.3. Relation to the Palm-Khinchin Functions 110
2.7.4. Definition of the Palm-Khinchin Functions for Stationary Renewal Streams 112

2.7.5. Basic Formulas for Renewal Processes 113
2.7.6. Statements of Some Theorems on Stationary Renewal Processes 115
2.8. Limit Theorems for Compound Streams 119
2.8.1. Statement of the Problem 119
2.8.2. Definitions and Notation 120
2.8.3. Statement of the Basic Result and a Proof of Necessity ... 121
2.8.4. Proof of Sufficiency 122
2.8.5. The Case of Stationary and Orderly Component Streams 125
2.8.6. Additional Remarks 126
2.9. Direct Probabilistic Methods 128
2.10. Limit Theorem for Thinning Streams 129
2.10.1. Statement of the Problem 129
2.10.2. Laplace Transform of Transformed Streams 130
2.10.3. Some Properties of the T-Operation 131
2.10.4. The T_q-Transformation for a Simple Stream 132
2.10.5. Rényi's Limit Theorem 132
2.11. Additional Limit Theorems for Thinning Streams 133
2.11.1. Belyaev's Theorem and its Generalizations 133
2.11.2. Rare Events in the Scheme of a Regenerative Process ... 134

Chapter 3. Some Classes of Stochastic Processes 136
3.1. Kendall's Method: Semi-Markov Processes 136
3.1.1. Semi-Markov Processes and Embedded Markov Chains 136
3.1.2. Some Results from the Theory of Markov Chains 139
3.1.3. Basic Relations for Semi-Markov Processes 141
3.1.4. Ergodic Properties of a Semi-Markov Process 143
3.1.5. Method of "Catastrophes" 144
3.2. Linear-Type Markov Processes 145
3.2.1. Definition 145
3.2.2. Basic Equations 145
3.2.3. The Ergodic Theorem for Lined Processes 148
3.2.4. The Method of Integrodifferential Equations 148
3.2.5. Lined Processes with a Fixed Remainder 150
3.2.6. Differential Equations 151
3.3. Piecewise-Linear Markov Processes 155
3.3.1. Method of Additional Variables 155
3.3.2. Piecewise-Linear Markov Process 156
3.3.3. Regularity Conditions 157
3.3.4. Two Reductions 157
3.3.5. Embedded Markov Chain 158
3.4. Other Important Classes of Random Processes 159

Chapter 4. Semi-Markov Models of Queueing Systems 162
4.1. Classification of Queueing Systems 162
4.2. $M|G|1$ System 163
4.2.1. Statement of the Problem, Notation 163
4.2.2. Embedded Markov Chain 164
4.2.3. Pollaczek-Khinchin Formula 167
4.2.4. Mathematical Law of a Stationary Queue 168
4.2.5. Virtual Waiting Time 169
4.2.6. The Limiting Distribution of the Waiting Time 171
4.2.7. The Case $\rho \geqslant 1$ 173
4.3. Nonstationary Characteristics of an $M|G|1$ System 175
4.3.1. The Busy Period 175
4.3.2. An Integral Equation 175
4.3.3. Functional Equation 176
4.3.4. Distribution of the Number of Customers Served During the Busy Period 178
4.3.5. Distribution of Time until the First Disengagement of the Server 178
4.3.6. Nonstationary Distribution of the Virtual Waiting Time . 179
4.3.7. Nonstationary Conditions of the Queueing System for a Simple Incoming Stream 180
4.4. A System of the $GI|M|m$ Type 181
4.4.1. Construction of an Embedded Markov Chain 181
4.4.2. Example 183
4.5. $M|G|1$ System with an Unreliable and "Renewable" Server 186
4.5.1. Possible Statements of the Problem 186
4.5.2. Failure During Idle Time 187
4.5.3. The General Case 189
4.5.4. The Influence of Partial Failure 191
4.6. Mixed Service Systems 197
4.6.1. Mixed System with Constant Service Rate 197
4.6.2. Condition for Ergodicity 197
4.6.3. Mixed System with Variable Service Rate 200
4.6.4. Example 205
4.6.5. $M|G|1|m$ System 205
4.7. Systems with Restrictions 208
4.7.1. Various Forms of Restrictions 208
4.7.2. Formulation of Restrictions 208
4.7.3. Existence of the Ergodic Distribution 209
4.7.4. Equation for the Stationary Distribution 211
4.7.5. Embedded Markov Chain 213
4.8. Priority Service 214
4.8.1. Assumptions and Notation 214
4.8.2. Service of Customers of the First Type 215
4.8.3. The Method of Investigation 215

4.8.4. Determination of the Function $\psi_\beta(s)$ 216
4.8.5. Determination of the Function $\Phi_2^*(s)$ 218
4.8.6. The Ergodicity Condition 218
4.9. The Generalized Scheme of Priority Service with a Limited Queue 219
4.9.1. Statement of the Problem 219
4.9.2. The Structure of the Process 220
4.9.3. Basic Equations 221
4.9.4. Remarks 224

Chapter 5. Application of More General Methods 225
5.1. The $GI|G|1$ System 225
5.1.1. Basic Recurrence Relations 225
5.1.2. The Integral Equation; The Existence of the Limiting Distribution 227
5.1.3. Analytic Methods 230
5.2. $GI|G|m$ Systems 231
5.2.1. Multidimensional Random Walk 231
5.2.2. Kiefer and Wolfowitz's Ergodic Theorem 232
5.3. The $M|G|m|0$ System 236
5.3.1. The Ergodic Theorem 236
5.3.2. Proof of Sevast'yanov's Formula 238
5.4. More Complex Systems with Losses 240
5.4.1. Reliability of a Renewable System 241
5.4.2. A Renewable System with a Variable Renewal Rate 242
5.4.3. Incompletely Accessible Service System 242
5.4.4. A Necessary and Sufficient Condition for Solvability of the State Equations of a System in Constants 244
5.4.5. Further Generalizations 247
5.4.6. The Problem of Redundancy with a Redistributed Load. . 250
5.5. Ergodic Theorems 253
5.5.1. Sevast'yanov's Theorem 253
5.5.2. Construction of Innovation Times 255
5.5.3. Stability of Queueing Systems 257
5.6. Heavily Loaded Queueing Systems 259
5.6.1. Limit Theorem for Distribution of Waiting Time in a $GI|G|1$ System 259
5.6.2. Utilization of the Invariance Principle 261
5.6.3. Borovkov's Theorem 262
5.6.4. Asymptotic Invariance 263
5.7. Underloaded Queueing Systems 263
5.7.1. Introductory Remarks 263
5.7.2. Statement of the Problem 264
5.7.3. Investigation of the Process $\lambda x(t)$ 264

5.8. Little's Theory and its Corollaries 268
5.8.1. General Statements 268
5.8.2. Little's Theorem 268
5.8.3. Notes 270

Chapter 6. Statistical Simulation of Systems 274
6.1. Principles of the Monte Carlo Method 274
6.1.1. Foundation of the Methodology 274
6.1.2. Weighted Simulation 275
6.2. Simulation of Some Classes of Random Processes 275
6.2.1. Preliminary Remarks 275
6.2.2. Simulation of Random Trials and Variables 276
6.2.3. Simulation of a Homogeneous Markov Chain 278
6.2.4. Simulation of a Markov Process with a Finite Set of States 278
6.2.5. Simulation of a Semi-Markov Process with a Finite Set of States 281
6.2.6. Simulation of a Piecewise-Linear Markov Process 281
6.3. Statistical Problems Associated with Simulation 282
6.4. Simulation of Queueing Systems 285
6.4.1. General Principles of Simulation of Systems 285
6.4.2. Block Principle of Simulation 285
6.4.3. Piecewise-Linear Aggregates 286
6.4.4. A Typical Element of the Model 287
6.4.5. Interpretation of Elements of Queueing Systems 288
6.5. Calculation of Corrections to Characteristics of Systems 289
6.5.1. Introductory Remarks 289
6.5.2. Statement of the Problem 290
6.5.3. Remark 292
6.5.4. Notes 293

Bibliography 294
Index 313

Introduction to the Second Edition

During the two decades since the publication of the first edition of our book, queueing theory has developed substantially. A large number of new service systems have been studied by the traditional methods of Markov processes and embedded Markov chains. Priority systems have been investigated intensively (mainly in connection with the problems of modern computational technology, teletraffic systems, systems of inventory control, and transportation and operations research). Along with Markovian methods, powerful ergodic methods have been developed and a stability theory of service systems has been worked out. The method of statistical simulation has become one of the most widespread computational tools for solving problems of queueing theory.

All of these developments necessitated substantial revision of the book. A new chapter (Chapter 6) dealing with statistical simulation was incorporated. Material on new qualitative methods—for example, the method of Borovkov's innovation moments, the method of marked point processes, and the method of consolidation of states of random processes—were added. However, the basic mathematical models adopted in this book remain at the level of Markov chains and processes. Thus, this book remains accessible to an engineer with a good mathematical background. In the framework of Markov models proofs of the theorems are given with sufficient mathematical rigor.

Selection of the material and references were influenced to some extent by our own scientific interests as well as those of our colleagues.

The authors of this book have always carried out their investigations in queueing theory and related areas in close cooperation with specialists in various fields of applications. The vivid interest of these specialists in queueing theory together with the attention received from our colleagues—mathematicians and cyberneticists—provided us with the resolve to complete this edition.

Substantial comments that resulted in improvements in this book were made at different times by G.P. Basharin, Yu. K. Belyaev, A.D. Solov'yev, V.V. Rykov, V.A. Ivnitskiy, and N. Yu. Kuznetsov. It is the authors' pleasant duty to express their sincere gratitude to all of them.

B.V. Gnedenko
I.N. Kovalenko

Introduction to the First Edition

At the beginning of the 20th century, the practical requirements of teletraffic, physics, and rational organization of "mass service" (theatre agencies, stores, automatic machines, etc.) gave rise to interesting, new mathematical problems. These problems dealt primarily with questions of priority service to telephone subscribers, regulation of inventory in stores to ensure an uninterrupted supply to customers, and determination of an adequate number of salespersons and cash registers in stores. The impetus to development of this theory was given by the well-known Danish scientist A.K. Erlang (1878–1929), for many years an employee of the Copenhagen Telephone Company. His basic research in the field dates from the period 1908–1922. From then on, interest in the models proposed by Erlang increased rapidly. Increasing numbers of mathematicians, engineers, and economists became interested in similar models and developed them. It turned out that models arising from teletraffic are also relevant in various other fields such as the natural sciences, engineering, economics, transportation, military problems, and organization of production.

Practical demands originated many new formulations of problems in queueing theory. Investigation of these problems is necessary if they are to be applied in practice and the real physical conditions are to be approximated. Moreover, these investigations are instructive for developing research methods and creating a systematic theory that would provide an immediate solution to particular problems. Among the most important tools of queueing theory are the theory of stochastic processes, especially Markov processes, and their various generalizations.

Before presenting a systematic account of the subject matter of this book, we shall consider briefly a few new fields of application.

Assume that calls arrive at a telephone exchange in random order. If a free line is available when a call arrives, the subscriber is connected, the conversation begins, and lasts as long as necessary. However, if all the lines are busy, various systems of service to the subscribers are possible. At present two systems have been investigated extensively: the system with waiting and the system with losses. In the first system, a call arriving at the exchange and finding all lines busy, joins a queue and waits until all earlier calls have been terminated. In the second system, if a call arrives when all the lines are busy, it is refused (the customer is lost) and the service continues as if the lost call had not arrived.

We note that the systems described above differ not only in their technical features but also in the nature of the problems involved in their investigation. Indeed, in estimating the quality of the service provided by a system with waiting, the mean waiting time until the service begins is of the greatest importance. In systems with losses, the waiting time is of neither technical nor mathematical interest. Here a different characteristic is important: the probability of refusal (loss of a customer). While the probability of refusal provides a complete picture of the functioning of the second system, the situation is more complicated in the first. The mean waiting time, although important, does not fully characterize the functioning of the system; the deviations of the actual waiting times from their mean play a very essential role. The mean queue length and its distribution, as well as the load on the servers, are also of interest.

It is obvious that the situation arisng at a theatre box-office, for example, is reminiscent of the description of the telephone service systems presented previously. While the original formulation refers to telephone lines, we are now interested in the box-office attendant's busy period. The desire to serve the customers, rationally leads us to investigate the regularities of the queue formation. This is necessary, in particular, to help determine an adequate number of cash registers in train stations or stores. The maintenance of each register involves a certain expense; on the other hand, refusal of customers also results in losses. Therefore, the problem is to find an optimal arrangement. The manner in which tools are issued to workers in large factories may be of even greater importance from an economical point of view. If only one distribution point is available, the workers may lose much time while waiting to receive their equipment, and while they wait, their machines are idle. On the other hand, if there are too many distribution points, the workers allocating the tools may not have enough work. It is evident that such problems are of general interest, and arise in planning the capacity of airports, approach roads, sluices, harbor quays, hospitals, and other installations.

Since the 1930s, because of automation in heavy industry, one worker may be assigned to a number of machines. For various reasons, the machines break down at random times and require the worker's attention. The length of a repair operation is, in general, a random variable. Thus, one is interested in the probability that at a particular instant of time (under specified operating conditions of machine and worker) a given number of machines will be in need of servicing. Other natural problems of practical importance are, for instance: what is the mean idle time of the machines allotted to the work? Under specified working conditions, how many machines should be allotted to one work? Is it better to assign n machines to one worker or ns machines to s workers? We shall not continue enumerating the additional problems that arise in a deeper analysis of servicing with several machines.

Geiger-Müller counters are used in many fields of modern science, in particular, in nuclear physics. A special feature of this device is that a particle entering it causes a discharge. For a first approximation, it may be assumed that this discharge has a definite duration τ, during which additional particles entering the counter are not recorded. Thus, the counter provides a slightly distorted picture of the phenomemon and the readings of the device require correction. First and foremost, one must calculate the probability of loss of a certain number of particles in a given time t. Another important problem in many specific applications is to reconstruct the actual particle flow entering the counter, based on the readings of the counter.

In many real-world problems encountered in science, industry, and economics, problems arise in which service times or waiting times are not unlimited. In fact, we often forego service if the waiting time is too long. If we see that there are more than, say, five customers in a queue at a store, we may depart and postpone the purchase. When ordering a long distance call, we often have a limited amount of time, and request the operator to cancel the call if it is not put through within a certain time. A somewhat different situation is when duration in a system is limited. This happens when perishable goods are sold; no more than a time τ may elapse from production to consumption, otherwise the goods lose their valuable properties and may even be harmful to the consumer's health.

Another real-world illustration of this formulation is treatment of casualties in a traffic accident. The total "holding" time of a casualty in the "service system" is limited by a random variable τ, since his or her ability to wait until the end of the service depends on the shock he or she has suffered and on his or her physical condition. Here the term "holding time in the service system" should be understood as the time from the accident to the recovery (waiting for the ambulance, transportation, doctor's examination, operation, and convalescence). Of course, not all casualties can wait for the completion of the "service"; some cannot even wait for its beginning.

The following formulations of similar problems are thus quite natural and comprise a large number of related problems. (1) Arriving customers join the queue if the number of earlier arrivals awaiting service does not exceed a given number k; otherwise, the customer is lost. (2) A customer remains in the system for a time no longer than τ, even if service has already begun. (3) Waiting time is limited by a random variable τ; but, if service begins before this time, it is completed. In all three cases the mean number of losses during a given time interval, the mean waiting time until service begins, and the time lost in waiting are of interest. In the second formulation the mean number of losses of customers whose service begins but is not completed should be noted.

In the previously described problems we have assumed that the servers are totally reliable and never break down; of course, such an assumption is an idealization. Therefore, a natural and important problem is to take into account the influence of the failure of the servers on the efficiency of the system. In this connection, there are innumerable problems of practical interest. For example, after each flight, airplanes undergo preventive inspection; they are either sent for repair with probability α or returned to service with probability $1 - \alpha$.

A related class of problems deals with so-called priority service; here, not one, but several streams of customers arrive at the serving system. The stream with the lowest serial number has claim to priority service; these customers are served out of turn in relation to customers with a higher serial number who arrived earlier. Here again, we consider two formulations: customers of higher rank may or may not interrupt service of customers of lower rank. In the first situation there are two possibilities: when a displaced customer returns to the serving system, the part of the service rendered before, either is lost and begins again from scratch, or it is taken into account. Both situations are relevant to the operation of computers, for instance. We may view a breakdown or malfunction of a computer as a priority customer. Then there are two possibilities: either the failure led to no errors in the previous run of the computer and the computation may be resumed, or the failure causes errors and the computation must begin anew. This example also shows that priority service may include failure of servers as a special case.

Problems of priority service arise constantly; a maintenance crew first deals with the emergency and then resumes its current repairs; dentists' patients with acute toothaches are treated out of turn; and so on.

Many more real-world problems can be treated by the mathematical methods of queueing theory. There is, however, no need for detailed enumeration, since no mathematical theory can claim to list all the practical problems (even the most important ones) to which it applies. The theory deals with general methods that are applicable not only to the solution of those specific problems which stimulate its elaboration, but also to other problems whose formulation may be very different from the initial ones.

Various aspects of queueing theory are discussed in numerous books. First, we shall mention the book by Khinchin (1963); next the books of Takács (1967), Syski (1960), Saaty (1961), Riordan (1962), Beneš (1963), and Le Gall (1952). The books by Syski, Saaty, and Le Gall present a most comprehensive treatment of the subject matter. Khinchin's, Takács', Beneš', and Le Gall's book are notable for their deep mathematical treatment. Among the monographs of a general nature published during the last 10–15 years, we mention books by Cohen (1969); Kingman (1966); König, Matthes, and Navrotzky (1967); König, Schmidt, and Stoyan (1976); Franken, König, Arnat, and Schmidt (1984); Gross and Harris (1974); and Cooper (1981). Newell's (1971) monograph is devoted to applications of queueing theory. Books by Syski (1960), Saaty (1961), and that of Bharucha-Reid (1960) contain comprehensive bibliographies for their time period. More current surveys are given in Gnedenko and König (1983/4) and in Cooper (1981).

Among Soviet works in queueing theory we note Klimov's (1966) monograph and Ivchenko, Kashtanov, and Kovalenko's (1982) textbook. The well-known textbook by Ventzel, *Theory of Probability*, the books by Ovcharenko, *Applied Problems of Queueing Theory*, and Rosenberg and Prokhorov's *What is Queueing Theory*, in which foundations of queueing theory are presented, substantially contributed to the development of the applications of queueing theory to operations research.

Many generalizing works directed toward specific problems have appeared in world and Soviet literature. We note the books by Kleinrock (1975); Artamonov and O.M. Brekhov (1979) dealing with models of computer operation; Barlow and Proschan (1981); Gnedenko, Belyayev, and Solov'yev (1965) on reliability theory; and Gnedenko et al. (1973) on priority systems. It should be mentioned that problem-oriented investigations, in addition to their main purpose, constitute a substantial contribution to the theory. Thus, teletraffic problems require the development of computational methods of solving systems of linear equations of large dimensions based on approximate consolidation of the states of a Markov process (Basharin, Kharkovich, and Shneps (1966). Problems of calculating highly reliable systems stimulated the development of an asymptotic method of analysis of the distribution of the moment of occurrence of a rare event in the model of a regenerative process (Solov'yev (1971)). Optimization methods of queueing theory were developed. We shall mention an application to the optimization of technical servicing of systems (Barzilovich, Belyayev, Kashtanov, and others (1983)).

In connection with the problems of queueing theory, a large number of classes of random processes of various generality was developed. Gikhman and Skorokhod (1971) developed the theory of jump Markov processes in an arbitrary measurable space of states. Along with Markov chains, semi-Markov, regenerative processes, renewal processes, and line Markov processes that have achieved a solid position as mathematical tools in queueing theory, there are, in the literature, many other classes of random processes that are interesting for this theory; Çinlar, Pyke, Shaffel, Sevast'yanov, Korolyuk, Ezhov, Turbin, Tomko, Franken, König, Brodi, Zakharin, and Shpak made substantial contributions to the development of the theory of these classes of processes. They serve as models for service processes with various special features. They are usually constructively defined and thus can be translated efficiently into the language of statistical simulation.

In recent years, general probabilistic and analytic methods were developed in queueing theory that allow us to carry out conclusions of an aggregate nature. Borovkov's (1972, 1980) monographs, which combine deep analytic results with a variety of probabilistic approaches and interpretations, are examples of such investigations. Systems that can be handled using Borovkov's method are far beyond the bounds of the classical "Markov" models; they are described by general transformations of stationary random sequences. Franken et al.'s (1984) monograph summarizes the results achieved in an important field—that of the theory of marked point processes. Utilization of this theory allows us to take a unifying view of the large number of ergodic relations between the characteristics of systems and achieve a natural maximal generality. We also note Shurenkov's (1981) monograph, which contains interesting ergodic theorems on queueing theory for strictly stationary processes.

One of the main foundations of queueing theory is the random walk theory. Classical results in this theory are due to Pollaczek (1957, 1961) and Spitzer; Borovkov, Korolyuk, Rogozin, Presman, Ezhov, Gusak, Narisova, and other scientists extended further the classical methods and originated a theory of analytic representation of the characteristics of random walks that are interpreted in terms of service systems.

We note that the investigation of stable service systems represents an important aspect of queueing theory. This problem was developed substantially in the works of Borovkov, Prokhorov, Zolotarev, Stoyan, Kalashnikov, and others.

One cannot imagine utilizing queueing theory in practice without an appropriate level of development of numerical methods. First and foremost among these is the statistical simulation of systems for which theoretical foundations as well as powerful software were developed (SIMUL, GPSS, SIMSCRIPT, NEDIS, and other systems). We mention as an important problem the development of special software for calculations associated with rare events in service systems.

1 Problems of Queueing Theory under the Simplest Assumptions

1.1. Simple Streams

1.1.1. *Historical Remarks*

The first task facing any serious study of the theory of queueing processes or its specific applications is the investigation of the stream of customers arriving at a queueing system. Thus, in order to calculate particle loss in a counter, we must know how the particles enter the counters. Similarly, when the operation of a telephone exchange is implemented, we must take into account the special features of the stream of subscriber-to-exchange calls.

Most of the literature of queueing theory, both what served as the foundation for the theory and current literature, deals with the *simple case of streams.* Here the probability that k customers arrive in a time interval of length t is given by

$$P_k(t) = \frac{(\lambda t)^k}{k!} e^{-\lambda t},$$

where $\lambda > 0$ is a constant whose menaing will be studied subsequently. Here the incoming stream is assumed to possess the following property: Given any finite group of nonoverlapping time intervals, the numbers of customers arriving during each interval are mutually independent random variables.

Attempts to state sufficiently general conditions under which such a stream actually occurs were made a long time ago. Thus, in Sections 81 and 82 of Fry's well-known book (1965) the concepts of randomness in the individual and collective sense of the term were presented. Fry has shown that when these conditions are simultaneously satisfied the queue must have a Poisson input of the type indicated above. Fry's arguments cannot be considered exhaustive; nevertheless they provided practitioners with conditions broad enough to indicate when a simple stream is the only possibility. Somewhat different conditions had been considered earlier by Smoluchowski and Einstein (1936) in their papers dealing with the theory of Brownian motion. Khinchin, in his monograph (1963) reduced the number of these conditions to three: the stream must be stationary, without aftereffects, and orderly. A fourth condition (see p. 11), often given in textbooks on probability theory, was shown by Khinchin to be a corollary of the above three conditions. This approach will be considered in this section. A Poisson stream is not only confirmed statistically in the course of observing numerous real-world streams, but it also appears as a limiting case in various probabilistic models.

1.1.2. *The Notion of a Stream of Homogeneous Events*

Definition. A finite or countable sequence $\{\tau_n\}$ of random variables defined on the same probability space under the condition that in any fixed time interval (a, b) with probability 1 a finite number of these variables falls in it is called a *stream of homogeneous events.*

If the given t coincides with r elements of the sequence $\{\tau_n\}$ we say that, at time t, r events of the stream occur. If the τ_n are ordered so that

$$\tau_n < a \leqslant \tau_{n+1} \leqslant \tau_{n+2} \leqslant \cdots \leqslant \tau_{n+k} < b \leqslant \tau_{n+k+1},$$

then τ_{n+i} is called the *arrival time* of the ith event of a stream of homogeneous events in the half-interval $[a, b)$. Chapter 2 is devoted to the study of properties of streams of homogeneous events. We shall only note one property that will be required in this section.

Lemma. *The number $\nu(a, b)$ of events of a stream in the half-interval $[a, b)$ is a random variable.*

PROOF. We have

$$\{\nu(a, b) \geqslant k\} = \bigcup_{n_1 < \cdots < n_k} \{a \leqslant \tau_{n_1} < b, \ldots, a \leqslant \tau_{n_k} < b\};$$

and we observe that a finite or countable union of random events is a random event. □

1.1.3. *Qualitative Assumptions and Their Analysis*

We shall define the properties of stationarity, absence of aftereffects, and orderliness of a stream of homogeneous events.

Stationarity of a stream means that the probability that $k_1, k_2, \ldots, k_n$ customers, respectively, will arrive in each interval of any finite group of n nonoverlapping time intervals depends only on the k_i $(i = 1, \ldots, n)$ and on the lengths of the time intervals, but not on their positions on the time axis. In particular, the probability that k customers will arrive in a time interval $(T, T + t)$ is independent of T and is a function of the variables k and t only.

The *absence of aftereffects* means that the probability of k customers arriving during a time interval $(T, T + t)$ does not depend on the number and type of customers arriving before T. Thus, the conditional probability of k customers arriving during a time interval $(T, T + t)$, under any assumption about the customers arriving prior to T, is equal to the unconditional probability of this event. In particular, the absence of aftereffects expresses the mutual independence of various numbers of customers arriving at the service system during nonoverlapping time intervals.

The orderliness of the stream expresses the practical impossibility that two or more customers will arrive at the same instant of time. This condition is expressed more precisely as follows: Let us denote the probability of two or more customers arriving in a time intervl of length h by $P_{>1}(h)$. The condition for orderliness is that

$$\frac{P_{>1}(h)}{h} \to 0,$$

as $h \to 0$, or, as we shall write in what follows,

$$P_{>1}(h) = o(h).$$

A stream of homogeneous events satisfying these three conditions is called a *simple* stream. We now proceed to a brief analysis of the three conditions mentioned previously, paying special attention to their intuitive, physical meaning. Such an analysis is essential, especially if we consider the importance of the theoretical conclusions and practical applications that serve as the basis for these assumptions.

Experimental verification in various fields—physics, teletraffic, reliability theory (failure of elements of a system), transportation, commerce, and so forth—have shown that simple streams occur less frequently than was originally supposed. It is obvious that such a conclusion was to be expected even before experimental investigation. Indeed, the assumption of stationarity in specific situations is a rather severe abstraction. In fact, it is invalid for many different reasons. During radioactive decay we must take into account that the decaying mass decreases with time, so that a stationary state does not exist in the strict sense of the term. A stream of calls entering a telephone exchange cannot be considered completely stationary, since in the course of 24 hours the operating conditions of the exchange vary substantially. The stream of calls to a first aid station also varies considerably over a full 24 hours. However, if we consider these phenomena over comparatively short time intervals, the stationarity assumption is satisfactory as a first approximation.

Neither is the hypothesis of absence of aftereffects justified in many cases. In numerous cases, one event leads to others. Thus, one telephone call may result in a large number of calls to other subscribers. Radioactive decay is another example: If a large amount of undecayed matter is present, the disintegration of one atom may lead to the disintegration of others, resulting in a chain reaction. However, if only a small amount of matter is present, the hypothesis of absence of aftereffects is sufficiently justified. A chain reaction of telephone calls affects the operation of the exchange only negligibly, because there are many other subscribers.

The assumption of orderliness is also far from being rigorously satisfied in numerous cases. It is well known that batches of customers appear suddenly at stores, ticket booths, etc. Not only individual barges appear at a river harbor for unloading, but convoys of barges are also brought in by tugboats. Besides a single ship, a tugboat with barges also arrives at a sluice. Similar group (batch) arrivals may be observed in numerous physical phenomena.

Regardless of the fact that the three conditions discussed previously, are not strictly fulfilled, they serve as a good starting point for the study of real-world queueing streams. We shall subsequently clarify the influence of each one of these conditions on the nature of the stream.

1.1.4. *Derivation of Equations for Simple Streams*

We denote by $P_k(t)$ the probability that customers will arrive for servicing in a time interval of length t. Since the stream is simple, this probability depends neither on the choice of the time origin nor on the previous history of the stream. The conditions defining the simple stream enable us to find a simple formula for $P_k(t)$, uniquely determined up to a parameter.

To simplify the argument we shall assume an additional property that

$$P_1(h) = \lambda h + o(h), \tag{1}$$

where λ is a constant. Later we shall prove that this assumption follows from the previous three conditions.

First we shall determine the probability that during a time interval of length $t + h$ exactly k customers arrive. This event may occur in $k + 1$ different ways, corresponding to the following disjoint possibilities:

1. During the time interval t, all k customers arrive, and in interval h, no customer arrives.
2. During the time interval t, $k - 1$ customers arrive, and during h, one customer arrives.

⋮

$k + 1$. During the time time interval t, no customer arrives, and during h, all k customers arrive.

We shall use the total probability formula; since there are no aftereffects, we have

$$P_k(t + h) = \sum_{j=0}^{k} P_j(t)P_{k-j}(h).$$

We introduce the notation

$$R_k = \sum_{j=0}^{k-2} P_j(t)P_{k-j}(h)$$

and estimate this sum, observing that the probability $P_k(t)$ never exceeds unity. Thus,

$$R_k \leqslant \sum_{j=0}^{k-2} P_{k-j}(h) = \sum_{s=2}^{k} P_s(h).$$

Extending the summation of the right-hand side to infinity only strengthens the inequality:

$$R_k \leqslant \sum_{s=2}^{\infty} P_s(h) = P_{>1}(h).$$

Since the stream is orderly,

$$P_{>1}(h) = o(h).$$

We thus obtain

$$P_k(t+h) = P_k(t)P_0(h) + P_{k-1}(t)P_1(h) + o(h). \tag{2}$$

In this equality we can substitute $\lambda n + o(h)$ for $P_1(h)$ in view of the additional condition (1), which we shall discard later.

Moreover, it is clear that

$$P_0(h) = 1 - \sum_{s=1}^{\infty} P_s(h) = 1 - P_1(h) - \sum_{s=2}^{\infty} P_s(h).$$

Hence

$$P_0(h) = 1 - \lambda h + o(h).$$

Now (2) may be rewritten as

$$P_k(t+h) = P_k(t)(1 - \lambda h) + P_{k-1}(t)\lambda h + o(h),$$

so that

$$\frac{R_k(t+h) - P_k(t)}{h} = -\lambda P_k(t) + \lambda P_{k-1}(t) + o(1).$$

Now let $h \to 0$. Since the right-hand side tends to a limit, so does the left-hand side. Thus, passing to the limit, we arrive at

$$\frac{dP_k(t)}{dt} = -\lambda P_k(t) + \lambda P_{k-1}(t). \tag{3}$$

In deriving this equation, we have assumed that $k \geqslant 1$. Letting k take all possible values, we obtain an infinite number of system equations for the infinite number of unknown probabilities $P_k(t)$. Thus, Eq. (3) is actually an infinite system of differential-difference equations. We need one more equation, satisfied by the function $P_0(t)$. In view of the conditions defining the simple stream, we have

$$P_0(t + h) = P_0(t)P_0(h).$$

Based on the preceding arguments, this equation may be replaced by the following equivalent one:

$$P_0(t + h) = P_0(t)[1 - \lambda h + o(h)].$$

Approaching the limit, we obtain the equation for $P_0(t)$:

$$\frac{dP_0(t)}{dt} = -\lambda P_0(t). \tag{4}$$

1.1.5. *Solution of the Equations*

As is easily seen, the last equation has the solution

$$P_0(t) = Ce^{-\lambda t}. \tag{5}$$

In order to determine the constant C we use (1), which implies that

$$P_0(0) = 1,$$

while according to (5),

$$P_0(0) = C.$$

Comparison of the last two equalities yields

$$P_0(t) = e^{-\lambda t}. \tag{6}$$

Substitution of $P_0(t)$ into the equation for $P_1(t)$ results in

$$P_1(t) = \lambda t e^{-\lambda t}.$$

By successive substitution of the previously determined probabilities into (3) we obtain the probabilities $P_k(t)$ for arbitrary k. A simple calculation shows that for any $k \geqslant 0$

$$P_k(t) = \frac{(\lambda t)^k}{k!} e^{-\lambda t}. \tag{7}$$

The numerical solution of our system of equations is simplified if we set

$$P_k(t) = e^{-\lambda t} v_k(t).$$

In terms of the functions $v_k(t)$, Eqs. (3) and (4) become

$$v_k'(t) = \lambda v_{k-1}(t), \quad k = 1, 2, \ldots,$$

and

$$v_0'(t) = 0.$$

The initial conditions are

$$v_0(0) = P_0(0) = 1,$$

$$v_k(0) = P_k(0) = 0, \quad k = 1, 2, \ldots.$$

Taking into account the initial conditions for $v_k(t)$, we obtain

$$v_0(t) = 1, \qquad v_1(t) = \lambda t, \qquad v_2(t) = \frac{(\lambda t)^2}{2!},$$

and, in general, for arbitrary $k \geqslant 0$,

$$v_k(t) = \frac{(\lambda t)^k}{k!}.$$

A return to functions $P_k(t)$ yields (7).

1.1.6. *Derivation of the Additional Assumption from the Other Three Assumptions*

In deriving equations describing the general form of a simple stream, we have temporarily used condition (1). Our immediate task is to derive this condition from the other three. To do this we consider a time interval of unit length and denote the probability that no customer arrives during this period by θ. Thus,

$$\theta = P_0(1).$$

Subdivide the time interval under consideration into n equal parts. In order that no customer arrives in the whole time interval it is necessary and sufficient that none arrive in any of these n subintervals. Using this fact, the stationarity of the stream, and the absence of aftereffects in it, we obtain

$$\theta = \left[P_0\left(\frac{1}{n}\right)\right]^n.$$

Hence

$$P_0\left(\frac{1}{n}\right) = \theta^{1/n}.$$

Thus, the probability of the absence of customers in a time interval of length k/n is

$$P_0\left(\frac{k}{n}\right) = \theta^{k/n}.$$

Let t be a nonnegative number. For any t there exists an integer k such that for a given n the following inequalities hold:

$$\frac{k-1}{n} \leqslant t < \frac{k}{n}.$$

Since $P_0(t)$ is a nonincreasing function of time,

$$P_0\left(\frac{k-1}{n}\right) \geqslant P_0(t) \geqslant P_0\left(\frac{k}{n}\right).$$

Thus, $P_0(t)$ satisfies the inequalities

$$\theta^{(k-1)/n} \geqslant P_0(t) \geqslant \theta^{k/n}.$$

Now let n tend to infinity; then

$$\lim_{n\to\infty} \frac{k}{n} = \lim_{n\to\infty} \frac{k-1}{n} = t$$

and thus from the preceding discussion

$$P_0(t) = \theta^t.$$

Since $P_0(t)$, as a probability, satisfies the inequality

$$0 \leqslant P_0(t) \leqslant 1,$$

there are three possibilities: (1) $\theta = 0$, (2) $\theta = 1$, (3) $0 < \theta < 1$. The first two cases are not interesting. In the first case we have $P_0(t) = 0$ for any t, and this means that the probability that at least one customer will arrive is equal to 1. In other words, infinitely many customers arrive with probability 1 in a time interval of arbitrary length. In the second case, $P_0(t) = 1$ for any $t > 0$, and hence there is no stream of customers. Only the third case is of interest for theoretical and practical purposes. Here we put $\theta = e^{-\lambda}$, where λ is a positive number.

Using stationarity and the absence of aftereffects we have thus obtained

$$P_0(t) = e^{-\lambda t}. \tag{8}$$

Note that in deriving this formula we did not assume that the stream is orderly.

Equation (8) is interpreted as follows: The probability that the interarrival time in a stationary stream without aftereffects exceeds 1, and is given by $e^{-\lambda t}$. Hence, the distribution function of the interarrival times is

$$F(t) = 1 - e^{-\lambda t}. \tag{9}$$

Since in any time interval t some number of customers will arrive, we have

$$P_0(t) + P_1(t) + P_{>1}(t) = 1.$$

For small t it follows from the orderliness and from (8) that

$$P_0(t) = 1 - \lambda t + o(t).$$

This equality and the orderliness condition give

$$P_1(t) = \lambda t + o(t), \tag{9'}$$

that is, we deduce the condition (1), which is what we wished to demonstrate.

1.1.7. *Distribution of Time of Events of a Stream*

Consider the subintervals

$$\begin{aligned} A_1 &= [0, t_1 - h_1), & B_1 &= [t_1 - h_1, t_1 + k_1), \\ A_2 &= [t_1 + k_1, t_2 - h_2), & B_2 &= [t_2 - h_2, t_2 + k_2), \\ &\vdots & &\vdots \\ A_n &= [t_{n-1} + k_{n-1}, t_n - h_n), & B_n &= [t_n - h_n, t_n + k_n). \end{aligned}$$

The quantities h_i and k_i are viewed as arbitrarily small. In view of the preceding equations, the probability that no events of the stream occur in the intervals $A_1, \ldots, A_n$, and that exactly one event occurs in each one of the intervals $B_1, \ldots, B_n$ is equal to

$$\begin{aligned} & e^{-\lambda(t_1 - h_1)}[\lambda(h_1 + k_1) + o(h_1 + k_1)]e^{-\lambda(t_2 - h_2 - t_1 - k_1)}[\lambda(h_2 + k_2) \\ & \quad + o(h_2 + k_2)] \cdots e^{-\lambda(t_n - h_n - t_{n-1} - k_n)}[\lambda(h_n + k_n) + o(h_n + k_n)] \\ & = \prod_{i=1}^{n} [e^{-\lambda(t_i - t_{i-1})}\lambda(h_i + k_i) + o(h_i + k_i)], \end{aligned}$$

where $t_0 = 0$. This is also the probability that

$$t_1 - h_1 \leqslant \tau_1 < t_1 + k_1, \ldots, t_n - h_n \leqslant \tau_n < t_n + k_n$$

(the τ_i are the arrival times of events). Hence, the random vector $(\tau_1, \ldots, \tau_n)$ has a density of the form

$$\lambda^n e^{-\lambda(t_n - t_0)} = \prod_{i=1}^{n} [\lambda e^{-\lambda(t_i - t_{i-1})}].$$

This means that $\tau_1, \tau_2 - \tau_1, \ldots, \tau_n - \tau_{n-1}$ are independent random variables exponentially distributed with parameter λ.

In this discussion we can assume that $\tau_1, \tau_2, \ldots,$ are times of successive events of a simple stream starting from any fixed time t_0, if the time origin is located at t_0.

As a corollary we shall prove the property that Fry ((1965), p. 172) called *randomness in the individual sense.*

Theorem. *Under the condition that the number of events of a simple stream in the interval (a, b) equals n, the times of these events are independent and uniformly distributed on the interval (a, b).*

PROOF. First, assume that the times of events are located in increasing order. In accordance with the above

$$\begin{aligned} & P\{\Delta(a, b) = n; t_i \leqslant \tau_i < t_i + dt_i, 1 \leqslant i \leqslant n\} \\ & \quad = P\{t_i \leqslant \tau_i < t_i + dt_i, 1 \leqslant i \leqslant n; \tau_{n+1} > b\}, \end{aligned}$$

where τ_i is the ith time of the events of the stream starting with time a. This probability equals $\lambda^n e^{-\lambda(b-a)}\,dt_1 \cdots dt_n$: The probability of the event associated with $\tau_1, \ldots, \tau_n$ is multiplied by the probability $e^{-\lambda(b-t_n-dt_n)} \sim e^{-\lambda(b-t_n)}$ of the absence of an event in the interval $(t_n + dt_n, b)$. We see that the random vector $(\tau_1, \ldots, \tau_n)$ is uniformly distributed in the region $\{a < t_1 < \cdots < t_n < b\}$. If nothing is known about the mutual disposition of the times of events and it is assumed that any system of inequalities $\tau_{i_1} < \tau_{i_2} < \cdots < \tau_{i_n}$ has the probability $(n!)^{-1}$, we then obtain a uniform distribution in the hypercube $\{a \leqslant \tau_i \leqslant b, 1 \leqslant i \leqslant n\}$, which is the required property. □

1.1.8. *The Intensity and Parameter of a Stream*

We now perform some simple calculations. First, it is easy to calculate that for a simple stream the mean number of customers arriving in a time interval t is

$$\mathsf{M}\mu(t) = \sum_{k=1}^{\infty} kP_k(t) = e^{-\lambda t} \sum_{k=1}^{\infty} k\frac{(\lambda t)^k}{k!} = \lambda t.$$

Here $\mu(t)$ is the actual number of customers arriving during a time interval of length t.

The mathematical expectation of the number of customers arriving in unit time is called the *intensity of the stream.* We shall denote the intensity by μ. For a simple queue

$$\mu = \lambda.$$

λ is called the *parameter of the stream.*

The last equation shows that *in a simple stream the intensity and the parameter of the stream are equal.*

We denote by $\pi_1(t)$ the probability that at least one customer arrives in a time interval t, that is, set

$$\pi_1(t) = \sum_{k=1}^{\infty} P_k(t) = 1 - P_0(t).$$

For a simple stream the following is valid:

$$\lim_{t \to 0} \frac{\pi_1(t)}{t} = \lambda. \tag{10}$$

This will be regarded as the definition of the parameter of a stream.

The inequality

$$\mu \geqslant \lambda$$

holds for any stationary stream provided the limit (10) *exists.*

Indeed, for a stationary stream

$$\mathsf{M}\mu(t) = \mu t = \sum_{k=1}^{\infty} kP_k(t) \geqslant \sum_{k=1}^{\infty} P_k(t) = \pi_1(t).$$

Thus, for any t we have

$$\mu \geqslant \frac{\pi_1(t)}{t}.$$

Clearly this inequality proves our assertion.

1.2. Service with Waiting

1.2.1. *Statement of the Problem*

We shall consider the classical problem of queueing theory here as considered and solved by Erlang. A simple stream of customers of intensity λ arrives at m "uniform" servers. If at the instant a customer arrives at least one server is free, it immediately serves a customer. On the other hand, if all the servers are busy, a newly arriving customer must wait in the queue behind all earlier arrivals who have not yet been served. Any server who becomes free proceeds immediately to serve the next customer, provided there is a queue. Each customer is served by one server only and, at any one time, each server is servicing only one customer. The service time is a random variable with a well-defined probability distribution $F(x)$. We shall assume that for $x \geqslant 0$

$$F(x) = 1 - e^{-\mu x}, \tag{1}$$

where $\mu > 0$ is a constant.

The problem just described is of considerable practical interest and the results to be presented here are widely used in practice. Subsequently we shall consider a schematic example of this type. There is no need to emphasize here that similar problems arise in many specific situations. As already mentioned in the Introduction, Erlang solved this problem with the questions of teletraffic in mind.

Our choice of distribution (1) for describing the service time is not arbitrary. The point is that under this assumption there is a simple solution that describes the process with a precision adequate for practical purposes. As we shall see, distribution (1) plays a very crucial role in queueing theory. This is due to the following property of the distribution:

Let the service time be an exponential distribution, and consider the time required at some instant to complete the service; then the distribution of this time is independent of the past duration of the service.

Indeed, let $f_a(t)$ denote the probability that the service which was in progress for a time a is prolonged at least by t. Assuming that the service time is exponentially distributed, $f_0(t) = e^{-\mu t}$. It is thus clear that

$$f_0(a) = e^{-\mu a} \quad \text{and} \quad f_0(a + t) = e^{-\mu(a+t)}.$$

And since we always have

$$f_0(a + t) = f_0(a) f_a(t),$$

it follows that

$$e^{-\mu(a+t)} = e^{-\mu}f_a(t),$$

and consequently,

$$f_a(t) = e^{-\mu t} = f_0(t).$$

QED.

It is evident that in specific situations exponential service time is, as a rule, only a crude approximation to reality. Thus, service time often cannot be less than some definite value. One assumption implicit in (1) is that a large proportion of the customers require only a short service time, close to zero. One of our subsequent problems will be to get rid of the excessive restrictions imposed by (1). This problem was already evident to Erlang, who, in a number of papers, tried to find another distribution suitable for the service time. In particular, he proposed the so-called *Erlang distribution*, whose density is

$$\varphi_k(t) = \begin{cases} 0 & \text{for } t \leqslant 0, \\ \mu \dfrac{(\mu t)^{k-1}}{(k-1)!} e^{-\mu t} & \text{for } t > 0, \end{cases}$$

where $\mu > 0$ and k is a positive integer.

The Erlang distribution is the distribution of the sum of k independent random variables, each one of which has distribution (1).

We denote by η the time for servicing one customer when the distribution is (1). Then the mean service time is

$$\mathsf{M}\eta = \int_0^\infty x\,dF(x) = \mu \int_0^\infty xe^{-\mu x}\,dx = \frac{1}{\mu}.$$

This equality allows us to evaluate the parameter μ based on experimental data. It is easy to see that the variance of the service time is equal to

$$\mathsf{D}\eta = \mathsf{M}\eta^2 - (\mathsf{M}\eta)^2 = \frac{1}{\mu^2}.$$

1.2.2. *The Servicing Process as a Markov Process*

Under the foregoing assumptions on the stream of customers and the service time, the problems of queueing theory acquire certain features that facilitate their study. We have already noted their computational simplicity. Now we shall mention a more basic consideration as applied to our problem.

At each instant the system under consideration can be in one of the following states: at the instant t, k customers ($k = 0, 1, 2, \ldots$) are in the system. If $k \leqslant m$, k customers are in the system and being served, while $m - k$ servers are idle. If $k > m$, m customers are being served and $k - m$ customers are waiting their turn. Let E_k denote the state when k customers are in the system.

Thus, the system may be in states $E_0, E_1, E_2, \ldots$. Denote by $P_s(t)$ the probability that the system is in state E_s at time t.

We shall now state the special features of our problems under the given assumptions. Let the system be in state E_i at time t_0. We shall prove that the subsequent course of the servicing process is independent (in the probability-theoretic sense) of its previous history. Indeed, the subsequent course of the process is completely determined by the following three factors:

1. The times of termination of the individual services that are being performed at the instant t_0.
2. The arrival times of new customers.
3. The service times for customers arriving after t_0.

Owing to the exponential distribution, the service not yet performed does not depend on its course prior to the instant t_0. Since the stream of customers is simple, the past has no effect on the number of customers arriving after t_0. Finally, the service time for customers appearing after t_0 does not depend on what and how the customers have been serviced prior to t_0.

As is well known, stochastic processes, whose future development depends only on the state arrived at at the given instant and not on their past history, are called *Markov processes* or *processes without aftereffects*. We have thus proved that a system with waiting, in the case of a simple stream and exponential service time, is a Markov process. This will facilitate our further arguments.

1.2.3. *Construction of Equations*

The problem now is to find the equations satisfied by the probabilities $P_k(t)$. One of the equations is obvious and holds for all t:

$$\sum_{k=0}^{\infty} P_k(t) = 1. \tag{2}$$

First we find the probability that at the time $t + h$ all the servers are free. This may occur in the following ways:

At time t all the servers are free and no new customers arrive during the time h.

At time t one server is busy and all the others are free; during the time h the service of the customer is completed and no new customers arrive.

It is easily seen that the probability of the remaining possibilities—such as that two or three servers are busy and their service is completed during the time h—is $o(h)$.

The probability of the first of the preceding events is

$$P_0(t)e^{-\lambda h} = P_0(t)[1 - \lambda h + o(h)],$$

while that of the second is

$$P_1(t)e^{-\lambda h}(1 - e^{-\mu h}) = P_1(t)\mu h + o(h).$$

Thus,

$$P_0(t + h) = P_0(t)(1 - \lambda h) + \mu h P_1(t) + o(h).$$

Hence, in an obvious manner we arrive at the equation

$$P_0'(t) = -\lambda P_0(t) + \mu P_1(t). \tag{3}$$

We now proceed to set up the equation for $P_k(t)$ when $k \geqslant 1$. We must consider two different cases: $1 \leqslant k < m$ and $k \geqslant m$. First let $1 \leqslant k < m$. We shall enumerate only the essential situations leading to the state E_k at the time $t + h$. These are:

At time t the system is in the state E_k; during the time h no new customers arrive and no server completes its service. The probability of this event is

$$P_k(t)e^{-\lambda h}(e^{-\mu h})^k = P_k(t)[1 - \lambda h - k\mu h + o(h)].$$

At time t the system is in the state E_{k-1}; during the time h a new customer arrives, but the service of no earlier customer is completed. The probability of this event is

$$P_{k-1}(t)(1 - e^{-\lambda h})(e^{-\mu h})^{k-1} = \lambda h P_{h-1}(t) + o(h).$$

At time t the system is in the state E_{k+1}; during the time h no new customers arrive, but one customer is served. The probability of this event is

$$P_{k+1}(t)e^{-\lambda h}C_{k+1}^1(e^{-\mu h})^k(1 - e^{-\mu h}) = P_{k+1}(t)(k + 1)\mu h + o(h).$$

The probability of all other conceivable possibilities is $o(h)$. Combining these probabilities we obtain

$$P_k(t + h) = P_k(t)(1 - \lambda h - k\mu h) + \lambda h P_{k-1}(t) + (k + 1)\mu h P_{k+1}(t) + o(h).$$

A simple transformation of this equality leads to

$$P_k'(t) = -(\lambda + k\mu)P_k(t) + \lambda P_{k-1}(t) + (k + 1)\mu P_{k+1}(t), \tag{4}$$

for $1 \leqslant k < m$.

Analogous arguments for $k \geqslant m$ yield

$$P_k'(t) = -(\lambda + m\mu)P_k(t) + \lambda P_{k-1}(t) + m\mu P_{k+1}(t). \tag{5}$$

We have thus obtained an infinite system of differential equations [(2)–(5)] for the probabilitis $P_k(t)$. Clearly the solution involves substantial technical difficulties.

1.2.4. *Determination of the Stationary Solution*

In queueing theory usually only the stationary solution for $t \to \infty$ is considered. The existence of such solutions is established by so-called ergodic theorems, some of which we shall prove later. It turns out that these limiting probabilities (or stationary probabilities, as they are usually called) do exist in our case. We shall denote them by P_k. Additionally, we state (although we shall not prove this now) that $P_k'(t) \to 0$ as $t \to \infty$. The preceding implies that (3), (4), and (5) become for the stationary probabilities:

$$-\lambda P_0 + \mu P_1 = 0; \tag{6}$$

for $1 \leqslant k < m$,

$$\lambda P_{k-1} - (\lambda + k\mu)P_k + (k+1)\mu P_{k+1} = 0; \tag{7}$$

and for $k \geqslant m$,

$$\lambda P_{k-1} - (\lambda + n\mu)P_k + m\mu P_{k+1} = 0. \tag{8}$$

The normalizing condition

$$\sum_{k=0}^{\infty} P_k = 1 \tag{9}$$

is added to these equations.

In order to solve the infinite algebraic system obtained, we introduce the notation: for $1 \leqslant k < m$

$$z_k = \lambda P_{k-1} - k\mu P_k,$$

and for $k \geqslant m$

$$z_k = \lambda P_{k-1} - m\mu P_k.$$

Using this notation the system (6)–(8) becomes

$$z_1 = 0, \quad z_k - z_{k+1} = 0 \quad \text{for } k \geqslant 1.$$

Hence for all $k \geqslant 1$

$$z_k = 0,$$

that is, for $1 \leqslant k < m$

$$k\mu P_k = \lambda P_{k-1} \tag{10}$$

and for $k \geqslant m$

$$m\mu P_k = \lambda P_{k-1}. \tag{11}$$

For convenience we introduce the notation

$$\rho = \lambda/\mu.$$

Equation (10) yields for $1 \leqslant k < m$

$$P_k = \frac{\rho^k}{k!} P_0. \tag{12}$$

For $k \geqslant m$ we obtain from (11)

$$P_k = \left(\frac{\rho}{m}\right)^{k-m} P_m,$$

and, consequently, for $k \geqslant m$

$$P_k = \frac{\rho^k}{m!\, m^{k-m}} P_0. \tag{13}$$

It remains for us to determine P_0. To this end we substitute the expressions for P_k from (12) and (13) in (9). The result is

$$P_0 \left[\sum_{k=0}^{m} \frac{\rho^k}{k!} + \frac{m^m}{m!} \sum_{k=m+1}^{\infty} \left(\frac{\rho}{m} \right)^k \right] = 1.$$

Since the infinite sum in square brackets converges only for

$$\rho < m, \tag{14}$$

assuming that this is true, we obtain

$$P_0^{-1} = \sum_{k=0}^{m} \frac{\rho^k}{k!} + \frac{\rho^{m+1}}{m!\,(m-\rho)}. \tag{15}$$

If (14) is not satisfied, that is, if $\rho \geqslant m$, the series in square brackets diverges; this means that P_0 must be zero. But it follows from Eqs. (12) and (13) that for all $k \geqslant 1$, $P_k = 0$ also. The methods of the theory of Markov chains allow us to conclude that whenever $\rho \geqslant m$, the queue tends to ∞ in probability as the time increases.

This result will be illustrated by some specific examples. These examples show that in practice, calculations based on purely arithmetic considerations, without allowing for the specifics of the random fluctuations in the arrival of customers, may lead to serious errors.

Suppose a physician can satisfactorily examine a patient and fill in his case history in 15 min. The planning administrators usually come to the conclusion that the physician should receive 16 patients in a four hour working session. However, the patients arrive at random times. Thus, if this calculation of the physician's capacity is used, a queue will accumulate, since here ρ is assumed to be 1. The same conclusions apply to the planning of the number of beds in a hospital, the number of cash registers in a store, the number of waiters in a restaurant, etc. Unfortunately, some economists make the same error when designing unloading equipment in mines, assigning the number of service personnel in elevators, the number of moorings in sea ports, and so on.

In what follows, we shall always assume that (14) holds.

1.2.5. *Some Preliminary Results*

We have already stated in the Introduction that in the problem with waiting the main characteristic of the quality of service is the waiting time until the commencement of service. The waiting time (in a queue) is a random variable, which will be denoted by γ. For the time being, we shall consider only the problem of determining the probability distribution of the waiting time in a stationary serving process. We further denote by $P\{\gamma > t\}$ the probability that the waiting time exceeds t, and by $P_k\{\gamma > t\}$ the conditional probability of this inequality under the condition that the customer to whom the waiting time refers will find k customers ahead of him or her in the queue. By the total probability formula we have

$$\mathsf{P}\{\gamma > t\} = \sum_{k=m}^{\infty} P_k P_k\{\gamma > t\}. \tag{16}$$

Before transforming this expression into a form suitable for applications, we need some preliminary results. First we shall find simple formulas for P_0 when $m = 1$ and $m = 2$. These formulas are easy to derive: for $m = 1$

$$P_0 = 1 - \rho, \tag{17}$$

and for $m = 2$

$$P_0 = \frac{2 - \rho}{2 + \rho}. \tag{18}$$

Now we shall calculate the probability that all the servers are busy at an arbitrary, randomly selected time. Clearly this probability is

$$\pi = \sum_{k=m}^{\infty} P_k = \frac{m^m}{m!} \sum_{k=m}^{\infty} \left(\frac{\rho}{m}\right)^k P_0 = \frac{\rho^m P_0}{(m-1)!\,(m-\rho)}. \tag{19}$$

For $m = 1$ this formula is very simple:

$$\pi = \rho, \tag{20}$$

and for $m = 2$

$$\pi = \frac{\rho^2}{2 + \rho}. \tag{21}$$

Recall that in (19) ρ can take any value from 0 to m (exclusively). Thus, in (20) $\rho < 1$ and in (21) $\rho < 2$.

1.2.6. *The Distribution Function of the Waiting Time*

If at the instant a customer arrives there are already $k - m$ customers in the queue, that customer must wait until $k - m + 1$ customers are served, since the customers are served according to their order of arrival. Let $q_s(t)$ be the probability that in a time interval of length t after the arrival of a specific customer the service of exactly s customers is completed. It is evident that for $k \geqslant m$ the equality

$$\mathsf{P}_k\{\gamma > t\} = \sum_{s=0}^{k-m} q_s(t)$$

is valid.

Since we have assumed the distribution of the service time is exponential and independent of the number of customers in the queue or the service time for other customers, the probability that during the time t the service of no customer will be completed (i.e., the probability that not a single server is freed) is equal to

$$q_0(t) = e^{-m\mu t}.$$

If all servers are busy and there is still a sufficiently long queue of customers, the stream of customers departing after service will be simple. Indeed, in this case all three conditions (stationarity, absence of aftereffects, and orderliness) are satisfied. The probability that exactly s servers will be freed in the time interval t is (as is easily shown)

$$q_s(t) = e^{-m\mu t}\frac{(m\mu)^s}{s!}.$$

Thus,

$$\mathsf{P}_k\{\gamma > t\} = \sum_{s=0}^{k-m} e^{-m\mu t}\frac{(m\mu)^s}{s!}$$

and hence

$$\mathsf{P}\{\gamma > t\} = P_m e^{-m\mu t}\sum_{k=m}^{\infty}\left(\frac{\rho}{m}\right)^{k-m}\sum_{s=0}^{k-m}\frac{(m\mu)^s}{s!}.$$

But the probabilities P_k are known:

$$P_k = \left(\frac{\rho}{m}\right)^{k-m} P_m,$$

and therefore

$$\mathsf{P}\{\gamma > t\} = P_m e^{-m\mu t}\sum_{k=m}^{\infty}\left(\frac{\rho}{m}\right)^{k-m}\sum_{s=0}^{k-m}\frac{(m\mu)^s}{s!}.$$

The right-hand side becomes, after obvious transformations,

$$\begin{aligned}
\mathsf{P}\{\gamma > t\} &= P_m e^{-m\mu t}\sum_{s=0}^{\infty}\frac{(m\mu)^s}{s!}\sum_{k=m+s}^{\infty}\left(\frac{\rho}{m}\right)^{k-m} \\
&= P_m e^{-m\mu t}\sum_{s=0}^{\infty}\frac{(m\lambda t)^s}{m^s s!}\sum_{k=m+s}^{\infty}\left(\frac{\rho}{m}\right)^{k-m-s} \\
&= \frac{P_m}{1-\frac{\rho}{m}} e^{-m\mu t}\sum_{s=0}^{\infty}\frac{(\lambda t)^s}{s!} = \frac{P_m}{1-\frac{\rho}{m}} e^{-(m\mu-\lambda)t}.
\end{aligned}$$

It follows from (13) and (19) that

$$P_m = \pi\left(1 - \frac{\rho}{m}\right);$$

therefore, for $t > 0$

$$\mathsf{P}\{\gamma > t\} = \pi e^{-(m\mu-\lambda)t}. \tag{22}$$

Clearly, for $t < 0$

$$\mathsf{P}\{\gamma > t\} = 1.$$

The function $P\{\gamma > t\}$ has a discontinuity at $t = 0$; the jump is equal to the probability of finding all the servers occupied.

1.2.7. *The Mean Waiting Time*

Using (22) we can determine all the required numerical characteristics of the waiting time. In particular, the mathematical expectation of the waiting time (usually called the mean waiting time) is equal to

$$a = \mathsf{M}\gamma = -\int_0^\infty t\,d\mathsf{P}\{\gamma > t\} = \pi\int_0^\infty t(m\mu - \lambda)e^{-(m\mu-\lambda)t}\,dt.$$

Simple integration yields

$$a = \frac{\pi}{\mu(m-\rho)}. \tag{23}$$

The variance of the variable γ is equal to

$$\mathsf{D}\gamma = \mathsf{M}\gamma^2 - (\mathsf{M}\gamma)^2 = \frac{\pi(2-\pi)}{\mu^2(m-\rho)^2}.$$

Formula (23) gives the mean waiting time for one customer. We shall now calculate the mean loss of time for the customers arriving at the service system during a time interval of length T. In a time interval T an average of λT customers arrive in the system: the (mean) total loss of time for these customers is equal to

$$a\lambda T = \frac{\pi\lambda T}{\mu(m-\rho)} = \frac{\pi\rho T}{m-\rho}. \tag{24}$$

Some simple arithmetical calculations will demonstrate how rapidly the total time loss increases as ρ changes. We shall confine ourselves to the case $T = 1$ and consider only small values of m: $m = 1$ and $m = 2$. For $m = 1$ we obtain from (20)

$$a\lambda = \frac{\rho^2}{1-\rho}.$$

For $\rho = 0.1, 0.3, 0.5$, and 0.9 the values of $a\lambda$ are approximately equal to 0.011, 0.267, 0.500, 1.633, and 8.100, respectively.

For $m = 2$ we have, in view of (21),

$$a\lambda = \frac{\rho^3}{4 - \rho^2}.$$

For $\rho = 0.1$, 1.0, 1.5, and 1.9 the values of $a\lambda$ are approximately 0.0003, 0.333, 1.350, and 17.587, respectively.

These values clearly illustrate a well-known phenomenon: a high sensitivity to an increase in the load of a service system that is already substantially loaded. A customer is immediately aware of the considerable increase in the waiting time. This fact must definitely be taken into account when designing the load on equipment in queueing systems.

1.2.8. *Example*

We shall now present an application of our results in a small example. This example is theoretical in nature, but there is no difficulty in applying it to real-world situations.

When a sea port is being planned there are several possibilities: (1) constructing two harbors and installing one quay in each, assigning an equal number of ships to each harbor; (2) constructing one harbor with two quays; and, finally, (3) constructing one harbor with one quay and concentrating all loading and unloading installations in it. Which one of the above possibilities is optimal with respect to minimization of time loss due to waiting for loading and unloading?

Processing actual data from several ports has shown that both assumptions—that the ships arrive at the harbor in a simple stream and that the loading and unloading time are exponentially distributed—are valid in the first approximation with sufficient accuracy (Gnedenko and Zubkov (1964)).

Let the intensity of the whole queue of ships be 2λ. In the first plan streams of intensity λ arrive at each of the two harbors; in the other two, streams of intensity 2λ enter the harbors. If the parameter characterizing the unloading rate for the first and second plans equals μ, it is 2μ for the third plan.

Table 1 presents the results of the calculations. The ratio $\rho = \lambda/\mu$ was used as a parameter. Versions 1–3 are shown in the corresponding rows of the table, which indicate the average time loss for all ships arriving in a unit time.

TABLE 1

	ρ					
Version	0.1	0.5	0.6	0.7	0.8	0.9
1	0.0222	1.0000	1.8000	3.2660	6.4000	16.2000
2	0.0020	0.3333	0.6750	1.3451	2.8444	7.6737
3	0.0111	0.5000	0.9000	1.6330	3.2000	8.1000

Comparison shows that the first version is the worst and the second the best. Comparison of the second and third versions must be continued, since building costs must be taken into account.

1.3. Birth and Death Processes

1.3.1. *Definition*

All the problems considered in the previous sections of this book result in strikingly similar systems of differential equations, and the methods of derivation of these equations are almost identical. It is therefore natural to suspect that we have considered two special cases of one general theory. Indeed, probability theory studies a class of Markov processes that include the problems just studied, as well as many others. This class of processes was first studied in connection with the biological formulations of problems of population size, spreading of epidemic diseases, etc. Thus, they were called "birth and death processes." Since the mathematical model used in their investigation is of sufficiently general character, birth and death processes have found wide application in many applied problems outside the field of biology. In particular, numerous problems of reliability theory, for example, redundancy theory, are often considered from the point of view of these processes.

Assume that, at any instant of time, the system under consideration can be in one of finitely or denumerably many states $E_0, E_1, E_2, \ldots$. For various reasons the state of the system changes with time, and during the time interval h the system in state E_n at the instant t passes into state E_{n+1} with probability $\lambda_n h + o(h)$ and into E_{n-1} with probability $\mu_n h + o(h)$. The probability that the system will pass into state E_{n+k} or E_{n-k} (for $k > 1$) during the time interval $(t, t + h)$ is infinitely small compared to h. Hence, the probability that the system will remain in state E_n during the same time interval is $1 - \lambda_n h - \mu_n h + o(h)$. The constants λ_n and μ_n are assumed to be functions of n but independent of t and the route by which the system has reached this state. The theory presented below may actually be extended to the case where λ_n and μ_n are also functions of t.

Random processes of the type just described are *birth and death processes*. If we view E_n as the event that a population consists of n individuals, then the transition $E_n \to E_{n+1}$ means that the population increases by one unit. On the other hand, the transition $E_n \to E_{n-1}$ represents the death of one individual.

If for any $n \geqslant 1$ we have $\mu_n = 0$, that is, if only transitions $E_n \to E_{n+1}$ are possible, the process is called a (pure) *birth process*. If, on the other hand, all $\lambda_n = 0$ $(n = 0, 1, 2, \ldots)$, then the transitions $E_n \to E_{n+1}$ are impossible and we have a *death process*.

The process considered in Section 1.1 is a birth process; there $\lambda_n = \lambda$ for all $n \geqslant 0$.

The service process with waiting, considered in Section 1.2, is a birth and death process. There $\lambda_n = \lambda$ for $n \geqslant 0$, $\mu_k = k\mu$ for $1 \leqslant k \leqslant m$, and $\mu_k = m\mu$ for $k \geqslant m$.

1.3.2. *Differential Equations for the Process*

We denote by $P_k(t)$ the probability that the system is in state E_k at time t. Using familiar arguments we derive the following differential equations:

$$P_0'(t) = -\lambda_0 P_0(t) + \mu_1 P_1(t), \tag{1}$$

and for $k \geqslant 1$

$$P_k'(t) = -(\lambda_k + \mu_k)P_k(t) + \lambda_{k-1}P_{k-1}(t) + \mu_{k+1}P_{k+1}(t). \tag{2}$$

Our notation is somewhat unsatisfactory, since we have not specified the initial state E_i of the system. A comprehensive notation would be $P_{ij}(t)$, that is, the probability that the system is in state E_j at time t, having been in state E_i at time 0. In the problems discussed in Sections 1.1 and 1.2 we assumed that the initial state was E_0.

Equations (2) and (1) are very simple when, for all $k \geqslant 1$, $\mu_k = 0$ (pure birth processes). In this special case we can find all the functions by successive integration. A general solution is easy to derive, and it can be proved that the functions $P_k(t)$ are nonnegative for arbitrary k and t. However, if λ_k increases very rapidly with k increasing, it may be that $\sum_{k=0}^{\infty} P_k(t) < 1$.

1.3.3. *Proof of Feller's Theorem*

The solutions $P_k(t)$ of the equations for a pure birth process satisfy the condition

$$\sum_{k=0}^{\infty} P_k(t) = 1,$$

for all t, if and only if, the series

$$\sum_{k=0}^{\infty} \lambda_k^{-1} \tag{3}$$

diverges.

PROOF. Consider the partial sum

$$S_n(t) = P_0(t) + \cdots + P_n(t). \tag{4}$$

The birth equation implies

$$S_n'(t) = -\lambda P_n(t).$$

Thus,

$$1 - S_n(t) = \lambda_n \int_0^t P_n(t)\, d\lambda. \tag{5}$$

[If instead of the initial condition $P_0(0) = 1$ we assume that for some i $P_i(0) = 1$, this equality applies for $n \geqslant i$.]

Since each term in (4) is nonnegative, the sum $S_n(t)$ does not decrease with the increase of n for any fixed t. Thus, the limit

$$\lim_{n\to\infty} [1 - S_n(t)] = \mu(t) \tag{6}$$

exists.

We find from (5) that

$$\lambda_n \int_0^t P_n(t)\, dt \geqslant \mu(t).$$

Hence, it is evident that

$$\int_0^t S_n(z)\, dz \geqslant \mu(t)\left(\frac{1}{\lambda_0} + \cdots + \frac{1}{\lambda_n}\right).$$

Since for any t and n the inequality $S_n(t) \leqslant 1$ is valid, we have

$$t \geqslant \mu(t)\left(\frac{1}{\lambda_0} + \cdots + \frac{1}{\lambda_n}\right).$$

If the series (3) is divergent, it follows from the last inequality that $\mu(t) = 0$ must be satisfied for all t. Taking (6) into account, we see that the divergence of (3) implies

$$\sum_{k=0}^{\infty} P_k(t) = 1.$$

It follows from (5) that

$$\lambda n \int_0^t P_n(t)\, dt \leqslant 1,$$

and hence,

$$\int_0^t S_n(t)\, dt \leqslant \frac{1}{\lambda_0} + \cdots + \frac{1}{\lambda_n}.$$

In the limit as $n \to \infty$ we obtain

$$\int_0^t [1 - \mu(t)]\, dt \leqslant \sum_{n=0}^{\infty} \lambda_n^{-1}.$$

If $\mu(t) = 0$ for all t, the left-hand side of the inequality equals t, and since t is arbitrary, the series on the right-hand side is divergent. The theorem is thus proved. □

In the theorem in Section 1.1 we dealt with the simplest problem of pure birth; there we had $\lambda_n = \lambda$ for all $n \geqslant 0$. Of course, the series (3) diverges in that case.

It follows from the theorem that when $\lambda_n = n^2, n \geqslant 0, \sum_{n=0}^{\infty} P_n(t) < 1$ necessarily holds. The sum $\sum_{n=0}^{\infty} P_n(t)$ can be viewed as the probability that infinitely many changes of state occur during the time interval t. In radioactive decay such a situation is called an explosion. This theorem is due to Feller (1950).

We now present another proof of Feller's theorem, introducing at the same time the notion of the Laplace-Stieltjes transform, one of the most important analytical tools of queueing theory.

Definition. Let $F(t)$ be a distribution function of a nonnegative random variable ξ. A function of the complex variable s given by

$$\varphi(s) = \int_0^{\infty} e^{-st}\, dF(t)$$

is called the *Laplace-Stieltjes transform* of the function $F(t)$ (or of the random variable ξ).

If the distribution is continuous, that is, if $F(t) = \int_0^t p(x)\,dx$, where $p(x)$ is the probability density, then $\varphi(s) = \int_0^{\infty} e^{-st} p(t)\,dt$. If $F(t)$ is a discrete distribution, namely, $dF(t) = \sum p_k \delta(t - x_k)\,dt$, where $\delta(t)$ is a delta function, then $\varphi(s) = \sum p_k e^{-sx_k}$. Many random variables in queueing theory (for example, waiting time) take on value 0 with probability $F(+0)$ and values in the interval $(x, x + dx)$ with probability $p(x)\,dx$ for $x > 0$. In such a case

$$\varphi(s) = F(+0) + \int_0^{\infty} e^{-st} p(t)\, dt.$$

We now present the basic properties of Laplace-Stieltjes transforms.

1. The function $\varphi(s)$ is defined for any complex number s with a nonnegative real part and is continuous in s for $\operatorname{Re} s \geqslant 0$.
2. In the region $\{\operatorname{Re} s > 0\}$ $\varphi(s)$ is an analytic function.
3. $\varphi(0) = 1$; $|\varphi(s)| \leqslant 1$ for $\operatorname{Re} s \geqslant 0$.
4. $|\varphi(s)| < 1$ for $\operatorname{Re} s > 0$ if $F(+0) < 1$; $\varphi(s) \to 0$ as $\operatorname{Re} s \to \infty$ if $F(+0) = 0$.
5. If $\xi_1, \ldots, \xi_n$ are independent random variables with transforms $\varphi_1(s), \ldots, \varphi_n(s)$, then the transform of $\xi_1 + \cdots + \xi_n$ is $\varphi_1(s) \cdots \varphi_n(s)$.
6. If $M|\xi|^k < \infty$, then

$$\mathsf{M}\xi^k = (-1)^k \varphi^{(k)}(0),$$

   where the derivative is meant to be in the direction of any ray going out from the origin and located on the right half-plane. [The derivative in the usual sense may not exist at the point $s = 0$; for example, it is possible to have $\varphi(-\varepsilon) = \infty$ for any $\varepsilon > 0$.]

Let $s > 0$. We have

$$F(t) = \int_0^t dF(t) \leqslant \int_0^t e^{-s(x-t)}\, dF(x) \leqslant e^{st}\varphi(s). \tag{7}$$

Inequality (7) is the key to the proof of Feller's theorem.

Let t_n be the instant at which the birth process arrives at the state n, $t^* = \lim_{n\to\infty} t_n$. Then $t_n = z_0 + \cdots + z_{n-1}$ where z_k are independent random variables; $\mathsf{P}\{z_k > t\} = e^{-\lambda_k t}$, $t \geqslant 0$. The Laplace–Stieltjes transform of z_k will be

$$\varphi_k(s) = \lambda_k \int_0^\infty e^{-st-\lambda_k t}\, dt = 1/(1 + s/\lambda_k).$$

Whence setting $s = 1$, we obtain from (7)

$$\mathsf{P}\{t^* < t\} \leqslant \mathsf{P}\{t_n < t\} \leqslant e^t \Big/ \prod_{k=0}^{n-1}\left(1 + \frac{1}{\lambda_k}\right) < e^t \Big/ \sum_{k=0}^{n-1} \frac{1}{\lambda_k}.$$

The last expression tends to zero as $n \to \infty$, provided the series (3) diverges, thus $\mathsf{P}\{t^* \leqslant t\} = 0$ for any t. It remains to note that

$$\mathsf{P}\{t^* < t\} = 1 - \sum_{k=0}^{\infty} P_k(t).$$

Now if the series (3) is convergent, then, noting that

$$\mathsf{M}t_n = \frac{1}{\lambda_0} + \cdots + \frac{1}{\lambda_{n-1}},$$

we arrive at

$$\mathsf{M}\{t_n - t_N\} = \frac{1}{\lambda_N} + \cdots + \frac{1}{\lambda_{n-1}} < \varepsilon^2,$$

provided N is sufficiently large. Hence,

$$\mathsf{P}\{t_n - t_N > \varepsilon\} \leqslant \mathsf{M}\{t_n - t_N\}/\varepsilon = \varepsilon$$

and by the continuity axiom

$$\mathsf{P}\{t^* - t_N \geqslant \varepsilon\} \leqslant \varepsilon.$$

Clearly, $\mathsf{P}\{t_N < x\} > 0$ for any $x > 0$ since t_N is a sum of independent random variables with positive densities; whence

$$\mathsf{P}\{t^* < t\} \geqslant \mathsf{P}\{t_N < t - \varepsilon, t^* - t_N \leqslant \varepsilon\} \geqslant \mathsf{P}\{t_N < t - \varepsilon\}(1 - \varepsilon) > 0.$$

1.3.4. *Passive Redundancy without Renewal*

Suppose we have a system consisting of one main element and n identical passive redundant elements. The main element is loaded and during the time interval $(t, t + h)$ may fail with probability $\lambda h + o(h)$. As soon as the main element fails, it is replaced by one of the (identical) redundant elements. The system as a whole breaks down when all the elements (the main and all redundant ones) have failed.

We denote by E_k the event that k elements of the system have failed. At time $t = 0$ the system is in state E_0. We have to determine the probability of state

E_k at time t. The probability of state E_{n+1} at time t is the probability that the system has failed prior to t.

Note that here we have a pure birth process, where $\lambda_k = \lambda$ for $0 \leqslant k \leqslant n$, and $\lambda_k = 0$ for $k > n$. Series (4) is divergent since there are $\lambda_k = 0$. Thus, in this case (3) necessarily holds.

Equations (2) and (3) are in this case of the form:

$$P_0'(t) = -\lambda P_0(t),$$

for $1 \leqslant k \leqslant n$

$$P_k'(t) = -\lambda P_k(t) + \lambda P_{k-1}(t),$$

and for $k = n + 1$

$$P_{n+1}'(t) = \lambda P_n(t).$$

By successive integration and using the initial conditions we arrive at

$$P_0(t) = e^{-\lambda t},$$

$$P_1(t) = \lambda t e^{-\lambda t},$$

$$P_2(t) = \frac{(\lambda t)^2}{2!} e^{-\lambda t},$$

$$\vdots$$

$$P_n(t) = \frac{(\lambda t)^n}{n!} e^{-\lambda t},$$

$$P_{n+1}(t) = 1 - \sum_{k=0}^{n} \frac{(\lambda t)^k}{k!} e^{-\lambda t}.$$

We denote by ξ_k the service life of the kth element under working conditions. It is evident that the service life of the whole system is

$$\xi_1 + \xi_2 + \cdots + \xi_{n+1}.$$

Since the mean sevice life of one element is

$$\int_0^\infty t\lambda e^{-\lambda t}\, dt = \frac{1}{\lambda},$$

the mean service life of the redundant system is $(n + 1)/\lambda$, that is, it is proportional to the total number of elements in the system.

1.3.5. *Active Redundancy without Renewal*

We shall consider yet another simple problem in redundancy theory. We again have n redundant elements, but they are all active, that is, in the same state as the main element. Each of the elements breaks down during the time interval $(t, t + h)$ with probability $\lambda h + o(h)$. The whole system breaks down when all the elements fail. This is again a pure birth process, with $\lambda_k = (n+1-k)\lambda$ for $0 \leqslant k \leqslant n$, $\lambda_{n+1} = 0$.

The equations of the problem are of the form:

$$P_0'(t) = -(n+1)\lambda P_0(t),$$

for $1 \leqslant k \leqslant n$

$$P_k'(t) = -(n+1-k)\lambda P_k(t) + (n-k+2)\lambda P_{k-1}(t),$$

and for $k = n+1$

$$P_{k+1}'(t) = \lambda P_n(t).$$

The solution of these equations yields

$$P_0(t) = e^{-(n+1)\lambda t},$$

$$P_1(t) = (n+1)e^{-n\lambda t}(1 - e^{-\lambda t}).$$

$$\vdots$$

$$P_n(t) = (n+1)e^{-\lambda t}(1 - e^{-\lambda t})^n,$$

$$P_{n+1}(t) = (1 - e^{-\lambda t})^{n+1}.$$

The mean service life of the redundant system is determined as follows: the instants of successive breakdowns are plotted on the time axis, $t_1, t_2, \ldots, t_{n+1}$. Denote $\tau_1 = t_1$, $\tau_2 = t_2 - t_1, \ldots, \tau_{n+1} = t_{n+1} - t_n$. Since $n+1$ elements are operating in the first interval, the probability that a single one of them fails during time t is given by $P_0(t) = e^{-(n+1)\lambda t}$. In the second interval, only n elements are operating. In view of the exponential distribution, the probability that all the elements function without fail for a time t starting from the instant t_1 is $e^{-n\lambda t}$. Finally, in the last interval, only one element functions. The probability that this element functions after the instant of time t_n for a time t is $e^{-\lambda t}$. The mean service life of the system is thus

$$T_2 = \mathsf{M} \sum_{k=1}^{n+1} \tau_k = \sum_{k=1}^{n+1} \mathsf{M}\tau_k = \frac{1}{\lambda}\left(1 + \frac{1}{2} + \cdots + \frac{1}{n+1}\right).$$

For large n we can approximate

$$1 + \frac{1}{2} + \cdots + \frac{1}{n+1} \approx \ln(n+1) + C,$$

where $C = 0.5772157\ldots$ is Euler's constant.

A comparison of the formulas for the mean duration of operation without failure for systems with active and passive redundancy yields

$$\frac{T_1}{T_2} \approx \frac{n+1}{\ln(n+1)}.$$

This increases as the number of redundant elements increases. Thus for $n = 2$, $T_1/T_2 \approx 2.72$, and for $n = 4$, $T_1/T_2 \approx 3.12$.

1.3.6. *Existence of Solutions for Birth and Death Equations*

For pure birth processes, the system (1)–(2) could be solved easily by successive integration, since the differential equations are of the form of recurrence relations. The structure of the general equations for birth and death processes is different, and a successive determination of the functions $P_n(t)$ is not possible. The conditions for the existence and uniqueness of solutions were determined by Feller (1940) and (1957), and Karlin and McGregor (1957a, b) among others. It turns out that the equality

$$\sum_{k=0}^{\infty} P_k(t) = 1$$

is not always satisfied. A sufficient condition for this to be valid is that the series

$$\sum_{k=1}^{\infty} \prod_{i=1}^{k} \frac{\mu_i}{\lambda_i} \tag{8}$$

diverges. If, in addition, the series

$$\sum_{k=1}^{\infty} \prod_{i=1}^{k} \frac{\lambda_{i-1}}{\mu_i} \tag{9}$$

is convergent, the following limits exist:

$$P_k = \lim_{t \to \infty} P_k(t) \qquad (k = 0, 1, \ldots). \tag{10}$$

In particular, this condition is satisfied if starting with some j on, we have

$$\frac{\lambda_k}{\mu_{k+1}} \leqslant \alpha < 1.$$

As a rule, this inequality holds in queueing theory, in particular, the problems with waiting, discussed previously.

These conditions are intuitively obvious; they indicate that the arrival rate of customers in a system should not grow too rapidly in comparison with the increase in the service rate. We have already discussed this in a particular case in Section 1.2.4.

To determine the limits (10) it is sufficient to solve the algebraic system obtained from (1)–(2) setting $P_i'(t) = 0$ and substituting P_i for $P_i(t)$. The resulting system is of the form:

$$\begin{gathered} -\lambda_0 P_0 + \mu_1 P_1 = 0, \\ -(\lambda_k + \mu_k)P_k + \lambda_{k-1} P_k + \mu_{k+1} P_{k+1} = 0 \qquad (k \geqslant 1). \end{gathered} \tag{11}$$

We introduce the notation

$$z_k = -\lambda_k P_k + \mu_{k+1} P_{k+1} \qquad (k = 0, 1, \ldots).$$

Using this notation (11) becomes

$$z_0 = 0, \qquad z_k - z_{k-1} = 0.$$

Hence, for all $k \geqslant 0$

$$z_k = 0.$$

We obtain

$$P_k = \frac{\lambda_{k-1}}{\mu_k} P_{k-1} = \prod_{i=1}^{k} \frac{\lambda_{i-1}}{\mu_i} P_0. \tag{12}$$

The normalizing condition $\sum_{k=0}^{\infty} P_k = 1$ determines P_0:

$$P_0 = \left[1 + \sum_{k=1}^{\infty} \prod_{i=1}^{k} \frac{\lambda_{i-1}}{\mu_i}\right]^{-1}. \tag{13}$$

These calculations were carried out for a special case in Section 1.2.

1.3.7. *Backward Equations*

We mentioned in Section 1.3.2 that the notation $P_k(t)$ for the probability of the state at the instant t is not satisfactory since it does not specify the state of the system at the instant $t = 0$. It would be natural to introduce a more complete notation $P_{ij}(t)$, from which it would follow immediately that the transition probability during time t from state E_i to state E_j equals $P_{ij}(t)$. For brevity, as long as there is no danger of confusion, we have used the simplified notation. If the state at time $t = 0$ is known, $P_{ij}(t)$ represents the absolute probability that the system is in state E_j at some later time t. If we know only the probability distribution π_i of the initial state E_i, the probability of the system being in state E_j at time t is given by the total probability formula as

$$P_j(t) = \sum_i \pi_i P_{ij}(t).$$

The initial conditions for the system of equations (1)–(2) are

$$P_i(0) = \pi_i, \qquad i = 0, 1, 2, \ldots$$

if only the initial distribution of the states of the system is known.

In particular, if the system is initially in state E_i, then

$$P_{ij}(0) = \begin{cases} 1 & \text{for } j = i, \\ 0 & \text{for } j \neq i. \end{cases}$$

Equations (1)–(2) were obtained by comparing the probabilities at times t and $t + h$. We have proceeded, so to speak, from the past to the future. In some cases a different problem is of interest: the state of the system at time t is known. What is the probability that the system reached this state from state E_i?

It is not difficult to set up a new system of equations if we note the corresponding rules of constructing these equations: we fix t and compare the modes of transition from various states to state E_j at the times $t = h$ and $t = 0$.

We present these equations, leaving their derivation as a simple exercise for the reader:

$$P'_{0j}(t) = -\lambda_0 P_{0j}(t) + \lambda_0 P_{1j}(t), \tag{14}$$

for $i \geqslant 1$

$$P_{ij}(t) = -(\lambda_i + \mu_i)P_{ij}(t) + \lambda_i P_{i+1,j}(t) + \mu_i P_{i-1,j}(t). \tag{15}$$

For comparison we rewrite equations (1)–(2), using the new notation:

$$P'_{i0}(t) = -\lambda_0 P_{i0}(t) + \mu_1 P_{i1}(t), \tag{1'}$$

for $i \geqslant 1$

$$P'_{ij}(t) = -(\lambda_j + \mu_j)P_{ij}(t) + \lambda_{j-1} P_{i,j-1}(t) + \mu_{j+1} P_{i,j+1}(t). \tag{2'}$$

These equations are a special case of the well-known (Chapman–) Kolmogorov equations, which govern continuous Markov processes.

As an example, consider a simple problem: $\lambda_i = \lambda$ for $i \geqslant 0$, $\mu_i = 0$; at time t there are n customers in the system. It is necessary to find the probability that at time 0 a certain number of customers are in the system. This is a pure birth process, therefore i can take only the values $0, 1, 2, \ldots, n$. System (14)–(15) becomes

$$P'_{nn}(t) = -\lambda P_{nn}(t),$$

for $0 \leqslant i \leqslant n-1$

$$P'_{in}(t) = -\lambda P_{in}(t) + \lambda P_{i,n-1}(t).$$

The initial conditions of the problem are: $P_{nn}(0) = 1$, $P_{in}(0) = 0$, and $i \neq n$. The solution is given by

$$P_{nn}(t) = e^{-\lambda t}, \qquad P_{n-1,n}(t) = \lambda t e^{-\lambda t}, \ldots.$$

1.4. Applications of Birth and Death Processes in Queueing Theory

1.4.1. *Systems with Losses*

There are m servers, each accessible to any arrival when free and capable of serving only one customer at a time. The service time is random and is exponentially distributed with parameter μ. The customer stream is simple and its parameter is λ. A customer is served immediately, provided that at least one server is free. When all the servers are busy, the customer is refused service and lost. As was already stated in the Introduction, the main characteristic of the quality of service in a system with refusals is the probability of refusal (loss of a customer).

If we denote by E_k the state of the system when it contains k customers, the system can only be in states $E_0, E_1, \ldots, E_m$. The transition probability from state E_k to E_{k+1} (for $k < m$) in time h is $\lambda h + o(h)$, while the probability of remaining in the state E_k (for $k \leqslant m$) during the same time interval is $1 - \lambda h - kh\mu + o(h)$. Finally, the transition probability from state E_k to state E_{k-1} (for $k > 0$) is $k\mu h + o(h)$. These are conditions for a birth and death process where $\lambda_k = \lambda$ for $k < m$ and $\lambda_k = 0$ for $k \geqslant m$, $\mu_k = 0$ for $k = 0$ and $k > m$, and $\mu_k = k\mu$ for $1 \leqslant k \leqslant m$.

It is easy to construct differential equations for the problem by substituting these values for λ_k and μ_k into Eqs. (1)–(2) of Section 1.3.6. The stationary solution is obtained from equation (12) of Section 1.3.6. It is

$$P_k = \frac{\lambda^k}{k!\,\mu^k} P_0 \qquad (1 \leqslant k < m), \tag{1}$$

the value of P_0 is determined by

$$\sum_{k=0}^{m} P_k = 1.$$

Substituting the values of P_k from (1) here, we obtain

$$\begin{aligned} P_0 &= \left[1 + \frac{\lambda}{\mu} + \frac{1}{2!}\left(\frac{\lambda}{v}\right)^2 + \cdots + \frac{1}{m!}\left(\frac{\lambda}{v}\right)^n\right]^{-1} \\ &= \left[1 + \rho + \frac{1}{2}\rho^2 + \cdots + \frac{1}{m!}\rho^n\right]^{-1}, \end{aligned}$$

where $\rho = \lambda/\mu$.

Thus, for $0 \leqslant k \leqslant m$

$$P_k = \frac{\dfrac{1}{k!}\rho^k}{1 + \rho + \dfrac{1}{2}\rho^2 + \cdots + \dfrac{1}{m!}\rho^m}. \tag{2}$$

These formulas were obtained by Erlang and are known as the *Erlang formulas*. Since Erlang's work became known, many attempts have been made in various directions to generalize his results. Later we shall prove a general result due to Sevast'yanov and developed by many researchers, according to which (2) retains its form for problems with losses of any distribution of service time, if only its mean value equals $1/\mu$.

For $k = m$, (2) gives the probability that at the given time all the servers are busy, each customer arriving at this time is refused service. Thus, the probability of refusal is

$$P_m = \frac{\dfrac{1}{m!}\rho^m}{\displaystyle\sum_{k=0}^{m} \frac{1}{k!}\rho^k}. \tag{3}$$

Note that if there are infinitely many servers in the system, then

$$P_0 = e^{-\rho}$$

and hence,

$$P_k = \frac{1}{k}\rho^k e^{-\rho}. \tag{4}$$

This equation is useful for computing the probabilities P_k for large n and for ρ not too large. Note that in (2) and (3) ρ may take any value $\geqslant 0$.

Formulas (2) enable us to obtain the average number of busy servers

$$a_m = \sum_{k=0}^{m} kP_k = \rho(1 - P_m). \tag{5}$$

We illustrate the rate of increase in the probability of losses with an increase in the load (i.e., in ρ) in the following short tables. We shall deal only with the cases of $m = 2$ and $m = 4$, and select values of ρ corresponding to streams of the same intensity arriving at one server:

$m = 2$

ρ	0.1	0.3	0.5	1.0	2.0	3.0	4.0
P_m	0.0045	0.0335	0.0769	0.2000	0.4000	0.5294	0.6054

$m = 4$

ρ	0.2	0.6	1.0	2.0	4.0	6.0	8.0
P_m	0.0001	0.0030	0.0154	0.0952	0.3107	0.4696	0.5746

We observe from these tables that the probabilities of losses are substantially decreased when a large number of servers is lightly loaded. Thus, for $m = 2$ and $\rho = 0.3$ the probability of loss is 0.0335, while for $m = 4$ and $\rho = 0.6$ the probability of loss is only 0.0030, and finally for $m = 6$ and $\rho = 0.9$ the same probability is but 0.0003. However, if the loads are large, the situation is almost the same in all cases and we obtain the following data:

$$m = 2, \quad \rho = 3, \quad P_m = 0.5294, \quad a_m = 0.9412;$$

$$m = 4, \quad \rho = 6, \quad P_m = 0.4696, \quad a_m = 2.1216;$$

$$m = 6, \quad \rho = 9, \quad P_m = 0.4405, \quad a_m = 3.3570.$$

1.4.2. *Systems with Limited Waiting Facilities*

Now assume that in our queueing system there are waiting facilities for customers who encountered all the servers busy. For these customers we have a limited number of waiting places, such as a few chairs in a barber shop or a limited storage space for components waiting their turn to be processed on a machine. Let the number of such facilities be r. If the customer finds at least one free server or a free waiting facility, it remains in the system; if not, it is lost. The other conditions are as in the preceding section.

The conditions here are again those of a birth and death process. Here $\lambda_k = \lambda$ for $0 \leqslant k < m + r$, $\lambda_k = 0$, for $k \geqslant m + r$, $\mu_0 = 0$, $\mu_k = k\mu$ for $1 \leqslant k \leqslant m$; $\mu_k = m\mu$ for $m \leqslant k \leqslant m + r$, $\mu_k = 0$ for $k > m + r$. Equation (12) of Section 1.3.6 yields the formulas:

$$P_k = \frac{\rho^k}{k!} P_0 \qquad (1 \leqslant k \leqslant m), \tag{6}$$

$$P_k = \frac{\rho^k}{m!\, m^{k-m}} P_0 \qquad (n \leqslant k \leqslant m + r), \tag{7}$$

$$P_0 = \left[\sum_{k=0}^{m} \frac{\rho^k}{k!} + \frac{\rho^m}{m!} \sum_{s=1}^{r} \left(\frac{\rho}{m}\right)^s\right]^{-1}. \tag{8}$$

It is easy to see that for $r = 0$ (6)–(8) result in the Erlang formulas, while for $r = \infty$ we obtain (12), (13), and (15) of Section 1.2.4. The probability

$$P_{m+r} = \frac{\rho^{m+r}}{m!\, m^r} \Bigg/ \left(\sum_{k=0}^{m} \frac{\rho^k}{k!} + \frac{\rho^m}{m!} \sum_{s=1}^{r} \left(\frac{\rho}{m}\right)^s\right)$$

is obviously the probability of loss of a customer. The mean number of busy servers in a stationary process is

$$a_{mr} = \sum_{k=1}^{m} kP_k + m \sum_{k=m+1}^{m+r} P_k = \frac{\displaystyle\sum_{k=0}^{m-1} \frac{\rho^k}{k!} + \frac{\rho^m}{(m-1)!} \sum_{s=1}^{r} \left(\frac{\rho}{m}\right)^s}{\displaystyle\sum_{k=0}^{m} \frac{\rho^k}{k!} + \frac{\rho^m}{m!} \sum_{s=1}^{r} \left(\frac{\rho}{m}\right)^s}. \tag{9}$$

The following short tables show the probabilities of customer losses for $m = 2$, $r = 1$ and $m = 4$, $r = 1$ for the same values of ρ as in the preceding subsection:

$m = 2, r = 1$

ρ	0.1	0.3	0.5	1.0	2.0	3.0	4.0
P_{2+1}	0.0002	0.0033	0.0188	0.0909	0.2857	0.4426	0.5477

$m = 4, r = 1$

ρ	0.2	0.6	1.0	2.0	4.0	6.0	8.0
P_{4+1}	0.0000	0.0004	0.0038	0.0454	0.2447	0.4133	0.5345

The tables show that even one waiting facility substantially decreases the probability of customer loss provided the load is not too heavy. It is easy to calculate that, when the waiting facilities are available, the mean load on the servers is increased.

1.4.3. *Distribution of the Waiting Time until the Commencement of Service*

Maintaining the notation of Section 1.2.6, the arguments may be carried over almost verbatim. The following formulas are obtained:

$$\begin{aligned}\mathsf{P}\{\gamma > t\} &= \sum_{k=m}^{m+r-1} P_k \mathsf{P}\{\gamma > t\} = \sum_{k=m}^{m+r-1} P_k \sum_{s=0}^{k-m} q_s(t) \\ &= \sum_{s=0}^{r-1} \sum_{k=s}^{r-1} P_{k+m} q_s(t) = e^{-m\mu t} P_m \sum_{s=0}^{r-1} \frac{(m\mu)^s}{s!} \sum_{k=s}^{r-1} \left(\frac{\rho}{m}\right)^k \\ &= \pi_{mr} e^{-m\mu t} \sum_{s=0}^{m-1} \frac{(m\mu)^s}{s!} \left(\left(\frac{\rho}{m}\right)^s - \left(\frac{\rho}{m}\right)^r\right),\end{aligned} \tag{10}$$

where

$$\pi_{mr} = \frac{P_m}{1 - \dfrac{\rho}{m}}.$$

For $r = \infty$ we again obtain equation (22) of Section 1.2.6.

Simple computations yield the mean waiting time:

$$\mathsf{M}\gamma = \int_0^\infty \mathsf{P}\{\gamma > t\}\, dt = \frac{m\mu P_m}{(m\mu - \lambda)^2}\left[1 - 2\left(\frac{\rho}{m}\right)^r + \left(\frac{\rho}{m}\right)^{r+1}\right]. \tag{11}$$

Note that the statement of the problem under consideration can be generalized naturally as follows: a customer is served on arrival if at least one server is free. If, on the other hand, at the time of arrival, all the servers are busy and there is a queue of $k - m$ customers, the new arrival remains in the queue with probability b_k which depends on the number of customers in the queue. In this form the problem is closer to real-life problems. All of us know from personal experience that we remain in the queue for a barber or in a store with a certain probability depending on the size of the queue. We shall not dwell on the solutions of these new problems here, since the principles involved in their solution are sufficiently clear from the preceding discussion.

1.4.4. *Team Servicing of Machines*

In the 1930s, when automatic machines were introduced, and the number of machines operated by a single worker increased, several interesting problems arose in queueing theory. Even today, these problems, under various conditions, are a topic of serious investigation. We shall consider here a solution of one of them under the assumption proposed by the Swedish researcher, Palm (1947).

A team consisting of r workers operates n identical machines or devices ($r \leqslant n$). Each of these machines may need the attention of a worker at random times. Thus, in the weaving industry there are special crews of repair personnel, each of which services a preassigned number of looms. Assume that each loom is operated by only one worker. The looms break down independently of each other. The probability that a loom functioning at time t requires attention before time $t + h$ is $\lambda h + o(h)$. The probability that a loom in repair at time t will again be operating at time $t + h$ is $\nu h + o(h)$. The parameters λ and ν depend neither on t nor on n nor on the number of looms being repaired.

Denote by E_k the event that at time t, k looms are out of service. Clearly our system can only be in states $E_0, E_1, E_2, \ldots, E_n$. It is easy to verify that this is again a birth and death process, with $\lambda_k = (n-k)\lambda$ for $0 \leqslant k < n$, $\lambda_n = 0$; $\mu_0 = 0$, $\mu_k = k\nu$ for $1 \leqslant k \leqslant r$; and $\mu_k = r\nu$ for $r \leqslant k \leqslant n$. Formulas (12) and (13) of Section 1.3.6 yield: for $1 \leqslant k \leqslant r$

$$P_k = \frac{n!}{k!\,(n-k)!}\left(\frac{\lambda}{\nu}\right)^k P_0, \tag{12}$$

for $r \leqslant k \leqslant n$

$$P_k = \frac{n!}{r^{k-r}\,r!\,(n-k)!}\left(\frac{\lambda}{\nu}\right)^k P_0, \tag{13}$$

and

$$P_0 = \left[\sum_{k=0}^{r} \frac{n!}{k!\,(n-k)!}\rho^k + \sum_{k=r+1}^{n} \frac{n!}{r!\,r^{k-r}(n-k)!}\rho^k\right]^{-1}. \tag{14}$$

In particular, for $r = 1$ and $1 \leqslant k \leqslant n$ we have

$$P_k = \frac{n!}{(n-k)!}\rho^k P_0, \tag{15}$$

$$P_0 = \left[\sum_{k=0}^{n} \frac{n!}{(n-k)!}\rho^k\right]^{-1}. \tag{16}$$

1.4.5. *A Numerical Example*

After having obtained general equations, we can solve numerous special problems that arise in organizing production processes. We shall consider one example.

Two workers service eight machines. How can the work be organized best? There are two possibilities: any free worker should service any of the machines when necessary, or each worker should be assigned four specific machines. We let $\rho = 0.2$. The results of the computation are presented in the following table, where $n = 8$, $r = 2$, and $\rho = 0.2$:

No. of machines out of service	No. of machines awaiting service	No. of idle workers	P_k	No. of machines out of service	No. of machines awaiting service	No. of idle workers	P_k
0	0	2	0.2048	5	3	0	0.0275
1	0	1	0.3277	6	4	0	0.0083
2	0	0	0.2294	7	5	0	0.0017
3	1	0	0.1417	8	6	0	0.0002
4	2	0	0.0687				

The mean number of machines that are idle because the workers are busy with other machines is

$$\sum_{k=2}^{8} (k-2)P_k = 0.3045.$$

In other words all the machines are idle (waiting for service) less than one-third of a working day. The average of all the machines that are idle in a working day, owing to both servicing and waiting for service, is

$$\sum_{k=2}^{8} kP_k = 1.6875.$$

We see that a comparatively small fraction of time is "lost time."

The mean idle time of the workers is

$$0.4096 + 0.3277 = 0.7373.$$

In other words, each worker is idle (not servicing the machines) 0.3686 of the working day. In the following table, $n = 4$, $r = 1$, and $\rho = 0.2$:

No. of machines out of service	No. of machines awaiting service	No. of idle workers	P_k	No. of machines out of service	No. of machines awaiting service	No. of idle workers	P_k
0	0	1	0.1914	3	2	0	0.0760
1	0	0	0.3189	4	3	0	0.0153
2	1	0	0.3984				

The mean loss of working time for each of the four machines (because they are waiting for service) is

$$1 \times 0.1914 + 2 \times 0.0760 + 3 \times 0.0153 = 0.3893.$$

The whole group of eight machines lose 0.7786 of the working day of one machine, so the loss of working time due to waiting increases by a factor of more than 2.5. The overall time loss for servicing and waiting, for each group of four machines, is on the average

$$1 \times 0.3189 + 2 \times 0.1914 + 3 \times 0.0760 + 4 \times 0.0153 = 0.9909.$$

All eight machines thus lose 1.9818 of the working day of one machine. The workers are idle for 0.3984 of the working day. Thus, with this method of organizing service, the machines are idle for a longer time and the workers are less busy.

We shall not deal here with other interesting problems, such as: What is the economically justifiable number of machines to be assigned to each worker; how should the repair of the machines be organized–in the order of machine breakdown or on the distance between the failed machine and the worker, etc?

1.4.6. *Duplicated Systems with Renewal (Passive Redundancy)*

Birth and death processes will now be applied to an important problem of reliability theory studied by Epstein and Hosford (1960). A unit may fail during an operation. To maintain continuous operation, an identical redundant (duplicated) unit is put into operation as soon as the main unit breaks down. The failed unit is immediately repaired, and is as good as new. When not in operation, the unit does not fail or age. Its uninterrupted service time is governed by the distribution $F(x) = 1 - e^{-\lambda x}$. The renewal time is a random variable with distribution $G(x) = 1 - e^{-\nu x}$. The question is, what is the distribution of uninterrupted service time of the duplicated (two-unit) system if the system as a whole breaks down only when both units fail?

The system may be in three states E_0, E_1, and E_2, corresponding to none, one, and two units out of service. The transition probabilities between the states in a time interval h are:

$$P\{E_0(t) \to E_1(t+h)\} = \lambda h + o(h),$$

$$P\{E_1(t) \to E_0(t+h)\} = \nu h + o(h),$$

$$P\{E_1(t) \to E_2(t+h)\} = \lambda h + o(h),$$

$$P\{E_2(t) \to E_1(t+h)\} = o(h).$$

Thus, in the birth and death process under consideration $\lambda_0 = \lambda_1 = \lambda$, $\mu_1 = \nu$, and $\mu_2 = 0$; all other λ_k and μ_k are 0. Equations (1)–(2) in Section 1.3.2 for this problem are

$$P_0'(t) = -\lambda P_0(t) + \nu P_1(t),$$
$$P_1'(t) = -(\lambda + \nu)P_1(t) + \lambda P_0(t),$$
$$P_2'(t) = \lambda P_1(t).$$

The initial conditions are $P_0(0) = 1$, $P_1(0) = 0$, $P_2(0) = 0$.

Substitution of $P_0(t)$ from the second equation of the system into the first yields the following equation for $P_1(t)$;

$$P_1''(t) + (2\lambda + \nu)P_1'(t) + \lambda^2 P_1(t) = 0.$$

The solution satisfying the initial conditions is of the form

$$P_1(t) = Ce^{-(\lambda+\nu/2)t}[e^{\sqrt{\lambda\nu+(\nu^2/4)}t} - e^{-\sqrt{2\nu+(\nu^2/4)}t}].$$

We now obtain

$$P_0(t) = Ce^{-(\lambda+\nu/2)t}\left[\left(\frac{\nu}{2} + \sqrt{\lambda\nu + \frac{\nu^2}{4}}\right)e^{\sqrt{\lambda\nu+(\nu^2/4)}t} + \left(\sqrt{\lambda\nu + \frac{\nu^2}{4}} - \frac{\nu}{2}\right)e^{-\sqrt{\lambda\nu+(\nu^2/4)}t}\right]$$

and

$$C = \frac{\lambda}{\sqrt{4\lambda\nu + \nu^2}}.$$

It is easy to see that the required probability of fail-free service is

$$R(t) = P_0(t) + P_1(t)$$
$$= e^{-(\lambda+\nu/2)t}\left[\cosh\frac{t}{2}\sqrt{4\lambda\nu + \nu^2} + \frac{2\lambda + \nu}{\sqrt{4\lambda\nu + \nu^2}}\sinh\frac{t}{2}\sqrt{4\lambda\nu + \nu^2}\right].$$

For $\nu = 0$ we obtain a system without renewal. In this particular case, as follows from the preceding formula,

$$R(t) = e^{-\lambda t}(1 + \lambda t).$$

This result is of course also derivable from the formulas presented in Section 1.3.4.

The mean duration of fail-free service of a duplicated system is

$$\int_0^\infty R(t)\,dt = \frac{2\lambda + \nu}{\lambda^2} = \frac{2}{\lambda} + \frac{\nu}{\lambda^2}.$$

The first term in this sum represents the mean fail-free service time for a two-unit system without renewal. The second term is an addition due to renewal. The larger the ν, that is, the higher the renewal rate, the greater its effect. Usually ν is substantially larger than λ, in other words, renewal is completed before the work is interrupted, and thus the efficiency due to renewal is usually large.

1.4.7. *Duplicated Systems with Renewal (Active Redundancy)*

We now consider an important and particular case of duplicating the system, in which the reserve unit is in the same state as the working unit. The other conditions are the same as in Section 1.4.6. Here we should set $\lambda_0 = 2\lambda$, $\lambda_1 = \lambda$, $\lambda_2 = 0$, $\mu_1 = \nu$, and $\mu_2 = 0$ in the birth and death equations. The birth and death equations thus become

$$P_0'(t) = -2\lambda P_0(t) + \nu P_1(t),$$

$$P_1'(t) = -(\lambda + \nu)P_1(t) + 2\lambda P_0(t).$$

The standard computations yield the solution

$$P_1(t) = \frac{4\lambda}{\sqrt{\lambda^2 + 6\lambda\nu + \nu^2}} {}^{-(3\lambda+\nu)(t/2)} \sinh\frac{t}{2}\sqrt{\lambda^2 + 6\lambda\nu + \nu^2}$$

and

$$P_0(t) = e^{-(3\lambda+\nu)(t/2)}\left[\cosh\frac{t}{2}\sqrt{\lambda^2 + 6\lambda\nu + \nu^2} + \frac{\nu - \nu\lambda}{\sqrt{\lambda^2 + 6\lambda\nu + \nu^2}}\sinh\frac{t}{2}\sqrt{\lambda^2 + 6\lambda\nu + \nu^2}\right].$$

Thus, the probability of fail-free service is

$$R(t) = e^{-((3\lambda+\nu)/2)t}\left[\cosh\frac{t}{2}\sqrt{\lambda^2 + 6\lambda\nu + \nu^2} + \frac{3\lambda + \nu}{\sqrt{\lambda^2 + 6\lambda\nu + \nu^2}}\sinh\frac{t}{2}\sqrt{\lambda^2 + 6\lambda\nu + \nu^2}\right].$$

In the special case of a system without renewal we obtain

$$R(t) = 2e^{-\lambda t} - e^{-2\lambda t}.$$

The mean fail-free service time is $3/2\lambda + \nu/2\lambda^2$. The second term represents the effect of renewal. For an active redundant unit, this is one-half the effect of a passive one.

1.4.8. *Duplicated Systems with Renewal (Partially Active Redundancy)*

Both types of redundancy may be combined by considering a partially active redundancy, in which the redundant unit may fail at a rate λ_1. If $\lambda_1 = 0$, the unit is passive. If $\lambda_1 = \lambda$, it is active. The parameters of the birth and death equations here are $\lambda_0 = \lambda + \lambda_1$, $\lambda_1 = \lambda$, $\lambda_2 = 0$, $\mu_1 = \nu$, and $\mu_2 = 0$. The differential equations are

$$P_0'(t) = -(\lambda + \lambda_1)P_0(t) + \nu P_1(t),$$
$$P_1'(t) = -(\lambda + \nu)P_1(t) + (\lambda + \lambda_1)P_0(t).$$

Solution of this system yields the following probability of fail-free service:

$$R(t) = e^{-(\lambda+(\lambda_1+\nu)/2)t}\left[\cosh\frac{t}{2}\sqrt{\lambda_1^2 + 2\nu(2\lambda + \lambda_1) + \nu^2}\right.$$
$$\left. + \frac{2\lambda + \lambda_1 + \nu}{\sqrt{\lambda_1^2 + 2\nu(2\lambda + \lambda_1) + \nu^2}}\sinh\frac{t}{2}\sqrt{\lambda_1^2 + 2\nu(2\lambda + \lambda_1) + \nu^2}\right].$$

Simple calculations show that the mean duration of fail-free service is

$$\int_0^\infty R(t)\,dt = \frac{2\lambda + \lambda_1 + \nu}{\lambda(\lambda_1 + \lambda)}.$$

These formulas comparise those obtained in Sections 1.4.6 and 1.4.7 as special cases.

Note that the problems considered in Sections 1.4.6–1.4.8 may be generalized to deal arbitrarily with many operating and redundant units. The solution of the generalized problem by the methods of birth and death processes are basically analogous to the preceding, but are technically more difficult.

1.5. Priority Service

1.5.1. *Statement of the Problem*

There are many real-world situations in which the stream consists of several types of customers. Customers of the first type are served out of turn, provided there is no queue of customers of this type. Customers of the second type have priority over customers of the third and the following types, and so on. A well-known example of such a service is at a telegraph office: urgent cables are transmitted before ordinary ones, even if the ordinary ones are brought in earlier. In the past there were three types of telegrams: express, urgent, and regular. Each of these types has priority over the succeeding one. In the same manner, a long-distance telephone call has priority over a local one in the sense that a long-distance call interrupts a local one.

A great variety of problems arise in investigations of servincing several streams. Indeed, in different real-world conditions there are different possible solutions when a customer of the first type finds all the servers busy. First, the "head-of-the-line" disciplines, when a customer of a higher rank does not interrupt customers already in service, but is placed ahead of all the customers of lower ranks were studied. Next, "preemptive-priority" disciplines, when service is interrupted, were studied. These were in turn subdivided into two categories: in the first one, when the interrupted service is resumed, the time previously spent for its servicing is taken into account (preemptive resume); in the second, this time is lost and the service starts from the beginning (preemptive repeat). Such is the situation when working with computers: here one type of failure only interrupts computation, and it is not necessary after the repair to recompute previous calculations; other failures introduce an error in previously obtained calculations and require carrying out all the calculations from the beginning.

Clearly, when serving several streams, it may be necessary to consider not only service with waiting time but also problems with losses. Recently, in problems of telephone service it became necessary to consider the following problem: two streams arrive at a device, one—the priority stream—is serviced in accordance with the scheme of systems with losses, while the other—without priority—in accordance with a waiting-service scheme. Clearly, all the previously considered statements of problems can be carried over to the case of a stream consisting of customers of different urgency. We are not going to enumerate all the possible formulations of problems and various types of efficiency characteristics of servicing customers belonging to different types.

Here we shall only touch on certain formulations, aiming to demonstrate an approach for their solution in the simplest situations. Analogous problems under more general assumptions are to be considered subsequently, where references will also be given.

1.5.2. *Problems with Losses*

Consider the following problem. Two independent simple streams of customers with parameters λ_1 and λ_2 arrive at m equivalent servers. Each customer that arrives at a time when at least one server is free is being serviced without delay. If a customer of the first type arrives at the system when all the servers are busy but some of them serve customers of type two, then one of the servers is immediately switched to servicing the newly arrived customer of the highest rank, and the customer from the second stream is lost. Thus, the second type of customer is lost, not only when all the servers are busy, but also when at the time of servicing customers of the second type all of the servers are busy and a customer of the first type appears. Customers of the first type can be refused only if all the servers are busy servicing customers of this same type.

Assume that for a customer of the first type the service time is exponential with parameter μ_1 and for the customers of the second type it is exponential with parameter μ_2.

Before proceeding to a formal solution, we note that the first stream is serviced as if the second stream does not exist. This observation allows us to immediately obtain the probability that a customer of the first type is lost. The probability is

$$p_{m0} = \frac{\rho_1^m/m!}{\sum_{i=0}^{m} \frac{\rho_1^i}{i!}}, \qquad \rho_1 = \frac{\lambda_1}{\mu_1}.$$

It is known that if two independent simple streams with parameters λ_1 and λ_2 arrive at the server, the combined stream is also simple with parameter $\lambda_1 + \lambda_2$. Indeed, the probability that, during time interval t, k customers of either type arrive, can be represented by the total probability formula as the sum of products of probabilities, and that during this time interval, s customers of the first type arrive and $k - s$ of the second, summed over all possible values of s. This probability is equal to

$$\sum_{k=0}^{k} \frac{(\lambda_1 t)^s}{s!} e^{-\lambda_1 t} \frac{(\lambda_2 t)^{k-s}}{(k-s)!} e^{-\lambda_2 t}.$$

In accordance with Newton's binomial formula, this sum is

$$\frac{[(\lambda_1 + \lambda_2)t]^k}{k!} e^{-(\lambda_1+\lambda_2)t}.$$

We denote by $p_{ij}(t)$ the probability that, at time t, i servers are occupied serving customers of the first type and j servers are busy with the customers of the second type. Since the total number of busy servers cannot exceed the total number of servers, we have $0 \leqslant i + j \leqslant m$. We set

$$p_{i\cdot}(t) = \sum_{j=0}^{m-i} p_{ij}(t) \qquad \text{and} \qquad p_{\cdot j}(t) = \sum_{i=0}^{m-j} p_{ij}(t).$$

Clearly, $p_{i\cdot}(t)$ and $p_{\cdot j}(t)$ are the probabilities that, at time t, i customers of the first type and j customers of the second type are served respectively.

Recall that p_{m0} is the probability that a customer of the first type who arrived at time t is lost. The sum

$$\sum_{i+j=m} p_{ij}(t)$$

is the probability that a customer of the second type arriving at time t is lost. Thus, the difference

$$\sum_{i+j=m} p_{ij}(t) - p_{m0}(t)$$

represents the probability of losing a customer of type two who is being serviced, provided a customer of the first type arrives at time t.

1.5.3. *Equations for* $p_{ij}(t)$

The derivation of equations for the probabilities $p_{ij}(t)$ will not be given here, since this procedure does not add anything new to what we already know from the theory of birth and death processes. Only the final results will be presented. They are

$$p'_{00}(t) = -(\lambda_1 + \lambda_2)p_{00}(t) + \mu_1 p_{10}(t) + \mu_2 p_{01}(t); \tag{1}$$

for $1 \leqslant i < m$

$$p'_{i0}(t) = -(\lambda_1 + \lambda_2 + i\mu_1)p_{i0}(t) + \lambda_1 p_{i-1,0}(t) + (i+1)\mu_1 p_{i+1,0}(t) + \mu_2 p_{i1}(t), \tag{2}$$

$$p'_{m0}(t) = -m\mu_1 p_{m0}(t) + \lambda_1[p_{m-1,0}(t) + p_{m-1,1}(t)]; \tag{3}$$

for $1 \leqslant j < m$

$$p'_{0j}(t) = -(\lambda_1 + \lambda_2 + j\mu_2)p_{0j}(t) + \lambda_2 p_{0,j-1}(t) + \mu_1 p_{1j}(t) + \mu_2(j+1)p_{0,j+1}(t), \tag{4}$$

$$p'_{0m}(t) = -(\lambda_1 + m\mu_2)p_{0m}(t) + \lambda_2 p_{0,m-1}(t); \tag{5}$$

for $i \geqslant 1, j \geqslant 1, i + j < m$

$$p'_{ij}(t) = -(\lambda_1 + \lambda_2 + i\mu_1 + j\mu_2)p_{ij}(t) + \lambda_1 p_{i-1,j}(t) + \lambda_2 p_{i,j-1}(t) + (i+1)\mu_1 p_{i+1,j}(t) + \mu_2 p_{i,j+1}(t); \tag{6}$$

for $i > 0, j > 0, i + j = m, i \neq m, j \neq m$

$$p'_{ij}(t) = -(\lambda_1 + i\mu_1 + j\mu_2)p_{ij}(t) + \lambda_1[p_{i-1,j}(t) + p_{i-1,j+1}(t)] + \lambda_2 p_{i-1,j}(t). \tag{7}$$

Summation of equations (1), (4), and (7) for all j from 0 to m, numerous cancellations, and the use of the notation introduced in subsection 2, yield

$$p'_{0\cdot}(t) = -\lambda_1 p_{0\cdot}(t) + \mu_1 p_{1\cdot}(t). \tag{8}$$

Summation of eqs. (2), (6), and (7) for all j from 0 to $m - i$, cancellations, and the use of the notation $p_{i\cdot}(t)$, yield the equations for $1 \leqslant i < m$:

$$p'_{i\cdot}(t) = -(\lambda_1 + i\mu_1)p_{i\cdot}(t) + \lambda_1 p_{i-1\cdot}(t) + (i+1)\mu_1 p_{i+1\cdot}(t). \tag{9}$$

Equation (3) can therefore be written in the form

$$p'_{m\cdot}(t) = -m\mu_1 p_{m\cdot}(t) + \lambda_1 p_{m-1\cdot}(t). \tag{10}$$

Equations (8)–(10) differ only in notation from the differential equations of the birth and death processes for the standard problem of losses when only a priority stream is present.

Summation of eqs. (1), (2), and (3) over i, yields the equation

$$p'_{\cdot 0}(t) = -\lambda_2[p_{\cdot 0}(t) - p_{m0}(t)] + \lambda_1 p_{m-1,1}(t) + \mu_2 p_{\cdot 1}(t).$$

Summation of eqs. (4), (6), and (7) over i, yields for $1 \leqslant j < m$,

$$p'_{\cdot j}(t) = -(\lambda_2 + j\mu_2)[p_{\cdot j}(t) - p_{m-j,j}(t)] + \lambda_2[p_{\cdot j-1}(t) - p_{m-j+1,j-1}(t)] + (j-1)\mu_2 p_{\cdot j+1}(t) - j\mu_2 p_{m-j,j}(t) - \lambda_1 p_{m-j,j}(t) + \lambda_1 p_{m-j-1,j+1}(t).$$

Equation (5) can be written in the form:

$$p'_{\cdot m}(t) = -(\lambda_1 + m\mu_2)p_{\cdot m}(t) + \lambda_2[p_{\cdot m}(t) - p_{1,m-1}(t)].$$

These equations show that the situation for the second queue is very different from that of the first: the first queue has a decisive influence on the state of the customers in the second queue. This conclusion is, of course, intuitively obvious.

1.5.4. *A Particular Case*

We shall now study the case $m = 1$ in more detail. Here the computations can be completed without difficulty. The system (1)–(7) in Section 1.5.3 reduces down to the following three equations:

$$p'_{00}(t) = -(\lambda_1 + \lambda_2)p_{00}(t) + \mu_1 p_{10}(t) + \mu_2 p_{01}(t),$$

$$p'_{10}(t) = -\mu_1 p_{10}(t) + \lambda_1[p_{00}(t) + p_{01}(t)],$$

$$p'_{01}(t) = -(\lambda_1 + \mu_2)p_{01}(t) + \lambda_2 p_{00}(t).$$

The solution of this system presents no difficulties; thus, we only present the final results. Here it is assumed that

$$p_{00}(0) = 1,$$

$$p_{10}(0) = 0,$$

$$p_{01}(0) = 0.$$

We have

$$p_{10}(t) = \frac{\lambda_1}{\lambda_1 + \mu_1}(1 - e^{-(\lambda_1+\mu_1)t}),$$

$$p_{01}(t) = \frac{\lambda_2\mu_1}{(\lambda_1 + \mu_1)(\lambda_1 + \lambda_2 + \mu_2)} + \frac{\lambda_1\lambda_2}{(\lambda_1 + \mu_1)(\lambda_2 + \mu_2 - \mu_1)}e^{-(\lambda_1+\mu_1)t}$$
$$- \left\{\frac{\lambda_2\mu_1}{(\lambda_1 + \mu_1)(\lambda_1 + \lambda_2 + \mu_2)} + \frac{\lambda_1\lambda_2}{(\lambda_1 + \mu_1)(\lambda_2 + \mu_2 - \mu_1)}\right\}e^{-(\lambda_1+\lambda_2+\mu_2)t},$$

$$p_{00}(t) = \frac{\mu_1(\lambda_1 + \mu_2)}{(\lambda_1 + \mu_1)(\lambda_1 + \lambda_2 + \mu_2)} + \frac{\lambda_1(\mu_2 - \mu_1)}{(\lambda_1 + \mu_1)(\lambda_2 + \mu_2 - \mu_1)}e^{-(\lambda_1+\mu_1)t}$$
$$+ \frac{\lambda_2}{\lambda_1 + \mu_1}\left(\frac{\mu_1}{\lambda_1 + \lambda_2 + \mu_2} + \frac{\lambda_1}{\lambda_2 + \mu_2 - \mu_1}\right)e^{-(\lambda_1+\lambda_2+\mu_2)t}.$$

Whence

$$p_{0\cdot}(t) = p_{00}(t) + p_{01}(t) = \frac{\mu_1}{\lambda_1 + \mu_1} + \frac{\lambda_1}{\lambda_1 + \mu_1} e^{-(\lambda_1+\mu_1)t},$$

$$p_{1\cdot}(t) = p_{10}(t) = \frac{\lambda_1}{\lambda_1 + \mu_1}(1 - e^{-(\lambda_1+\mu_1)t});$$

$$p_{\cdot 0}(t) = \frac{\lambda_1(\lambda_1 + \lambda_2 + \mu_2) + \mu_1(\lambda_1 + \lambda_2)}{(\lambda_1 + \mu_1)(\lambda_1 + \lambda_2 + \mu_2)} - \frac{\lambda_1\lambda_2}{(\lambda_1 + \mu_1)(\lambda_2 + \mu_2 - \mu_1)} e^{-(\lambda_1+\mu_1)t} + \frac{\lambda_2}{\lambda_1 + \mu_1}\left(\frac{\mu_1}{\lambda_1 + \lambda_2 + \mu_2} + \frac{\lambda_1}{\lambda_2 + \mu_2 - \mu_1}\right) e^{-(\lambda_1+\lambda_2+\mu_2)t},$$

$$p_{\cdot 1}(t) = p_{01}(t) = \frac{\lambda_2\mu_1}{(\lambda_1 + \mu_1)(\lambda_1 + \lambda_2 + \mu_2)} + \frac{\lambda_1\lambda_2}{(\lambda_1 + \mu_1)(\lambda_2 + \mu_2 - \mu_1)} e^{-(\lambda_1+\mu_1)t} - \frac{\lambda_2}{\lambda_1 + \mu_1}\left\{\frac{\mu_1}{\lambda_1 + \lambda_2 + \mu_2} + \frac{\lambda_1}{\lambda_2 + \mu_2 - \mu_1}\right\} e^{-(\lambda_1+\lambda_2+\mu_2)t}.$$

These formulas show that as t tends to infinity the solutions rapidly approach the limiting values

$$p_{00} = \mu_1(\lambda_1 + \mu_2)/[(\lambda_1 + \mu_1)(\lambda_1 + \lambda_2 + \mu_2)],$$

$$p_{01} = p_{\cdot 1} = \lambda_2\mu_1/[(\lambda_1 + \mu_1)(\lambda_1 + \lambda_2 + \mu_2)],$$

$$p_{10} = p_{1\cdot} = \lambda_1/(\lambda_1 + \mu_1),$$

$$p_{0\cdot} = \mu_1/(\lambda_1 + \mu_1),$$

$$p_{\cdot 0} = \frac{\lambda_1(\lambda_1 + \lambda_2 + \mu_2) + \mu_1(\lambda_1 + \mu_2)}{(\lambda_1 + \mu_1)(\lambda_1 + \lambda_2 + \mu_2)}.$$

Suppose a customer of the second type is being served at a given time. What is the probability that the service wil be completed? This occurs if, and only if, its service time ends before a customer of the first type arrives. If the service time is x, the service will be completed if no customer of the first type arrives during time x. The probability of this event is $\mu_2 e^{-\mu_2 x} e^{-\lambda_1 x}\, dx$, and using the total probability formula, we obtain the required probability:

$$\int_0^\infty \mu_2 e^{-(\lambda_1+\mu_2)x}\, dx = \frac{\mu_2}{\lambda_1 + \mu_2}.$$

Thus, the probability that a customer of the second type whose service has already started will not be served completely is $\lambda_1/(\lambda_1 + \mu_2)$.

1.5.5. *The Possibility of Failure of the Servers*

The problems described in this section may be interpreted in a slightly different manner. For numerous problems of practical importance we must take into account failure of the servers themselves. In a telephone exchange a line may be out of order and, to resume normal service, a certain time must be spent on repairs. Loading devices in a harbor may need renewal, and a certain amount of time, which generally depends on the individual case, may be required to prepare them for unloading a ship. In a complex electronic device the failure of one element may cause the whole instrument to fail; consequently, service to the arriving customers must be interrupted until the cause of the failure is discovered and repairs are carried out.

The importance of problems in queueing theory allowing for failure of the servers when a server itself requires service is obvious.

There are numerous formulations of these problems. In some cases, a customer waits for the completion of repairs no matter how long the waiting time is. In other cases, it is immediately lost when a server breaks down. Another possibility is that the holding time of the customer in the service system cannot exceed a certain fixed time τ. Different formulations may also result from other causes. Thus, for example, breakdown can occur either during service time only or during both service time and the idle period. Also, it is possible that the probability of breakdown is greater during the service time than in the idel period. It should be noted that a customer whose service is interrupted may be lost independently of whether or not other free servers are available in the system. Alternatively, a customer may switch from the failed server to any free server. Other situations need not be discussed here. The problem studied in this section may be viewed as a problem of failures in serving mechanisms. Indeed, the stream of failures may be regarded as a stream of the first type; hence, it is evidently a stream with priority. The failure is an emergency, and it cannot be deferred until the service in progress has been completed. Moreover, we have assumed that during the repair no new failures in the server being repaired can occur (a customer of the first type is lost only if the server is busy with another customer of the first type). This type of failure is equally possible in the busy and the free periods. In what follows, we shall return to these problems under more general assumptions.

1.6. General Principles of Constructing Markov Models of Systems

1.6.1. *Homogeneous Markov Processes*

Let X be a finite or countable set and $\xi(t)$ be a homogeneous Markov process with the set of states X and the infinitesimal matrix $\Lambda = \|\lambda_{ij}\|$, whose elements are determined by the relations

$$\mathsf{P}\{\xi(t+\Delta)=j|\xi(t)=i\}=\lambda_{ij}\Delta+o(\Delta),\quad j\neq i, \tag{1}$$

$$\mathsf{P}\{\xi(t+\Delta)=i|\xi(t)=i\}=1-\lambda_i\Delta+o(\Delta)=1+\lambda_{ii}\Delta+o(\Delta), \tag{2}$$

$$\lambda_i=-\lambda_{ii}=\sum_{j\neq i}\lambda_{ij}. \tag{3}$$

The quantity λ_{ij} for $j\neq i$ is called the *transition intensity* of the state i into the state j, and the quantity λ_{ij} is the *exit intensity* from the state i.

As a rule, in real-world cases one can postulate the regularity property of a Markov process: the trajectory of the process $\xi(t)$ is, with probability 1, a step function with a finite number of jumps on any finite interval and, for all t simultaneously, $\xi(t)$ coincides with $\xi(t-0)$ or $\xi(t+0)$, either with the left-hand- or right-hand-side limit. More often one can define the process $\xi(t)$ in such a manner that almost all its trajectories will be right continuous, $\xi(t)=\xi(t+0)$.

Let $\xi(t)$ be a regular Markov process continuous from the right with characteristics (1)–(3). Define $\gamma(t)$ as the number of jumps of the process in the interval $(0,t)$ and denote

$$\zeta(t)=(\xi(t),\gamma(t)).$$

Then clearly

$$p_j(t)=\mathsf{P}\{\xi(t)=j\}=\sum_{m=0}^{\infty}p_j^{(m)}(t), \tag{4}$$

where

$$p_j^{(m)}(t)=\mathsf{P}\{\xi(t)=j,\gamma(t)=m\}.$$

The process $\zeta(t)$ will never return to the initial state. In this sense it is similar to a pure birth process. Equations for its transition probabilities are in the form of recurrence formulas

$$\frac{d}{dt}p_j^{(m)}(t)+\lambda_j p_j^{(m)}(t)=\sum_{i\neq j}\lambda_{ij}p_i^{(m-1)}(t),\quad m\geqslant 1, \tag{5}$$

$$\frac{d}{dt}p_j^{(0)}(t)+\lambda_j p_j^{(0)}(t)=0 \tag{6}$$

and satisfy the initial conditions

$$p_j^{(m)}(0)=0,\quad m\geqslant 1;\qquad p_j^{(0)}(0)=p_j(0)=\mathsf{P}\{\xi(0)=j\}.$$

Equations (5) and (6) are easily solved:

$$p_j^{(m)}(t)=\sum_{i\neq j}\lambda_{ij}\int_0^t p_i^{(m-1)}(t-x)e^{-\lambda_j x}\,dx,\quad m\geqslant 1, \tag{7}$$

$$p_j^{(0)}(t)=p_j(0)e^{-\lambda_j t}. \tag{8}$$

Let t_m be the time of the mth jump of the process for $t > 0$, $\xi_m = \xi(t_m + 0)$, that is, let ξ_m be the state of the process directly after the mth jump. Equations (7) and (8) express the following important properties:

1. For a given state $i = \xi(0)$, the quantity t_1, which equals the duration time of the process in state i, is distributed exponentially with parameter λ_i.
2. If the values $\xi_m = i$ and $t_m = t$ are known, then $t_{m+1} = t + \gamma$, where γ is a random variable exponentially distributed with parameter λ_i, which is independent of the behavior of the process up to time t_m; ξ_{m+1} equals j with probability λ_{ij}/λ_i, $j \neq i$, independently of γ.

1.6.2. *Characteristics of Functionals*

When using Markov models of the service system, both the characteristics of instantaneous values of the process

$$\mathsf{P}\{\xi(t) = j\} = p_j(t), \qquad \mathsf{P}\{\xi(t) \in A\} = \sum_{j \in A} p_j(t),$$

$$\mathsf{M}f(\xi(t)) = \sum_j f(j) p_j(t),$$

as well as the characteristics of the functionals of trajectories are of interest. We shall present a few that are most commonly used.

Let $\tau_j(T)$ be the duration time of the process in the state j during the interval $(0, T)$. Then

$$\mathsf{M}\tau_j(T) = \int_0^T p_j(t)\,dt.$$

Let a Poisson stream with parameter λ be given such that the occurrence of an event in the interval $(t, t + \Delta)$ does not depend on $\xi(t)$, although after this event the state of $\xi(t)$ may change. Denote by $N_j(T)$ the number of events of the stream that encountered the process $\xi(t)$ in the state j. Then

$$\mathsf{M}N_j(T) = \lambda \int_0^T p_j(t)\,dt. \tag{9}$$

Indeed $N_j(T) = N_{jn}(T) + N'_{jn}(T)$, where $N_{jn}(T)$ is the number of events of the stream in the interval $(0, T)$ such that in any interval $((k-1)\Delta, k\Delta)$, where $\Delta = T/n$, up until the time of the event in the stream there were no jumps in the process; $N'_{jn}(T) = N_j(T) - N_{jn}(T)$. We have

$$\begin{aligned} e^{-\lambda\Delta} p_j((k-1)\Delta)(1 - e^{-\lambda\Delta}) &\leqslant \mathsf{M}\{N_{jn}(k\Delta) - N_{jn}((k-1)\Delta)\} \\ &\leqslant p_j((k-1)\Delta)\lambda\Delta. \end{aligned} \tag{10}$$

Summing up with respect to k, we easily obtain

$$\lim_{n\to\infty} \mathsf{M}N_{nj}(t) = \lambda \int_0^T p_j(t)\,dt. \tag{11}$$

Choosing $n = 2^k$, we find that $N'_{jn}(T) \downarrow 0$ with probability 1; since $\mathbf{M}N'_{jn}(T) \leqslant \lambda T$ however, it follows that $\mathbf{M}N'_{jn}(T) \to 0$. Formula (9) is thus completely verified.

Corollary. *The mean number of customers in the interval* (0, *T*) *lost in the system with m servers and r waiting locations equals*

$$\int_0^T p_{m+r}(t)\,dt.$$

If $p_{m+r}(t) \xrightarrow[t\to\infty]{} P_{m+r}$, then the mean value of the quantity over the interval $(0, T)$ tends to P_{m+r} as $T \to \infty$; this justifies the identification of this quantity with the probability of loss of a customer in a stationary mode.

Finally, let $N_j(T, \Delta)$ be a number of intervals in which the process is in the state j and included in the interval $(0, T)$ of a longer duration than Δ. Then

$$\mathsf{M}N_j(T, \Delta) = e^{-\lambda_j \Delta} \sum_i \lambda_{ij} \int_0^{T-\Delta} p_i(t)\,dt. \tag{12}$$

1.6.3. *A General Scheme for Constructing Markov Models of Service Systems*

1.6.3.1. *Choice of the Phase Space X.*

This space is chosen with the stipulation that, based on the value $i \in X$, one can determine which operations are carried out in a given state.

Denote by $|i|$ (the "rank" of i) the total number of operations in state i and call them $O_{i1}, \ldots, O_{i|i|}$. Along with actual operations, this number also includes fictitious ones—the expectation of a certain event. Thus, one can assume that an event of the simplest stream occurs at the time of completion of the "operation" of waiting for this event; the duration of such an operation is distributed exponentially.

1.6.3.2. *Determination of Parameters of Operations.*

Denote by μ_{ij} the parameter of executing the operation O. Thus, if at time t the state of the system is i, then during the time Δ the jth operation may be completed with probability $\mu_{ij}\Delta + o(\Delta)$.

1.6.3.3. *Determination of Possible Transitions after Completion of Operations.*

Denote by $p_{ik}^{(j)}$ the probability that, after completion of operation O_{ij}, the system moves into the state k. (If the transitions are deterministic, then $p_{im}^{(j)} = 1$, $p_{ik}^{(j)} = 0$ for $k \neq m$, where m is the state into which the process proceeds after the completion of operation O_{ij}.)

A Markov model of behavior of the system is thus constructed. Namely, a homogeneous Markov process with the set of states X and transition intensities

$$\lambda_{ik} = \sum_{j=1}^{|i|} \mu_{ij} p_{ik}^{(j)} \tag{13}$$

is determined.

Evidently, the model will be adequate only in the case when the probability of completion of an operation during the time Δ does not depend on the behavior of the system up to time t.

All the preceding examples fit into the scheme just described.

In Sections 1.7 and 1.8 systems with limitations will be considered. These are examples of systems with a simple stream and exponential service time, for which a discrete Markov model is not adequate: here a Markov process with a continuum of states is required.

1.6.4. *The HyperErlang Approximation*

A Markov model can also be used in the case when the defining random variables are distributed according the Erlang distribution, with the density

$$p_r(x; \lambda) = \lambda \frac{(\lambda x)^{r-1}}{(r-1)!} e^{-\lambda x}, \quad x > 0, \quad \text{for } r = 1, 2, \ldots, \lambda > 0.$$

Let, for example, the service activity η follow this distribution. We shall use the method of fictitious phases due to Erlang: we represent η as the sum $\eta_1 + \cdots + \eta_r$, where η_i are independent random variables exponentially distributed with parameter λ. Thus, η is subdivided into r service phases. Denoting by $\gamma(t)$ the number of phases that should take place for the completion of the service starting with time t, we obtain that $\gamma(t)$ is a Markov process with the states $0, 1, \ldots, r$ starting with state r and successively passing to the states $r - 1, \ldots, 1, 0$; the latter means that the service of the given customer has been completed. The transition intensity from k into $k - 1$ equals λ. If the distribution is hyperErlang, having density of the form

$$\lambda \sum_{r=1}^{n} p_r \frac{(\lambda x)^{r-1}}{(r-1)!} e^{-\lambda x}, \quad x > 0, p_r \geqslant 0,$$

then one can use the same method, choosing r randomly with probabilities p_r. If the actual distribution differs from the hyperErlang, it nevertheless could be approximated by this distribution. Various aspects of the hyperErlang approximation are discussed by Marshall and Harris (1976), and Neuts (1975).

1.7. Systems with Limited Waiting Time

1.7.1. *Statement of the Problem*

Assume that the service system consists of m identical servers. A simple stream of customers with parameter λ arrives at the system. The service times have the same probability distribution $H(x) = 1 - e^{-\mu x}$. Any customer arriving at the service system is either immediately served, if at least one server is free, or waits its turn. The waiting, however, is limited by some amount τ. If the customer's service has not begun within time τ after arrival, it is lost. The problem as stated here arises often in real-world situations. Calls arriving at a telephone exchange may have only a limited time available to be put through. Another example is perishable goods that can be stored for only a limited time τ. If they are not sold within this period, they become unfit for consumption. The problem is of even greater importance in the technological, economical, and military fields. As Barrer (1957*a*) points outs: "An attacking airplane engaged by anti-aircraft or guided missiles is available for 'service', i.e. is within range, for only a limited time."

It is evident that the basic characteristics of the quality of service in the problem under consideration are the mean number of losses (or, equivalently, the probability of loss) and the mean waiting time for those customers whose service commences.

In this section we shall study both the case discussed by Barrer (τ = const) and another case of practical interest: τ is a random variable with distribution $G(x) = P\{\tau < x\} = 1 - e^{-\nu x}$.

1.7.2. *The Stochastic Process Describing the State of a System for* τ = const

The method of birth and death processes used until now is not applicable to the problem at hand. Indeed, when τ = const, the number of customers in the system at a given time is no longer even a Markov process. If it is known that at time t there are k customers in the system, then the state of the system at time $t + h$ for any $h > 0$ depends not only on k and t, but also on how long the customers that arrived before time t are waiting. Here we shall use the method proposed by Kovalenko (1960), since Barrer's derivation is not sufficiently rigorous mathematically. Actually, the results obtained by Barrer are correct. The random process we will have to deal with is of a more complex structure.

Consider an m-dimensional stochastic process $\xi(t) = \{\xi_1(t), \dots, \xi_m(t)\}$, where $\xi_i(t)$ is the time that must elapse from the instant t until the ith server completes servicing customers arriving prior to t. If at the time t the ith server is free and there are no customers waiting in the system, then $\xi_i(t) = 0$.

The vector $\xi(t)$ changes with time as follows: all nonzero components are decreased by the amount equal to the time elapsing from t, provided that no new customer arrives and no difference becomes negative. If at time $t_1 > t$ some difference vanishes, then $\xi(t_1) = 0$.

Now consider the behavior of the vector $\xi(t)$ at time s when the first customer arrives after t. In the preceding paragraph we defined the vector $\xi(t)$ for all the values of the argument up to and including s. We shall now define

$\xi(s+0)$. For this purpose we consider the variable $\zeta(t) = \min_i \xi_i(t)$. If $\zeta(s-0) = 0$, the arriving customer is immediately served by a free server. (If more than one is free, the server is chosen at random.) The corresponding ξ_i jumps from 0 (at $s-0$) by the amount equal to the service time for the new arrival, this being defined as the service time at $s+0$. If $\zeta(s-0) \neq 0$, two cases should be considered; $\zeta(s-0) > \tau$ and $\zeta(s-0) \leqslant \tau$. In the first case the arrival of a new customer has no effect on the vector $\xi(s+0)$ since it leaves the system, not waiting to be served. In the second case the component for which $\xi_i(s-0) = \zeta(s-0)$ increases by an amount equal to the service time for the newly arrived customer.

Let the service time for a customer arriving at time s be η. Then all three cases can be combined in the formula

$$\xi_j(s+0) = \xi_j(s-0) + \tfrac{1}{2}[1 + \operatorname{sign}(\tau - \zeta(s-0))]\eta. \tag{1}$$

Indeed, if $\zeta(s-0) = 0$ or $0 < \zeta(s-0) \leqslant \tau$, then $\operatorname{sign}(\tau - \zeta) = 1$ and ξ_i increases by η. However, if $\zeta(s-0) > \tau$, then $\operatorname{sign}(\tau - \zeta(s-0)) = -1$, and in accordance with (1) the jump is zero.

To give a general idea of the process, we shall represent one of its components graphically. Let this component be $\xi_i(t)$. A customer arriving at time t selects the ith server if and only if

$$\xi_i(t-0) = \min_k \xi_k(t-0).$$

Assume that the customers arrive at the ith server at times $t_{i_1}, t_{i_2}, \ldots$, and the required service times are $\eta_{i_1}, \eta_{i_2}, \ldots$, respectively. For definiteness, suppose that at time $t = 0$ the ith server is free. The function $\xi_i(t)$ is zero up to the time t_{i_1}. At t_{i_1} it jumps by η_{i_1} and then it decreases by the amount of time elapsing, as long as the difference is positive or no new customer arrives. If at the instant of arrival of a new customer $\xi_i(t)$ is smaller than τ, it increases by a jump in the corresponding amount η. However, if $\xi_i(t)$ is greater than τ, the customer arriving at that instant is lost.

In Fig. 1 we sketch the process $\xi_i(t)$. At the points $t_{i_1}, t_{i_2}, t_{i_3}$, and t_{i_4} customers arrived for service. The second customer was lost since at the instant of its arrival $\xi_i(t) > \tau$ and thus the duration of waiting for the beginning of the service exceeds τ.

The preceding description shows that the state of the process $\xi(t)$ at time $s > t$ is completely determined by its state at time t. Thus $\xi(t)$ is a Markov process.

1.7.3. *System of Integro-differential Equations for the Problem*

Consider a system of integro-differential equations for $m = 2$. In this particular case the generl principle for derivation of these equations is quite clear, while

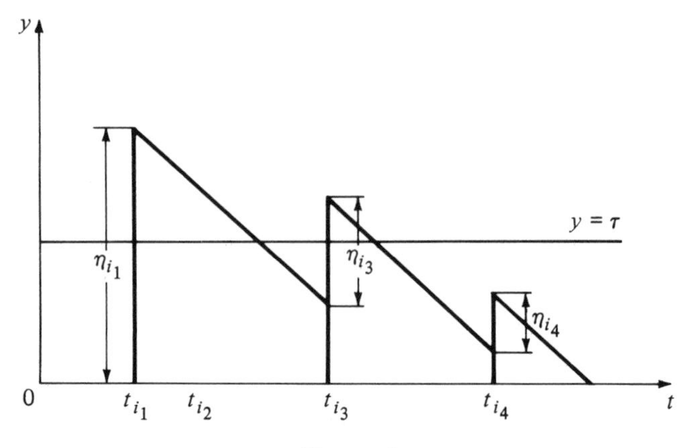

FIGURE 1

the notation is substantially simplified. For an arbitrary m these equations were derived in Kovalenko's paper (1961b).

We assume that the distribution of the process does not depend on t and denote

$$P_0 = \mathsf{P}\{\nu(0) = 0\},$$
$$P_1(x) = \mathsf{P}\{\nu(0) = 1, \xi_1(0) < x\},$$
$$P_2(x, y) = \mathsf{P}\{\nu(0) = 2, \xi_1(0) < x, \xi_2(0) < y\},$$

where $\nu(t)$ is the number of busy servers at time t. We shall assume that the function $P_2(x, y)$ is symmetric; this is justifiable if with the arrival of a customer into the state $\nu(t) = 1$, the enumeration of the variables ξ_1 and ξ_2 is selected randomly. Moreover, we shall assume that

$$\mathsf{P}\{x < \xi_j(0) < x + h\} \leqslant ch,$$
$$\mathsf{P}\{x < \xi_1(0) < x + h, y < \xi_2(0) < y + h\} \leqslant ch^2$$

for all $x \geqslant 0$, $y \geqslant 0$. In view of the above the equalities

$$P_0 = \mathsf{P}\{\nu(h) = 0\} = (1 - \lambda h)\mathsf{P}\{\nu(0) = 0\} + \mathsf{P}\{\nu(0) = 1, \xi_1(0) < h\} + o(h);$$

$$\begin{aligned} P_1(x) &= \mathsf{P}\{\nu(h) = 1, \xi_1(h) < x\} \\ &= (1 - \lambda h)\mathsf{P}\{\nu(0) = 1, h < \xi_1(0) < x + h\} \\ &\quad + 2\mathsf{P}\{\nu(0) = 2, \xi_1(0) < x + h, \xi_2(0) < h\} \\ &\quad + \lambda h \mathsf{P}\{\nu(0) = 0, \eta < x + \theta h\} + o(h) \\ &= (1 - \lambda h)[P_1(x + h) - P_1(h)] + 2P_2(x + h, h) \\ &\quad + \lambda h P_0 B(x + \theta h) + o(h); \end{aligned}$$

$$
\begin{aligned}
P_2(x, y) &= \mathsf{P}\{v(h) = 2, \xi_1(h) < x, \xi_2(h) < y\} \\
&= (1 - \lambda h)\mathsf{P}\{v(0) = 2, h < \xi_1(0) < x + h, h < \xi_2(0) < y + h\} \\
&\quad + \lambda h[\mathsf{P}\{v(0) = 2, \xi_1(0) < \xi_2(0) < y, \xi_1(0) < \tau, \xi_1(0) + \eta < x\} \\
&\quad + \mathsf{P}\{v(0) = 2, \xi_2(0) < \xi_1(0) < x, \xi_2(0) < \tau, \xi_2(0) + \eta < y\} \\
&\quad + \tfrac{1}{2}\lambda h[\mathsf{P}\{v(0) = 1, \xi_1(0) < x, \eta < y + \theta\} \\
&\quad + \mathsf{P}\{v(0) = 1, \xi_1(0) < y, \eta < x + \theta\}] \\
&\quad + \lambda h \mathsf{P}\{v(0) = 2, \tau < \xi_1(0) < x, \tau < \xi_2(0) < y\}] + o(h) \\
&= (1 - \lambda h)[P_2(x + h, y + h) - P_2(h, y) - P_2(x, h)] \\
&\quad + \lambda h\left[\iint\limits_{\substack{u<v<y\\ u<\tau}} (1 - e^{-\mu(x-u)})\, dP_2(u, v)\right. \\
&\quad \left. + \iint\limits_{\substack{u<v<x\\ u<\tau}} (1 - e^{-\mu(y-u)})\, dP_2(u, v)\right] \\
&\quad + \tfrac{1}{2}\lambda h[P_1(x)(1 - e^{-\mu y}) + P_1(y)(1 - e^{-\mu x})] \\
&\quad \lambda h \iint\limits_{\substack{\tau<u<x\\ \tau<v<y}} dP_2(u, v) + o(h)
\end{aligned}
$$

are fulfilled. When writing these relations using the boundedness of the density, we have neglected the probabilities of various "small angles" of the type $\{\xi_1(0) < h,\ \xi_2(0) < h\}$ or $\{x < \xi_1(0) < x + h\}$ in the case when a customer arrives during the time h. In the usual manner we obtain the system of integro differential equations

$$\lambda P_0 = P_1'(0);$$

$$P_1'(x) - P_1'(0) - \lambda P_1(x) + 2\left.\frac{\partial P_2(x, y)}{\partial y}\right|_{y=0} + \lambda P_0(1 - e^{-\mu x}) = 0;$$

$$
\begin{aligned}
&\left(\frac{\partial}{\partial x} + \frac{\partial}{\partial y}\right) P_2(x, y) - \left.\frac{\partial P_2(x, y)}{\partial x}\right|_{x=0} - \left.\frac{\partial P_2(x, y)}{\partial y}\right|_{y=0} - \lambda P_2(x, y) \\
&\quad + \lambda \iint\limits_{\substack{u<v<y\\ u<x\wedge\tau}} (1 - e^{-\mu(x-u)})\, dP_2(u, v) + \lambda \iint\limits_{\substack{u<v<x\\ u<y\wedge\tau}} (1 - e^{-\mu(y-u)})\, dP_2(u, v) \\
&\quad + \frac{\lambda}{2}[P_1(x)(1 - e^{-\mu y}) + P_1(y)(1 - e^{-\mu x})] = 0,
\end{aligned}
$$

where $a \wedge b = \min(a, b)$. Strictly speaking $(\partial/\partial x + \partial/\partial y)P_2$ is just

$$\lim_{h \to 0} \frac{1}{h}[P_2(x + h, y + h) - P_2(x, y)].$$

A solution (P_0, P_1, P_2) corresponds to this system, where

$$P_1(x) = \frac{\lambda}{\mu} P_0(1 - e^{-\mu x}),$$

$$P_2(x, y) = \frac{\lambda^2}{2} P_0 \iint\limits_{\substack{0<u<x \\ 0<v<y}} e^{-\mu(u+v)+\lambda(\tau \wedge u \wedge v)}\, du\, dv,$$

This can easily be verified by direct substitution.

For an arbitrary $m \geqslant 1$ the stationary solution is of the form

$$P_k(x_1, \ldots, x_k) = P_0 \frac{\rho^k}{k!} \prod_{i=1}^{k} (1 - e^{-\mu x_i}), \quad 0 \leqslant k \leqslant m - 1,$$

$$P_m(x_1, \ldots, x_m) = P_0 \frac{\lambda^m}{m!} \int\limits_{\substack{0<u_i<x_i \\ 1 \leqslant i \leqslant m}} \cdots \int e^{-\mu(u_1 + \cdots + u_m) + \lambda \min\{\tau, u_1, \ldots, u_m\}}\, du_1 \cdots du_m,$$

where $\rho = \lambda/\mu$. The constant P_0 is determined by the formula

$$P_0^{-1} = \begin{cases} \displaystyle\sum_{k=0}^{m-1} \frac{\rho^k}{k!} + \frac{\rho^m}{m!} \frac{\lambda e^{-\mu\tau(m-\rho)} - m\mu}{\lambda - m\mu} & \text{for } \lambda \neq m\mu, \\ \displaystyle\sum_{k=0}^{m-1} \frac{m^k}{k!} + \frac{m^m}{m!}(1 + \lambda\tau) & \text{for } \lambda = m\mu. \end{cases}$$

1.7.4. *Various Characteristics of Service*

The intensity of the stream of losses (the average number of losses per unit of time) is of special interest for the system under consideration. Denoting this quantity by λ_0 we obtain

$$\begin{aligned} \lambda_0 &= \lambda \mathsf{P}\{\xi_i > \tau, 1 \leqslant i \leqslant m\} \\ &= \lambda \int_\tau^\infty \cdots \int_\tau^\infty p_m(x_1, \ldots, x_m)\, dx_1 \cdots dx_m \\ &= P_0 \frac{\mu\rho^{m+1}}{m!} e^{-\mu\tau(m-\rho)}. \end{aligned}$$

This formula was derived by Barrer.

There is no difficulty in determining the distribution of the waiting time for the stationary process. First, it is clear that the waiting time is zero if at least one server is free. The probability of this event is

$$\mathsf{P}\{\gamma = 0\} = P_0 \sum_{k=0}^{m-1} \frac{\rho^k}{k!} = a.$$

It is also clear from the conditions of the problem that

$$\mathsf{P}\{\gamma > \tau\} = 0.$$

The probability that the waiting time is τ is equal to

$$\mathsf{P}\{\gamma = \tau\} = \mathsf{P}\{\xi_i > \tau, 1 \leqslant i \leqslant m\} = P_0 \frac{\rho^m}{m!} e^{-\mu\tau(m-\rho)}.$$

Furthermore, for $0 < x < \tau$

$$\mathsf{P}\{\gamma < x\} = a + \mathsf{P}\{\min\{\xi_1, \ldots, \xi_m\} < x\}.$$

Evaluation of the multiple integrals, yields for $0 < x < \tau$ the equalities

$$\mathsf{P}\{\gamma < x\} = \begin{cases} a + \dfrac{m\rho^m P_0}{m!} \dfrac{1 - e^{-\mu(m-\rho)x}}{m - \rho} & \text{for } \lambda \neq m\mu, \\[2ex] a + \dfrac{m^{m+1}}{m!} P_0 \mu x & \text{for } \lambda = m\mu. \end{cases}$$

From here we easily obtain the mean waiting time until the beginning of the service:

$$\mathsf{M}\gamma = \begin{cases} \dfrac{\rho^m}{m!} P_0 \dfrac{m\mu - e^{-(m\mu-\lambda)\tau}[m\mu + \lambda\tau(m\mu - \lambda)]}{(m\mu - \lambda)^2} & \text{for } \lambda \neq m\mu, \\[2ex] \dfrac{m^m}{m!} \tau P_0 \left(1 + \dfrac{\lambda\tau}{2}\right) & \text{for } \lambda = m\mu. \end{cases}$$

1.7.5. *Distribution of the Queue Length*

Let $\nu(t)$ be the number of customers in the system at time t. $\nu(t)$ is not a Markov process. To construct a Markov process additional components should be introduced: if $\nu(t) = m + k$, we denote by $\xi_j(t)$ the time since the arrival of the customer that occupies the jth position in the queue. Then the multidimensional process $\zeta(t) = \{\nu(t); \xi_1(t), \ldots, \xi_k(t)\}$ will be Markovian. This process was studied by Polyaev (1983).* Stationary distribution of the process $\zeta(t)$ is described by the functions

$$p_k = \mathsf{P}\{\nu(t) = k\}$$

and

---

* In Polyaev's paper a more general model is considered: the parameter of the incoming stream depends on the number of customers in the system, while each server has its own servicing parameter; occupancy of the free servers is equally probable.

$$p_{m+k}(x_1,\ldots,x_k)\,dx_1\cdots dx_k = \mathsf{P}\{v(t)=m+k; x_j<\xi_j(t)<x_j+dx_j, 1\leqslant j\leqslant k\}.$$

Polyaev observed that for known values of $v(t)=k>m$ and $\xi_1(t)$ the distribution of the random vector $(\xi_2(t),\ldots,\xi_{k-m}(t))$ can be evaluated from probabilistic considerations. Indeed, customers in the second, ..., $(k-m)$th position in the queue arrived during the time interval $(t-\xi_1(t),t)$ and had no effect on the service process before time t. This clever observation allows us to proceed immediately from a multidimensional Markov process to the process $(v(t),\xi_1(t))$ and thus obtain explicit solutions to the equations.

We present the final result

$$p_k = p_0\rho^k/k!,\quad 0\leqslant k\leqslant m;$$

$$p_{m+k} = \frac{p_0\rho^m\lambda^k}{m!}\prod_{i=1}^{m}(m\mu+c_i)^{-1},\qquad c_i=\frac{\tau^{i-1}e^{-m\mu\tau}}{\int_0^\tau x^{i-1}e^{-m\mu x}\,dx}.$$

This result was obtained earlier by Barrer (1957b) using a heuristic approach.

1.7.6. *Waiting Time Bounded by a Random Variable*

Now consider a somewhat different formulation of the problem. A customer arriving at the system and finding all the servers busy can wait, but the waiting time is bounded by a random variable with distribution $G(x)=P\{\tau<x\}=1-e^{-vx}$. We will show that the situation is that of birth and death processes. Indeed, since the function $G(x)$ possesses the "lack-of-memory" property, the fact that the system is in state E_k at some time t completely determines the probability of the system being in state E_j at time $t+h$. This probability is not affected by our knowing how long the customers in the system have waited prior to time t.

We shall determine λ_k and μ_k for our problem. Since we have here a simple stream, $\lambda_k=\lambda$. If we are in state E_k $(0<k\leqslant m)$, we can pass to state E_{k-1} in time h in only one way: one server completes its service in this time interval. Thus, for $0<k\leqslant m$ we have $\mu_k=k\mu$. If $k=0$, then clearly $\mu_0=0$. If $k>m$, the transition from state E_k to E_{k-1} in time h may proceed in two ways: during this time interval either one of the m servers completes its service or for one of the $k-m$ waiting customers, waiting time expires. Thus, $\mu_k=m\mu+(k-m)v$.

We obtain from (12) of Section 1.3.6. that for $k\leqslant m$

$$p_k=\frac{1}{k!}\left(\frac{\lambda}{\mu}\right)^k p_0,$$

and for $k>m$

$$p_k=\lambda^k p_0\bigg/\left(\mu^m m!\prod_{i=m+1}^{k}[(i-m)v+m\mu]\right)$$

and finally

$$p_0^{-1} = \sum_{k=1}^{m} \left(\frac{\lambda}{\mu}\right)^k + \frac{1}{\mu^m m!} \sum_{k=m+1}^{\infty} \lambda^k \Big/ \prod_{i=m+1}^{\infty} ((i-m)v + m\mu).$$

The system $M|M|m$ with waiting time bounded by an arbitrarily distributed variable was studied by Yurkevich (1970); see also Kovalenko and Yurkevich (1970).

1.8. Systems with Bounded Holding Times

1.8.1. *Statement of the Problem and Assumptions*

No customer arriving at the serving system can be in the system longer than time τ. Thus, there are three nonoverlapping possibilities: waiting time and service time are together less than τ (the customer is completely served); waiting time is less than τ but the time remaining up to τ is insufficient for completion of service and the customer is lost since its service is not completed; the third possibility is that the waiting time is greater than τ and a pure loss occurs with no time being used for service.

Such servicing systems occur quite frequently, but here we shall consider a purely qualitative example. Consider the effective region of some apparatus (e.g., a particle counter). Each customer passing through this region is serviced, provided that the server is free. Each customer stays in the region for a time τ. Beyond the bounds of the region, the server can no longer serve the customer. The server can serve only one customer at a time. After having served one customer, the server proceeds without time loss to another. Each customer is serviced by one server only.

If the customers are serviced in the order of their arrival, no pure losses can occur. Indeed, we have assumed the stream to be orderly and thus any two customers arrive at the system separately. Consider two customers arriving at times t_1 and t_2, $t_1 < t_2$. If $t_2 - t_1 > \tau$, the second customer arrives at the system only when the first has already departed. If, however, $t_2 - t_1 \leqslant \tau$, the first customer leaves the system no later than time $t_1 + \tau$; therefore, the second customer is served at least from time $t_1 + \tau$. (It must leave the system at time $t_2 + \tau$.)

Clearly all three of the previously mentioned cases are possible, if the customers are chosen randomly rather than serviced in the order of their arrival.

We shall consider the case of ordered servicing of a simple stream by m identical servers. The distribution of the waiting time is exponential with parameter μ. Two cases will be considered: $\tau = \text{const}$ and τ is a random variable with distribution $G(x) = 1 - e^{-vx}$.

1.8.2. *A Stochastic Process Describing the Service*

As in Section 1.7, the stochastic process to be considered here is more complicated than a birth and death process for $\tau = \text{const}$. This is because the number of customers in the system at a given time does not at all determine the state of the system at subsequent times. The fate of each customer is to a large extent determined by its time of arrival.

We denote by $\xi_i(t)$ the random function defined as follows: $\xi_i(t) = 0$ if the ith server is free at time t; otherwise, $\xi_i(t)$ is equal to the time that must elapse between t and the time the ith server completes servicing customers who have arrive prior to time t. From the statement of the problem it follows that for any t we have $\xi_i(t) \leqslant \tau$. We present a geometrical representation of the process $\xi_i(t)$. On the abscissa axis we plot the arrival times t and on the ordinate axis Oy the function $\xi_i(t)$. Let t_{i_s} denote the time at which the customers arrive at the ith server. Let the server be free until t_{i_1}. This means that, until t_{i_1}, $\xi_i(t) = 0$ and at t_{i_1}, the functin $\xi_i(t)$ jumps. If the necessary service time η_{i_1} is less or equal to τ, then $\xi_i(t_{i_1} + 0) = \eta_{i_1}$; while if $\eta_{i_1} > \tau$, then $\xi_i(t_{i_1} + 0) = \tau$. As the time increases, the function $\xi_i(t)$ decreases by the length of the period elapsed until it becomes zero or until the next customer arrives. Figure 2 depicts the behavior of the process $\xi_i(t)$.

Consider now the m-dimensional stochastic process

$$\xi(t) = \{\xi_1(t), \ldots, \xi_m(t)\}.$$

It is clear from the definitions that if a new customer arrives at time t, its waiting time until the beginning of service is equal to

$$\zeta(t) = \min_{1 \leqslant i \leqslant m} \xi_i(t).$$

Thus, the process $\xi(t)$ provides the necessary information about the waiting time of the arriving customers. If there are several servers for which $\xi_i(t - 0)$ is minimal, the customer may select any one of them at random.

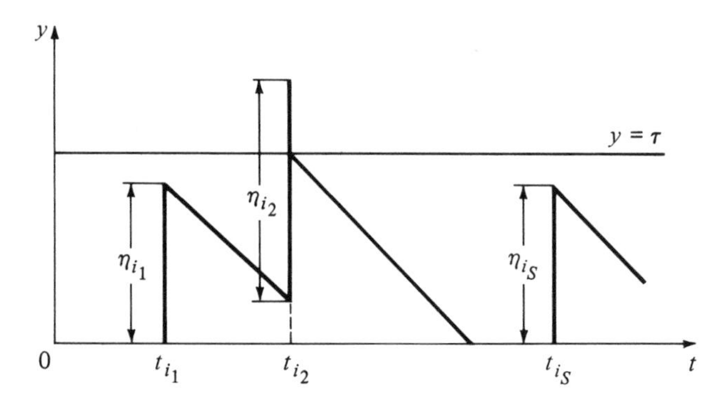

FIGURE 2

The processes $\xi(t)$, defined in Section 1.7 and here, are the same when described verbally. However, the specific features of each problem have a considerable effect on their behavior. This becomes evident on comparing Figs. 1 and 2: in Fig. 1 the function $\xi_i(t)$ can take on any nonnegative value, whereas in Fig. 2 it is bounded from above by τ.

1.8.3. *Stationary Distributions*

Equations for the stationary distribution of the random process $\zeta(t)$ will not be presented here since they are completely analogous to the corresponding equations for the case of a bounded waiting time. The same is valid concerning the method of their derivation. Stationary characteristics of the process $P_k(x_1, \dots, x_k)$, analogous to those introduced in Section 1.7, are of the form

$$P_k(x_1, \dots, x_k) = P_0 \frac{\lambda^k}{k!} \prod_{i=1}^{k} (1 - e^{-\mu x_i}), \quad 0 \leqslant k \leqslant m - 1;$$

$$P_m(x_1, \dots, x_m) = P_0 \frac{\lambda^m}{m!} \underset{\substack{0 < t_i < x_i \\ 1 \leqslant i \leqslant m}}{\int \cdots \int} e^{-\mu(t_1 + \cdots + t_m) + \lambda \min\{t_1, \dots, t_m\}} \, dt_1 \cdots dt_m$$

(in both cases $0 \leqslant x_i \leqslant \tau$).

Among the formulas for various stationary characteristics of servicing derived, based on the previously presented relations, we shall single out the formula for $f(x)$. Here $f(x)\,dx$ for $x > 0$ is the probability that a customer arriving for service will have to wait between x and $x + dx$ unit of time until the commencement of service. The formula is

$$f(x) = \frac{P_0 \lambda^m}{(m-1)!\, \mu^{m-1}} e^{(\lambda - \mu)x} (e^{-\mu x} - e^{-\mu\tau})^{m-1}.$$

The solution, using the method of integro-differential equations, was obtained by Kovalenko (1960). See also Barrer (1957a, 1957b).

1.8.4. *Holding Time in a System Bounded by a Random Variable*

As was mentioned in the Introduction, in many practical problems it is natural to view τ not as a constant (as we have done so far) but as a random variable. Under the assumption that $G(x) = \mathsf{P}\{\tau < x\} = 1 - e^{-\nu x}$, the problem reduces to a birth and death process where $\lambda_k = \lambda$ for $k \geqslant 0$, $\mu_k = k(\mu + \nu)$ for $1 \leqslant k \leqslant m$, and finally, $\mu_k = m\mu + k\nu$ for $k \geqslant m$. Formulas (12)–(13) of Section 1.3.6 yield: for $k \leqslant m$

$$p_k = \frac{1}{k!} \left(\frac{\lambda}{\mu + \nu} \right)^k p_0,$$

for $k \geqslant m$

$$p_k = \frac{\lambda^k p_0}{m!(\mu + \nu)^m \prod_{i=m+1}^{k} (i\nu + m\mu)}, \qquad \sum_{k=0}^{\infty} p_k = 1.$$

NOTES

Practical problems associated with Markov models of queueing systems result in the solution of systems of linear equations of large dimensions. Here general computational methods as well as methods specifically tailored to an analysis of queueing systems are used. See Faddeyev and Faddeyeva (1963), Shneps (1974), Basharin and Gromov (1978), Shwetlick (1979).

Additional literature on systems with restrictions includes Cohen (1968), Yurkevich (1970), Morozov (1977), Shwab (1973), and Kovalenko and Yurkevich (1970), (1972).

For Markov models of queueing systems, numerous optimization problems associated with strategies for controlling the stream of customers, the choice of intensity of service depending on the queue size, and choice of an optimal rule for priority services were solved. The mathematical basis is the theory of optimal control of Markov processes. See Rykov (1966), (1970), and (1975). Principles for solving similar problems by means of the linear programming methodology were detailed in the paper by Mova and Ponomarenko (1971).

2
The Study of the Incoming Customer Stream

2.1. Some Examples

2.1.1. *The Notion of the Incoming Stream*

An important notion in queueing theory now attracting increasing attention is the stream of customers arriving at the servicing system. In Chapter 1 we discussed simple streams, which until recently played the central role in queueing theory. From the viewpoint of a probabilist the simple stream is, although very important, only a particular type of discrete stochastic process. It also has recently become evident that in practical applications the simple stream is not the only one applicable. However, all streams studied until now in queueing theory are discrete stochastic processes of some kind assuming only nonnegative integral values. For the sake of brevity, in what follows we shall use the term *incoming stream*, which will express the law governing arrivals of customers at the servicing system over a period of time.

Denote by $x(t)$ the number of customers arriving at the system between the times 0 and t. If at the time t a new customer arrives at the system, the value of $x(t)$ does not take the customer into account. Thus, $x(t) = x(t-0)$. For each t, the number of customers arriving at time t is the difference $x(t+0) - x(t-0)$. The function $x(t)$ is a step function; its value for any t increases by the number of customers arriving at that time. A simple stream being orderly, it changes by one unit at each jump, and the distances between the jumps are independent random variables distributed according to the distribution $F(x) = 1 - e^{-\lambda x}$. In the general case, to define the stream $x(t)$ we must provide the distribution function of the vectors $(x(t_1), x(t_2), \ldots, x(t_n))$ for any sequence of values $t_1, t_2, \ldots, t_n$ $(n = 1, 2, \ldots)$. In our case, however, we can also proceed by indicating the distribution of other variables.

We assume that our process begins at time $t = 0$, so that $x(-0) = 0$. We denote the arrival times by $\tau_1, \tau_2, \tau_3, \ldots$, and the corresponding numbers of arriving customers by $\nu_1, \nu_2, \nu_3, \ldots$. For a simple stream $\nu_1 = \nu_2 = \nu_3 = \cdots = 1$, but in general they are random variables. Thus, when recording arrivals of boats at a river sluice gate, we must keep in mind not only single boats ($\nu_i = 1$), but also tugboats with barges. The number of barges may also vary, say from 1 to 6 (thus the corresponding v_i will take values between 2 and 7). Similarly, the number of victims of a street accident may be different. We set $z_0 = \tau_1$, $z_k = \tau_{k+1} - \tau_k$ for $k \geqslant 1$.

We shall verify that the incoming stream of customers can be defined in two equivalent ways: (1) as a random stream $x(t)$; (2) as a sequence of distributions of vectors $(z_0 v_1, z_1 v_2, \ldots, z_{n-1} v_n)$ for all $n \geqslant 1$. The latter method is especially appropriate for practical purposes. For streams which are orderly in the sense that at each time τ_k one and only one customer arrives, only the distribution of $(z_0, z_1, \ldots, z_{n-1})$ must be given.

As we know, for a simple stream the lengths of the intervals z_k are independent, and the probability that the length of such an interval exceeds t is given by the formula

$$\mathsf{P}\{z_k > t\} = e^{-\lambda t}.$$

Knowing this distribution, we can easily obtain the probability that k customers arrive in an interval $(t_0, t_0 + t)$. Indeed, under the stipulated conditions, the probability that customers arrive in the intervals $(t_1, t_1 + dt_1)$, $(t_2, t_2 + dt_2)$, ..., $(t_k, t_k + dt_k)$ and none arrive at other points of the interval $(t_0, t_0 + t)$ is

$$e^{-\lambda(t_1 - t_0)} \lambda\, dt_1 e^{-\lambda(t_2 - t_1)} \lambda\, dt_2 \cdots e^{-\lambda(t_k - t_{k-1})} \lambda\, dt_k e^{-\lambda(t_0 + t - t_k)}$$

up to the first-order of accuracy. Hence, the probability that k customers arrive at some instants during this time interval is

$$P_k(t_0, t_0 + t) = e^{-\lambda t} \lambda^k \int_{t_0}^{t_0+t} \int_{t_1}^{t_0+t} \cdots \int_{t_{k-1}}^{t_0+t} dt_k\, dt_{k-1} \cdots dt_1.$$

Simple calculations yield the well-known formula

$$P_k(t_0, t_0 + t) = \frac{(\lambda t)^k}{k!} e^{-\lambda t}.$$

Suppose that only one customer arrives at each time τ_k; since $x(t)$ denotes the number of customers arriving between 0 and t, the inequalities $\tau_k < t$ and $x(t) \geqslant k$ express the same event. Analogously, the events $\tau_r < u_r$ and $x(u_r) \geqslant r$ for $1 \leqslant r \leqslant k$ are equivalent for any k and arbitrary u_1, u_2, ... Thus, the distributions of the vectors $\{\tau_1, \tau_2, \ldots, \tau_k\}$ and $(x(u_1), x(u_2), \ldots, x(u_k))$ can be derived from each other. Since for any $k > 0$

$$\tau_k = \sum_{i=0}^{k-1} z_i,$$

the distribution of either of the vectors $(\tau_1, \tau_2, \ldots, \tau_k)$ and $(z_0, z_1, \ldots, z_{k-1})$ uniquely determines that of the other, and it is irrelevant which distribution of streams is given. In this manner, the equivalence of two approaches to the definition of streams is also checked in the general case.

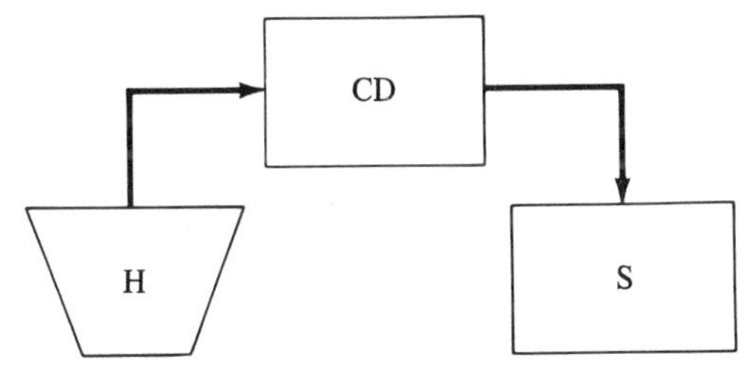

FIGURE 3

2.1.2. *Feed of Components from a Hopper*

Automation of technological processes in the machinery industry, instrument-making, and many other branches of industry has posed interesting problems involving the automatic feeding of component parts (blanks) into machines. One of the most widely and successfully applied solutions of the problem is the system of feeding from hoppers. A simpler version of this system is presented in Fig. 3. A large number of components are stored in bulk in a hopper H. These components must then proceed one by one in oriented positions to a charging device CD, which transfers them to a store S. The store has a definite capacity. When it is filled, it cannot receive more components. Only if the components are appropriately oriented can they enter the containers of the charging device. It may occur that when the container of the charging device passes underneath the output of the hopper, a component, not being in the correct position, does not fall into it. In this case the container will be empty when it reaches the store. The components in the hopper are systematically shaken up, thus changing their orientation, so that when the next container arrives the component will be correctly oriented and can leave the hopper. It may be assumed that there is a probability p that the container underneath the hopper will be filled with components. The problem is: What capacity of the store will ensure a continuous supply to the machine with a sufficiently high probability?

We have here a typical queueing problem that was considered in the previous chapter under different assumptions: A stream of customers arrives at m servers. If at the time a customer arrives at least one server is free, service begins. If, however, all the servers are busy, the new arrival is refused. The probability of refusal for an incoming stream and the servicing process satisfying certain conditions was obtained by Erlang. In this case, the server is a location in the store. If the store is not full, the component proceeds from the container of the charger into the store. If it is full, the new component is refused and remains in the container.

It should especially be noted that while the components arrive at the store in random order, the conditions here are not those of a simple stream. The containers of the charger pass at given times underneath the hopper, and at each passage they may receive a component with probability p. However, this simple process is complicated by the possibility that in the containers there may remain components that were "refused" by the store when it was full. This creates an aftereffect of the past state.

2.1.3. *A Regular Stream of Customers*

Another type of incoming stream occurs often in practice: The customers arrive at equal time intervals. For example, radar signals are transmitted according to this principle. Such a stream is called *regular*.

Assume that the arrival times are $0, h, 2h, \ldots$ and that the service time is exponentially distributed with parameter μ. Also assume that a customer is lost if it finds a busy server. If at some time ih an arriving customer is served, the service will be completed at time $(i + 1)h$ with probability $1 - e^{-\mu h}$. Hence, this is also the probability that a customer arriving at the time $(i + 1)h$ will be served. In this example, the probability that a server will be freed does not depend on the time at which service of the new customer begins. Another possibility is that at the time ih service of the earlier arrival is still in progress. It follows from a well-known property of the exponential distribution that the probability that service of this customer will not be completed at time $(i + 1)h$ is also $e^{-\mu h}$. Thus, the probability of losing a customer, under all conditions, is given by $e^{-\mu h}$. Let n customers appear in a time interval of length t. According to Bernoulli's formulas, the probability that m losses will occur during this time is equal to $P_m(t) = C_n^m e^{-m\mu h}(1 - e^{-m\mu h})^{n-m}$.

The mean number of losses during this time interval is

$$\mathsf{M}\varkappa = ne^{-\mu h},$$

where $\varkappa$ denotes the actual number of lost customers, and the variance is

$$\mathsf{D}\varkappa = ne^{-\mu h}(1 - e^{-\mu h}).$$

Assume now that the service time is a constant τ. It is clear that if $\tau < h$, no customers will be lost; if, however, $\tau > h$, systematical losses will occur. Under the assumptions that $(k - 1)h < \tau < kh$, only each kth customer will be served, and consequently, of each k customers arriving in the system, $k - 1$ will be lost and only one will be served.

2.1.4. *Streams of Customers Served by Successively Positioned Servers*

We shall now consider a stream of customers encountered in numerous specific problems, restricting ourselves to only general considerations. Assume that the service is organized as follows: A customer finding the first server in a sequence busy, proceeds to the second one; if this server is also busy, it goes on to the next server in the sequence. If all the servers are busy, the customer is lost. The simple stream arriving at the first server is thus not completely served. Some of the customers are lost to the first server and proceed to the second. The streams arriving at the second, third, and subsequent servers will no longer be simple. Indeed, suppose the stream arriving at the first server is simple with intensity λ and the service time for each customer is a constant τ. The probability that in time t no customer will arrive at the second server is (as we shall prove) equal to

$$\pi_0(t) = e^{-\lambda t} \sum_{k=0}^{n} \frac{\lambda^k(t - (k-1)\tau)^k}{k!}, \tag{1}$$

where n is determined by the inequality

$$(n-1)\tau \leqslant t \leqslant n\tau.$$

It is clear from (1) that the probability $\pi_0(t)$ cannot be represented for every t as $e^{-\lambda_1}$, where λ_1 is a constant. This implies that the stream arriving at the second server is no longer simple.

We shall prove (1). Assume, for definiteness, that the first server is free at the time $t = 0$. We start with the case $t \leqslant \tau$. Under this assumption $\pi_0(t)$ is obtained directly. Indeed, it is evident that no customer will arrive at the second server if and only if one or zero customers arrive at the first. In accordance with our assumption the probability of this event is

$$\pi_0(t) = e^{-\lambda t} + \lambda t e^{-\lambda t} = e^{-\lambda t}(1 + \lambda t).$$

Let $t \geqslant \tau$ now. One of the following two disjoint events must occur so that not a single customer will arrive at the second server during the time interval $t + h$:

1. There were no losses at the first server during time t and no arrivals during time h.
2. There were no losses at the first server during time $t - \tau$, no arrivals at the first server during the time interval $(t - \tau, t)$, and there was one arrival during the interval from t to $t + h$.

Thus, we obtain the equality

$$\pi_0(t+h) = \pi_0(t)(1 - \lambda h) + \pi_0(t - \tau)e^{-\lambda\tau}\lambda h + o(h),$$

which yields the following differential-difference equation:

$$\pi_0'(t) = -\lambda\pi_0(t) + \lambda e^{-\lambda\tau}\pi_0(t - \tau). \tag{2}$$

By substitution

$$\pi_0(t) = e^{-\lambda t}u(t)$$

eq. (2) is simplified:

$$u'(t) = \lambda u(t - \tau).$$

Since for $0 \leqslant t \leqslant \tau$

$$u(t) = 1 + \lambda t,$$

the function $u(t)$ satisfies the equation

$$u'(t) = \lambda[1 + \lambda(t - \tau)]$$

on the interval $\tau \leqslant t \leqslant 2\tau$.

Integration of this equation with respect to t from $t = \tau$ to t yields

$$u(t) = 1 + \lambda t + \frac{\lambda^2(t-\tau)^2}{2!} = \sum_{k=0}^{2} \frac{\lambda^k(t-(k-1)\tau)^k}{k!}.$$

By induction it is easy to prove that for any $n > 0$ the equation

$$u(t) = \sum_{k=0}^{n} \frac{\lambda^k(t-(k-1)\tau)^k}{k!}$$

holds on the interval $(n-1)\tau \leqslant t \leqslant n\tau$.

Thus, eq. (1) is verified.

2.1.5. *A Wider Approach to the Notion of the Incoming Stream*

In view of the preceding remarks, it is natural to consider the advisability of defining streams in a manner that would be applicable not only to events in time (to which we have restricted ourselves until now, and shall continue to do so in what follows), but also to more general situations.

Let a set of elementary events X be given together with its σ-algebra of measurable subsets $\mathfrak{U}_X$. A *random stream* $\eta(A)$ with the phase space $(X, \mathfrak{U}_X)$ is defined as a system of random variables $\eta(A)$ defined on the elements of the set $\mathfrak{U}_X$ $(A \in \mathfrak{U}_X)$ possessing the following two properties:

1. $\eta(A)$ is an absolutely additive set function.
2. $\eta(A)$ assumes only nonnegative integral values.

For example, in a stream of earthquakes the set X is the set of points (t, φ, θ, h), where $0 \leqslant t < \infty$, $0 \leqslant \varphi < 2\pi$, $0 \leqslant \theta < \pi$, $0 < h < 6000$ km.

An important particular type of random stream of events is the Poisson stream. In accordance with the general approach previously described, a *Poisson stream* must be defined as follows: Let $\eta(A)$ be an absolutely additive nonnegative set function defined on $A \in \mathfrak{U}_X$. For any system $A_1, A_2, \ldots, A_n$ of pairwise disjoint sets, the variables $\eta(A_1), \eta(A_2), \ldots, \eta(A_n)$ are mutually independent, and for any admissible set, the equalities

$$\mathsf{P}\{\eta(A) = k\} = \frac{[\Lambda(A)]^k}{k!} e^{-\Lambda(A)} \quad (k = 0, 1, 2, \ldots)$$

are valid.

Using the arguments of Section 1.1 it is easy to prove that a stream $\eta(A)$ is defined by

$$\mathsf{P}\{\eta(A)=k\}=\frac{(\lambda|A|)^k}{k!}e^{-\lambda|A|}\quad(k=0,1,2,\ldots)$$

(where $\lambda > 0$ is a constant and $|A|$ is the measure of the set A) if and only if the conditions of "stationarity," "orderliness," and "absence of aftereffects" are satisfied. Here a stationary stream is a stream in which the probabilities $P_k(A)$ do not depend on the position of the set A in the space X or on its "shape," but only on k and $|A|$. "Orderliness" means that the condition

$$\lim_{|A|\to 0}\frac{\mathsf{P}\{\eta(A)\geqslant 2\}}{|A|}=0$$

is fulfilled. Finally, absence of aftereffects means the independence of $\eta(A_1)$, $\eta(A_2), \ldots, \eta(A_n)$ for disjoint sets $A_1, A_2, \ldots, A_n$. Evidently, in the general case it is also natural to call such streams simple.

The preceding examples are sufficient to convince us that, aside from simple streams, streams of a more general nature should be studied. Much attention has been devoted recently to these problems. Here we shall present only a few results of theoretical and practical interest obtained recently.

2.1.6. *Marked Streams* [*Franken, Köning, Arndt, and Schmidt* (1984)]

Numerous practical problems where one should distinguish between types of occurring events of a stream lead us to the definition of a marked stream. Let $\{Z, \mathscr{B}\}$ be a measurable space, $\{t_n\}$ be a stream of homogeneous events, and $\{z_n\}$ be a collection of $\mathscr{B}$-measurable random variables with values in Z. Then the sequence of pairs (t_n, z_n) is called a *(random) marked stream*. We shall call t_n the *moment* of the nth event of the stream, and z_n is the *marking* of this event. For example, in priority systems $z_n = 1, 2, \ldots$ may indicate the degree of priority of customers; in a system with a bounded waiting time, one could set $z_n = (\eta_n, \tau_n)$ where η_n is the service time of the nth customer, and τ_n is the bound on its waiting time.

For any $A \in \mathscr{B}$ the sequence of t_n for which $z_n \in A$ represents a stream of homogeneous events in the usual sense.

The notion of a marked stream is very general. Thus, any random process $\xi(t)$ may be represented by a marked stream. To show this, consider an arbitrary stream of homogeneous events (for example, the simplest) associated with the process $\xi(t)$ or independent of it and define z_n as the interval of the trajectory of $\xi(t)$ for $t_n \leqslant t < t_{n+1}$. Then from the stream $\{(t_n, z_n)\}$ the trajectory of the process $\xi(t)$ can be reconstructed.

Based on the marked stream in the monography by Franken *et al.* cited previously, an ergodic theory of queueing systems was developed, and many formulas for the characteristics of specific systems were derived. In particular, a theory that connects stationary characteristics of queueing systems with characteristics determined at the times of a customer's arrival and other events in the systems has been fully developed.

2.2. A Simple Nonstationary Stream

2.2.1. *Definition of a Simple Nonstationary Stream*

In many problems the physical conditions of the arrival of customers are such that the assumptions of orderliness and absence of aftereffects are quite natural. At the same time, the assumption of stationariness is questionable and sometimes totally wrong. This section deals with orderly streams without aftereffects that are, however, nonstationary. Such streams will be called briefly *simple nonstationary streams.*

Since the stream is nonstationary, the probability that k customers arrive during a time interval of length t depends not only on t but also on the initial time t_0 of this interval. Subsequently, we shall denote this probability by $p_k(t_0, t_0 + t)$.

In particular, $p_0(t_0, t_1)$ is the probability that in the interval (t_0, t) no customer arrives, while $p_1(t_0, t)$ is the probability that exactly one customer arrives in (t_0, t).

Consider (as in Section 1.1) the functions

$$\pi_1(t_0, t) = \sum_{k=1}^{\infty} p_k(t_0, t) = 1 - p_0(t_0, t)$$

and

$$\pi_2(t_0, t) = \sum_{k=2}^{\infty} p_k(t_0, t) = 1 - p_0(t_0, t) - p_1(t_0, t).$$

The first defines the probability that at least one customer arrives in the time interval (t_0, t), and the second is the probability that at least two customers arrive in this interval.

The requirement of orderliness, expressed in quantitative—but not descriptive—terms, is that for any constant $t \geqslant 0$ the equality

$$\lim_{h \to 0} \frac{\pi_2(t, t + h)}{h} = 0 \tag{1}$$

is assumed to be valid.

We shall further assume that for any $t \geqslant 0$ the limit

$$\lim_{h \to 0} \frac{\pi_1(t, t + h)}{h} = \lambda(t) \tag{2}$$

exists. The limit is the *instantaneous value of the parameter.*

2.2.2. *Equations for the Probabilities* $p_k(t_0, t)$

To determine the general form of the stream, we must construct equations for the functions $p_k(t_0, t)$. The assumptions stated in the preceding section are quite sufficient for this purpose.

In a time interval $(t_0, t + h)$ no customer arrives if and only if no customers arrive either in (t_0, t) or in $(t, t + h)$. Because of the absence of aftereffects, the relation

$$p_0(t_0, t + h) = p_0(t_0, t)p_0(t, t + h) \tag{3}$$

is valid. In view of condition (2),

$$\pi_1(t, t + h) = \lambda(t)h + o(h)$$

holds for $t > 0$, whence

$$p_0(t, t + h) = 1 - \pi_1(t, t + h) = 1 - \lambda(t)h + o(h). \tag{4}$$

Now (3) may be written in the form

$$p_0(t_0, t + h) - p_0(t_0, t) = -\lambda(t)p_0(t_0, t)h + o(h). \tag{4'}$$

Dividing each term by h and approaching the limit as $h \to 0$ we obtain

$$\frac{\partial p_0(t_0, t)}{\partial t} = -\lambda(t)p_0(t_0, t). \tag{5}$$

Note that the existence of the derivative is ensured by the existence of the limit of the right-hand side of Eq. (4′).

By arguments analogous to those in Section 1.1 for the simple stream, we can derive equations for the probabilities $p_k(t_0, t)$ for $k > 0$. It is evident that under the conditions of the present section,

$$p_k(t_0, t + h) = \sum_{m=0}^{k} p_m(t_0, t)p_{k-m}(t, t + h). \tag{6}$$

Since it follows from (1) and (2) that for small h

$$\pi_2(t, t + h) = o(h)$$

and

$$p_1(t, t + h) = \lambda(t)h + o(h),$$

expression (6) can be rewritten as

$$p_k(t_0, t + h) = p_k(t_0, t)(1 - \lambda(t)h) + p_{k-1}(t_0, t)\lambda(t)h + o(h).$$

Hence

$$\frac{p_k(t_0, t + h) - p_k(t_0, t)}{h} = -\lambda(t)(p_k(t_0, t) - p_{k-1}(t_0, t)) + o(1).$$

Letting $h \to 0$ we arrive at

$$\frac{\partial p_k(t_0, t)}{\partial t} = \lambda(t)[p_{k-1}(t_0, t) - p_k(t_0, t)]. \tag{7}$$

We introduce the function $p_{-1}(t_0, t)$ defined by

$$p_{-1}(t_0, t) = 0.$$

Now relations (4) and (7) may be combined in one equation (7) for all $k \geqslant 0$.

2.2.3. *Solution of the System* (7)

The system (7) is easily solved by the method of generating functions.

Let the random variable ξ take on only the values 0, 1, 2, ..., with probabilities $p_0, p_1, p_2, \ldots$, respectively. The *generating function* of the random variable ξ is the series

$$\varphi(x) = p_0 + p_1 x + p_2 x^2 + \cdots = \sum_{k=0}^{\infty} p_k x^k.$$

This series is convergent for all complex-valued x satisfying $|x| \leqslant 1$; moreover, $\varphi(1) = 1$.

We introduce the notation

$$\varphi(t_0, t, x) = \sum_{k=0}^{\infty} p_k(t_0, t) x^k.$$

Multiply (7) by x^k and sum the resulting equalities over all k from 0 to ∞:

$$\sum_{k=0}^{\infty} x^k \frac{\partial p_k(t_0, t)}{\partial t} = \lambda(t) \sum_{k=0}^{\infty} x^k p_{k-1}(t_0, t) - \lambda(t)\varphi(t_0, t, x).$$

Since

$$\sum_{k=0}^{\infty} x^k p_{k-1}(t_0, t) = x\varphi(t_0, t, x)$$

and

$$\sum_{k=0}^{\infty} x^k \frac{\partial p_k(t_0, t)}{\partial t} = \frac{\partial \varphi(t_0, t, x)}{\partial t},$$

we have

$$\frac{\partial \varphi}{\partial t} = (x - 1)\lambda(t)\varphi.$$

This equation may be written in the form

$$\frac{\partial \ln \varphi(t_0, t, x)}{\partial t} = (x - 1)\lambda(t)\varphi(t_0, t, x).$$

Hence, integrating we have

$$\ln \varphi(t_0, t, x) - \ln \varphi(t_0, t_0, x) = (x - 1) \int_{t_0} \lambda(z)\, dz.$$

Since it follows from the definition of the generating function and the conditions of the problem that

$$\varphi(t_0, t, x) = p_0(t_0, t_0) = 1,$$

we arrive at

$$\ln \varphi(t_0, t, x) = (x - 1) \int_{t_0}^{t} \lambda(z)\, dz.$$

Introducing the notation

$$\Lambda(t_0, t) = \int_{t_0}^{t} \lambda(z)\, dz, \tag{8}$$

we obtain from the preceding that

$$\varphi(t_0, t, x) = e^{(x-1)\Lambda(t_0,t)} = e^{-\Lambda(t_0,t)} \sum_{k=0}^{\infty} \frac{[\Lambda(t_0, t)]^k}{k!} x^k.$$

Comparing this result with the definition of the function $\varphi(t_0, t, x)$, we find that for all $k \geqslant 0$

$$p_k(t_0, t) = \frac{[\Lambda(t_0, t)]^k}{k!} e^{-\Lambda(t_0,t)}. \tag{9}$$

We thus see that even in the nonstationary case an orderly stream without aftereffects is a Poisson process; however, here the parameter of the stream $\lambda(t)$ is no longer a constant.

It is easy to verify that the function $\Lambda(t_0, t)$ gives the mean number of customers arriving during the time interval (t_0, t).

2.2.4. *Instantaneous Intensity of a Stream*

The mathematical expectation of the number of customers arriving during (t_0, t) is equal to

$$\mu(t_0, t) = \sum_{k=0}^{\infty} k p_k(t_0, t) = \Lambda(t_0, t).$$

The mean intensity in the interval (t_0, t) is

$$\frac{\mu(t_0, t)}{t - t_0} = \frac{1}{t - t_0} \int_{t_0}^{t} \lambda(z)\, dz.$$

If the instantaneous parameter $\lambda(t)$ is a constant $\lambda(t) = \lambda$ in the interval (t_0, t), the mean intensity of the stream is also constant and equals λ.

We define the *instantaneous intensity* of the stream $\mu(t_0)$ as the limit

$$\mu(t_0) = \lim_{t \to t_0} \frac{\mu(t_0, t)}{t - t_0}.$$

By the well-known properties of a definite integral at the points of continuity of $\lambda(t)$:

$$\mu(t_0) = \lim_{t \to t_0} \frac{1}{t - t_0} \int_{t_0}^{t} \lambda(z)\, dz = \lambda(t_0).$$

Thus, *the instantaneous intensity of an orderly stream without aftereffects coincides with the instantaneous value of the parameter.* This result was obtained in Section 1.1 in the particular case where we assumed in addition that the stream is stationary.

If it is assumed that $\lambda(t)$ is only integrable, it follows from the theory of the Lebesgue integral that the last result is valid almost everywhere.

We shall make two additional remarks.

It follows from the definition of the function $\Lambda(t_0, t)$ that, if $t_0 < \tau < t$, then

$$\Lambda(t_0, t) = \Lambda(t_0, \tau) + \Lambda(\tau, t).$$

Without altering the calculations, we can generalize the theorem proved in Section 1.1.7. The following result holds under the conditions of the present section.

If it is known that in the interval (t_0, t) n customers of an orderly stream without aftereffects have arrived, then each of them is distributed over the interval independently of the others, and the probability that the customer will arrive in the interval (a, b) $(t_0 \leqslant a < b \leqslant t)$ is equal to $\Lambda(a, b)/\Lambda(t_0, t)$.

2.2.5. *Examples*

Very frequently, streams have pronounced periodicity. This occurs, for instance, in streams of calls at telephone exchanges, of cargo boats, and of calls to first aid stations. In the first case this periodicity is manifested at certain times of the day, in the second yearly, and in the third daily and during the week. Streams of customers at stores also have pronounced periodicity. From practical aspects it is very important to take these features into account, since this allows us a timely adoption of measures ensuring efficient service.

For example, let

$$\lambda(t) = 2\lambda \sin^2 at = \lambda(1 - \cos 2at).$$

As it is easily seen, the mean intensity of the stream in the interval (t_0, t) is

$$\mu(t_0, t) = \lambda \left[1 - \frac{\sin a(t - t_0)}{a(t - t_0)} \cos a(t + t_0) \right].$$

In this case the mean intensity tends to the constant λ as the length of the interval increases to infinity. This conclusion is valid for any periodic function $\lambda(t)$.

As a second example, consider a parameter $\lambda(t)$ that tends to a constant λ with an increase of t. Such streams occur frequently in practice. It should be noted that for these streams all the results of Chapter 1 are valid.

It is clear that the study of general, simple nonstationary streams involves substantial analytical difficulties. However, the system of differential equations with which we dealt in Chapter 1 may be constructed without difficulty for any simple nonstationary stream as well. We leave it to the reader to verify this on his or her own.

2.2.6. *The General Form of Poisson Streams without Aftereffects*

It is natural to consider the genera form of all *Poisson streams without aftereffects*. Can they be reduced to simple nonstationary streams? It is intuitively clear that this is not the case, since for such streams the function $\Lambda(t) \equiv \Lambda(0, t)$ is abolutely continuous (as, the integral of a function). It is natural to expect that an arbitrary Poisson stream without aftereffects will have the following gneeral form: Let $\Lambda(t)$ be a nondecreasing function of t [$\Lambda(t) = 0$ for $t \leqslant 0$]. The probability of the arrival of k ($k \geqslant 0$) customers in a time interval $\alpha \leqslant t < \beta$ is calculated by the formula

$$p_k(\alpha, \beta) = \frac{[\Lambda(\beta) - \Lambda(a)]^k}{k!} e^{-[\Lambda(\beta) - \Lambda(\alpha)]}.$$

The structure of Poisson streams without aftereffects was investigated by Khinchin (1963). Here we shall only state his result.

Denote the discontinuity points of the function $\Lambda(t)$ by $t_1, t_2, \ldots$. These points will play an essential role in what follows. A stream is called a *singular Poisson stream* if it has the following properties:

1. The events of the stream may occur only at preassigned times t_i ($i = 1, 2, \ldots$) and the number of events occurring at different times t_i are mutually independent random variables.
2. The probability of the occurrence of k events at time t_i is equal to

$$e^{-\alpha_i} \frac{\alpha_i^k}{k!} \quad (\alpha_i > 0, k = 0, 1, 2, \ldots).$$

3. The series

$$c(t) = \sum_{0 \leqslant t_i < t} \alpha_i$$

is convergent for any $t > 0$.

It is easy to verify that the generating function of this stream is of the form

$$\Gamma(x;\alpha,\beta) = \exp[c(\beta) - c(\alpha)](x-1).$$

For a given function $\Lambda(t)$ we determine the discontinuity points $t_1, t_2, \ldots$, the jumps $\alpha_i = \Lambda(t_i + 0) - \Lambda(t_i - 0)$, and the function

$$c(t) = \sum_{0 \leqslant t_i < t} \alpha_i; \qquad \Lambda_1(t) = \Lambda(t) - c(t).$$

The function $\Lambda_1(t)$ is continuous for all t. We construct a stochastic process with the generating function

$$\Phi_1(x;\alpha,\beta) = \exp[\Lambda_1(\beta) - \Lambda_1(\alpha)](x-1).$$

This is a Poisson process without aftereffects. It is orderly in the sense that for any $t > 0$ and $\varepsilon > 0$ a value of $\delta > 0$ can be found such that, for $\alpha < \beta < \alpha + \delta < t$, the inequality

$$\pi_2(\alpha,\beta) \leqslant \varepsilon\pi_1(\alpha,\beta)$$

is valid. Khinchin called such processes *regular processes.*

Thus, we see that any Poisson process without aftereffects is the sum of two independent Poisson streams without aftereffects, one regular and the other singular. As Khinchin has shown, this property characterizes Poisson processes.

We present a method for constructing a nonstationary Poisson process from a stationary simple process using time substitution. First, let $\Lambda(t)$ be a strictly increasing continuous function. We associate variable t with variable s by the relation $s = \Lambda(t)$, $t = \Lambda^{-1}(s)$ and consider in the s-time the simple stream $\{s_n\}$ with parameter 1. Then, as it is easy to see, the stream $\{t_n\}$ with $t_n = \Lambda^{-1}(s_n)$ is a Poisson stream with the mathematical expectation $\Lambda(t)$ of the number of events in the interval $(0, t)$. This conclusion is valid also for an arbitrary nondecreasing function $\Lambda(t)$ by setting

$$t_n = \sup\{t\colon \Lambda(t) \leqslant s_n\}.$$

Nonstationary simple streams appear in many models associated with random walks. Thus, Gergely and Ezhov (1975) observe that if $\{\xi_n\}$ are independent random variables with the distribution function $1 - \exp\{-\Lambda(t)\}$, where $\Lambda(t)$ is a continuous function, then the set of values of ξ_n each one of which is larger than $\xi_1, \ldots, \xi_{n-1}$ form a stream of this type with the leading function $\Lambda(t)$.

2.2.7. *A System with Infinitely Many Servers*

Utilizing the properties of a simple stream, one can study the characteristics of a queueing system with infinitely many servers with a general distribution function $B(x)$ of duration of service.

Let $\nu(t)$ be the number of customers in the system at time t, and $\nu_T(t)$ be the analogous variable in the case when $n = \lambda T$ customers appear independently at the times that are uniformly distributed on the interval $(0, T)$ where λT is an integer. It is easily seen that $\nu_T(t)$ is the number of successes in n independent trials with the probability of success

$$p = \frac{1}{T}\int_0^t \bar{B}(x)\,dx$$

[Here dx/T is the probability that a customer arrives in the interval $(t - x - dx, t - x)$.] From the Poisson limit theorem

$$\mathsf{P}\{\nu_T(t) = k\} \sim e^{-np}\frac{(np)^k}{k!} \xrightarrow[T\to\infty]{} e^{-\lambda\int_0^t \bar{B}(x)dx}\frac{\left(\lambda\int_0^t \bar{B}(x)\,dx\right)^k}{k!}.$$

It is easy to verify that this limit is actually $\mathsf{P}\{\nu(t) = k\}$. If

$$\tau = \int_0^\infty \bar{B}(x)\,dx < \infty,$$

then

$$\lim_{t\to\infty} \mathsf{P}\{\nu(t) = k\} = e^{-\lambda\tau}\frac{(\lambda\tau)^k}{k!}.$$

Systems with infinite numbers of servers, which are approximations of real-world systems, are discussed in Franken and Kerstan (1968), Smith (1972), Annayev and Manilov (1975), Daley (1976), and Veretennikov (1977).

2.3. A Property of Stationary Streams

2.3.1. *Existence of the Parameter*

It was proved in Section 1.1 that

$$p_0(t) = 1 - \lambda t + o(t)$$

(where λ is a constant) holds for any stationary stream. Hence, it follows that for any stationary stream without aftereffects, the limit

$$\lim_{t\to 0}\frac{\pi_1(t)}{t} = \lambda$$

exists. This limit will be called the *parameter of a stream.*

It was shown by Khinchin (1955) that the assumption of the absence of aftereffects is redundant for the existence of the parameter of a stream; we used this assumption because of the method of derivation and not because of the essence of the problem.

In this section we shall prove the following theorem.

Theorem. *For any stationary stream, the limit*

$$\lim_{t \to 0} \frac{\pi_1(t)}{t} = \lambda > 0$$

exists, where the case $\lambda = +\infty$ is also possible.

The proof is based on an elementary analytical result which we shall now state and prove.

2.3.2. *A Lemma*

Lemma. *Let a function $f(x)$ be nonnegative and nondecreasing in the interval $0 \leqslant x \leqslant a$, and furthermore, let*

$$f(x + y) \leqslant f(x) + f(y) \tag{1}$$

for all x and y belonging, together with the sum $x + y$, to the interval $0 < x \leqslant a$. Then, as $x \to 0$, $f(x)/x$ tends either to infinity or to a finite limit. The limit is zero only if $f(a) = 0$.

PROOF. It follows from (1) that for any positive integer m and an arbitrary x $(0 \leqslant x \leqslant a)$:

$$f(x) \leqslant mf\left(\frac{x}{m}\right). \tag{2}$$

In particular, for $x = a$

$$\frac{f\left(\frac{a}{m}\right)}{\frac{a}{m}} \geqslant \frac{f(a)}{a} \quad (m = 1, 2, 3, \ldots).$$

If $f(a) \neq 0$, then

$$\alpha = \overline{\lim_{x \to 0}} \frac{f(x)}{x} \geqslant \frac{f(a)}{a} > 0,$$

where clearly the case $\alpha = +\infty$ is also possible.

First consider the case $\alpha < +\infty$. Then, for a given $\varepsilon > 0$ there exists a $c > 0$ such that the inequality

$$\frac{f(c)}{c} > \alpha - \varepsilon$$

is fulfilled.

Now let x be in the interval $0 < x < c$. Define $m \geqslant 2$ to be an integer such that

$$\frac{c}{m} \leqslant x < \frac{c}{m - 1}.$$

In view of the monotonicity of the function $f(x)$ and the choice of m,

$$\frac{f(x)}{x} \geqslant \frac{f\left(\frac{c}{m}\right)}{\frac{c}{m-1}}.$$

However, since by (2)

$$f(x) \leqslant mf\left(\frac{c}{m}\right),$$

we have

$$\frac{f(x)}{x} \geqslant \frac{m-1}{m}\frac{f(c)}{c}.$$

Thus, in view of the choice of c,

$$\frac{f(x)}{x} \geqslant \left(1 - \frac{1}{m}\right)(\alpha - \varepsilon).$$

Since ε can be selected arbitrarily small and $m \to \infty$ as $x \to 0$, we obtain

$$\lim_{x \to 0} \frac{f(x)}{x} = \alpha.$$

Consider now the case $\alpha = +\infty$. Let A be an arbitrarily large number. Select c so that

$$\frac{f(c)}{c} > A.$$

It can be proved by similar arguments that

$$\frac{f(x)}{x} \geqslant \frac{m-1}{m}A.$$

Hence,

$$\frac{f(x)}{x} \to +\infty$$

as $x \to 0$. The lemma is proved. □

2.3.3. *Proof of Khinchin's Theorem*

To prove Khinchin's theorem it is sufficient to verify that the function $\pi_1(t)$ satisfies the conditions of the lemma.

It is evident that $\pi_1(t) \geqslant 0$ and it cannot decrease as t increases. If we disregard those streams in which no customers arrive [and, consequently,

$p_0(t) = 1$ and $\pi_1(t) = 0$ for all t], then $\pi_1(a) > 0$ must hold for some $a > 0$. If in the interval $(0, t_1 + t_2)$ at least one customer arrives, then at least one customer must have arrived in one of the intervals $(0, t_1)$ and $(t_1, t_1 + t_2)$, thus

$$\pi_1(t_1 + t_2) \leqslant \pi_1(t_1) + \pi_2(t_2) \quad (t_1 > 0, t_2 > 0, t_1 + t_2 > 0).$$

We see that the function $\pi_1(t)$ satisfies all the conditions of the lemma; the theorem is thus proved.

In Chapter 1 (Section 1.6) we have shown that for any stationary stream without aftereffects

$$p_0(t) = 1 - e^{-\lambda t},$$

and hence

$$\pi_1(t) = 1 - e^{-\lambda t}.$$

If we exclude streams in which, with probability one, in any time interval either no customers or infinitely many of them arrive, the value of λ in the last formula is neither 0 nor $+\infty$. The discarded cases are of no interest, and we shall not consider them.

2.3.4. *An Example of a Stationary Stream with Aftereffects*

If aftereffects are allowed in a stationary stream, the parameter of the stream may be infinite. To prove this assertion, consider an example.

Let the parameter z of a simple stream be a random variable with the distribution function $F(x)$ $[F(+0) = 0, F(+\infty) = 1]$. If the parameter takes on the value x, the conditional probability that, in a time interval of length t, k customers arrive is

$$p_{k,x}(t) = \frac{(xt)^k}{k!} e^{-xt}.$$

Consequently, the total probability of k customers arriving during the time interval t is

$$p_k(t) = \int_0^\infty \frac{(xt)^k}{k!} e^{-xt}\, dF(x).$$

Evidently

$$\sum_{k=0}^\infty p_k(t) = \sum_{k=0}^\infty \int_0^\infty \frac{(xt)^k}{k!} e^{-xt}\, dF(x) = \int_0^\infty dF(x) = 1.$$

Since

$$\pi_1(t) = \int_0^\infty (1 - e^{-xt})\, dF(x),$$

we have

$$\frac{\pi_1(t)}{t} = \int_0^\infty \frac{1 - e^{-xt}}{t} dF(x).$$

Assuming that

$$\int_0^\infty x\, dF(x) < +\infty,$$

the inequality

$$1 - e^{-xt} \leqslant xt$$

implies that

$$\frac{\pi_1(t)}{t} \leqslant \int_0^\infty x\, dF(x).$$

Thus, under the stipulated assumption, the parameter of the stream is finite.

Now let

$$\int_0^\infty x\, dF(x) = +\infty.$$

For an arbitrarily small $\varepsilon > 0$ there exists an A such that

$$\int_0^A x\, dF(x) > \frac{1}{\varepsilon}.$$

However,

$$\frac{\pi_1(t)}{t} \geqslant \int_0^A \frac{1 - e^{-xt}}{t} dF(x)$$

and

$$\lim_{t\to 0} \int_0^A \frac{1 - e^{-xt}}{t} dF(x) = \int_0^A x\, dF(x),$$

therefore for any $\varepsilon > 0$,

$$\lim \frac{\pi_1(t)}{t} \geqslant \frac{1}{\varepsilon}.$$

Hence,

$$\lim_{t\to 0} \frac{\pi_1(t)}{t} = +\infty.$$

Under the new assumption the parameter of the stream is infinite.

It is easy to see that the parameter λ of this stream is, in both cases, equal to

$$\lambda = \int_0^\infty x\, dF(x).$$

2.4. General Form of Stationary Streams without Aftereffects

2.4.1. *Statement of the Problem*

In this section the propabilities $\pi_k(t)$ of arrivals during a time interval t of at least k streams for $k = 0, 1, 2, \ldots$ will play an important role. It will be shown that for any stationary stream without aftereffects, the limits

$$\lim_{t \to 0} \frac{\pi_k(t)}{t}$$

exist for all $k > 0$. For a simple stream

$$\lim_{t \to 0} \frac{\pi_1(t)}{t} = \lambda, \qquad \lim_{t \to 0} \frac{\pi_k(t)}{t} = 0 \quad \text{for } k > 1.$$

The contents of this section are taken from Khinchin (1955). Another characterization of stationary streams without aftereffects was given by Redheffer (1953).

Lemma. *If a function $f(x)$ is nonnegative and nondecreasing in the interval $0 < x \leqslant a$, then the ratio $f(x)/x$ is bounded in this interval, and if for any integer n and some constant $c > 0$ the inequality*

$$f(nx) \leqslant nf(x) + cn^2x^2 \tag{1}$$

is satisfied, then the limit

$$\lim_{x \to 0} \frac{f(x)}{x} = l \geqslant 0$$

exists.

PROOF. Denote

$$\overline{\lim_{x \to 0}} \frac{f(x)}{x} = l.$$

It can be assumed that $l > 0$ since, if $l = 0$, the assertion of the lemma is trivial.

Set $z = nx$ in (1); then

$$f(z) \leqslant nf\left(\frac{z}{n}\right) + cz^2.$$

Hence,

$$\frac{f\left(\frac{z}{n}\right)}{\frac{z}{n}} \geqslant \frac{f(z)}{z} - cz.$$

Select $\varepsilon > 0$ sufficiently small and $z < \varepsilon/c$ such that the inequality

$$\frac{f(z)}{z} > l - \varepsilon$$

is satisfied. Let $0 < x < z$. Define an integer $n > 1$ by the inequality

$$\frac{z}{n} \leqslant x < \frac{z}{n-1}.$$

Since by the assumption on the function $f(x)$,

$$f(x) \geqslant f\left(\frac{z}{n}\right),$$

we have

$$\frac{f(x)}{x} \geqslant \frac{f\left(\frac{z}{n}\right)}{\frac{z}{n-1}} = \frac{n-1}{n} \frac{f\left(\frac{z}{n}\right)}{\frac{z}{n}}$$

and, hence, for x sufficiently small

$$\frac{f(x)}{x} \geqslant \frac{n-1}{n}\left[\frac{f(z)}{z} - cz\right] = \left(1 - \frac{1}{n}\right)\frac{f(z)}{z} - cz$$

$$\geqslant \left(1 - \frac{1}{n}\right)(l - \varepsilon) - \varepsilon > l - 3\varepsilon.$$

However, for any $\varepsilon > 0$ and x sufficiently small, the inequality

$$\frac{f(x)}{x} < l + \varepsilon$$

is satisfied.

This implies that

$$\lim_{x \to 0} \frac{f(x)}{x} = l.$$

The lemma is thus proved. □

2.4.2. *The Existence of the Limits* $\lim_{t \to 0} \frac{\pi_k(t)}{t}$

We shall now prove the assertion stated previously that the limits

$$\lim_{t \to 0} \frac{\pi_k(t)}{t}$$

exist for any $k \geqslant 1$. It is sufficient to verify that the function $\pi_k(t)$ satisfies all the conditions of the lemma. It follows from the definition of the functions

$\pi_k(t)$ that for any $k \geqslant 1$ they are nonnegative and nondecreasing. Thus, since

$$\pi_k(t) \leqslant \pi_1(t)$$

and

$$\lim_{t \to 0} \frac{\pi_1(t)}{t} = \lambda,$$

we obtain that $\pi_k(t)/t$ is bounded in any interval $0 < t < a$. It remains to verify that $\pi_k(t)$ satisfies (1).

Consider the least upper bound g_k of $\pi_k(t)/t$ in the region $0 < t < \infty$ and the number

$$A_k = \sum_{i=1}^{k} g_i g_{k-i}.$$

We shall prove that for any t and integers n

$$\pi_k(nt) \leqslant n\pi_k(t) + A_k \frac{n(n-1)}{2} t^2. \tag{2}$$

This will complete the proof that the conditions of the lemma are satisfied.

For $n = 1$ the inequality (2) is trivial. Assume now that it holds for some n. In order that at least k customers arrive in the interval $(0, (n+1)t)$ of length $(n+1)t = nt + t$, it is necessary that at least i $(i = 0, 1, 2, \ldots, k)$ customers arrive in the interval $(0, t)$ and at least $k - i$ customers in the interval $(t, (n+1)t)$. We thus have the inequality

$$\pi_k((n+1)t) \leqslant \sum_{i=0}^{k} \pi_i(t)\pi_{k-i}(nt) = \pi_0(t)\pi_k(nt) + \pi_k(t)\pi_0(nt) + \sum_{i=1}^{k-1} \pi_i(t)\pi_{k-i}(nt)$$

$$\leqslant \pi_k(nt) + \pi_k(t) + nt^2 \sum_{i=1}^{k-1} g_i g_{k-i} = \pi_k(nt) + \pi_k(t) + A_k nt^2.$$

However, by the induction hypothesis

$$\pi_k(nt) \leqslant n\pi_k(t) + A_k \frac{n(n-1)}{2} t^2,$$

and, therefore,

$$\pi_k((n+1)t) \leqslant (n+1)\pi_k(t) + A_k \frac{n(n+1)}{2} t^2.$$

Thus, we have shown that the functions satisfy (2). Consequently, the conditions of the lemma are fulfilled and the limits

$$\lim_{t \to 0} \frac{\pi_k(t)}{t}$$

exist.

Since

$$\lim_{t \to 0} \frac{\pi_1(t)}{t} = \lambda$$

there also exist the limits

$$\lim_{t \to 0} \frac{\pi_k(t)}{\pi_1(t)}.$$

From the above it also follows that the limits

$$a_k = \lim_{t \to 0} \frac{p_k(t)}{\pi_1(t)} \quad (k = 0, 1, 2, \ldots)$$

exist where $p_k(t) = \pi_k(t) - \pi_{k+1}(t)$.

The ratio $p_k(t)/\pi_1(t)$ is the probability that k customers arrive if it is known that at least one customer arrived in this interval. The limit of this ratio for $t \to 0$ can be naturally regarded as the probability that k customers arrive at some instant of time, given that at this time at least one customer arrived. This simple interpretation of a_k will be useful below. Note that

$$\sum_{k=1}^{\infty} a_k = 1.$$

2.4.3. *Equations for the General Stationary Stream without Aftereffects*

We now write the equations defining the functions $p_k(t)$ in the general case of a stationary stream without aftereffects. We know that for such streams, for any $t > 0$ and $h > 0$, the equalities

$$p_k(t + h) = \sum_{i=0}^{k} p_i(t) p_{k-i}(h) \quad (k = 0, 1, 2, \ldots)$$

are valid.

Since for streams without aftereffects

$$p_0(t) = e^{-\lambda t},$$

we have as $h \to 0$

$$p_0(h) = 1 - \lambda h + o(h).$$

Thus, for $k > 0$

$$p_k(t + h) = (1 - \lambda h) p_k(t) + \sum_{i=0}^{k-1} p_i(t) p_{k-i}(h) + o(h)$$

and, hence,

$$\frac{p_k(t + h) - p_k(t)}{h} = -\lambda p_k(t) + \sum_{i=0}^{k-i} p_i(t) \frac{p_{k-i}(h)}{h} + o(1). \tag{1}$$

We know that as $h \to 0$

$$\frac{p_{k-i}(h)}{h} = \frac{\pi_1(h)}{h} \frac{p_{k-i}(h)}{\pi_1(h)} = \lambda a_{k-i}. \tag{2}$$

Hence, for $k \geqslant 1$, the equalities

$$p_k'(t) = -\lambda p_k(t) + \lambda \sum_{i=0}^{k-1} a_{k-i} p_i(t) \tag{3}$$

are valid. The existence of the derivatives $p_k'(t)$ is self-evident. We now add to the system (3) an equation defining $p_0(t)$:

$$p_0'(t) = -\lambda p_0(t). \tag{4}$$

The two equations [(3) and (4)] uniquely determine the probabilities $p_k(t)$.

2.4.4. *Solution of Systems* (3) *and* (4)

The substitution

$$p_k(t) = e^{-\lambda t} u_k(t)$$

converts the system of equations under consideration into a simpler form. Direct substitution shows that the equations for the functions $u_k(t)$ are as follows:

$$u_0'(t) = 0$$

and for $k \geqslant 1$

$$u_k'(t) = \lambda(a_1 u_{k-1}(t) + a_2 u_{k-2}(t) + \cdots + a_k u_0(t)). \tag{5}$$

It follows from

$$p_0(t) = e^{-\lambda t}$$

that

$$u_0(0) = p_0(0) = 1.$$

Since for any $t \geqslant 0$

$$\sum_{k=0}^{\infty} p_k(t) = 1,$$

we evidently have for all $k \geqslant 1$

$$u_k(0) = p_k(0) = 0.$$

We write down a few of the first equations of system (5):

$$u_1'(t) = \lambda a_1 u_0(t),$$
$$u_2'(t) = \lambda(a_1 u_1(t) + a_2 u_0(t)),$$
$$u_3'(t) = \lambda(a_1 u_2(t) + a_2 u_1(t) + a_3 u_0(t)).$$

Substituting $u_0(t) = 1$ into the first equation, and taking into account the initial condition on u_1, yields

$$u_1(t) = \lambda a_1 t.$$

Substituting $u_0(t)$ and $u_1(t)$, and taking into account the initial condition on u_2, results in

$$u_2(t) = \frac{(\lambda a_1 t)^2}{2!} + \lambda a_2 t.$$

Analogously,

$$u_3(t) = \frac{(\lambda a_1 t)^3}{3!} + (\lambda a_1 t)(\lambda a_2 t) + \lambda a_3 t,$$

and the expression for $u_4(t)$ is:

$$u_4(t) = \frac{(\lambda a_1 t)^4}{4!} + \frac{(a_1 \lambda t)^2}{2!} a_2 \lambda t + \frac{\lambda a_1 t}{1!} a_3 \lambda t + \frac{(a_2 \lambda t)^2}{2!} + a_4 \lambda t.$$

A general formula for $u_k(t)$ will not be given, since it is rather complicated. The required probabilities are:

$$p_0(t) = e^{-\lambda t},$$

$$p_1(t) = a_1 \lambda t e^{-\lambda t},$$

$$p_2(t) = a_1^2 \frac{(\lambda t)^2}{2!} e^{-\lambda t} + a_2 \lambda t e^{-\lambda t},$$

$$p_3(t) = a_1^3 \frac{(\lambda t)^3}{3!} e^{-\lambda t} + 2a_1 a_2 \frac{(\lambda t)^2}{2!} e^{-\lambda t} + a_3 \lambda t e^{-\lambda t},$$

$$p_4(t) = a_1^4 \frac{(\lambda t)^4}{4!} e^{-\lambda t} + 3a_1^2 a_2 \frac{(\lambda t)^3}{3!} e^{-\lambda t} + (2a_1 a_3 + a_2^2) \frac{(\lambda t)^2}{2!} e^{-\lambda t} + a_4 \lambda t e^{-\lambda t}.$$

These formulas have an intuitive probabilistic interpretation. For example, consider the probability $p_4(t)$. In a time interval of length t, four customers may arrive in the following disjoint ways:

1. The customers arrive at four distinct instants, one customer at each instant;
2. The customers arrive at three instants, one customer at each of the two instants and two customers at the third;
3. The customers arrive at two instants, at one instant one customer and at the other three;
4. The customers arrive at two instants, two at each instant;
5. All four customers arrive at a single instant.

2.4.5. *A Special Case*

Let

$$a_1 = p, \qquad a_2 = 1 - p = q, \qquad a_k = 0 \quad \text{for } k \geqslant 3.$$

This case may also be of direct practical interest, e.g., in analyzing the physical phenomena of spontaneous particle splitting, in economic operations, and in industry.

The equations for the function $u_k(t)$ for $k \geqslant 2$ are quite simple:

$$u_k'(t) = \lambda(a_1 u_{k-1}(t) + a_2 u_{k-2}(t)).$$

Direct substitution shows that the function

$$u_k(t) = \sum_{i=0}^{[k/2]} C_{k-i}^{i} a_1^{k-2i} a_2^{i} \frac{(\lambda t)^{k-i}}{(k-i)!}$$

satisfies the differential equation as well as the initial condition. Hence, the required formula is:

$$p_k(t) = \sum_{i=0}^{[k/2]} C_{k-i}^{i} a_1^{k-2i} a_2^{i} \frac{(\lambda t)^{k-i}}{(k-i)!} e^{-\lambda t}.$$

The number of times at which customers arrive constitutes a simple stream with parameter λ; at each of these times one or two customers arrive with probabilities p and q, respectively. The nature of the stream in this special case is quite evident.

2.4.6. *The Generating Function of the Stream*

For a complete solution of the systems (4) and (5) we shall employ generating functions. Set

$$F(t, x) = \sum_{k=0}^{\infty} p_k(t) x^k.$$

Multiplying relations (3) and (4) by the corresponding powers x^k and summing with respect to k from 0 to ∞, we obtain

$$\frac{\partial F}{\partial t} = -\lambda F + \lambda \sum_{k=1}^{\infty} x^k \sum_{i=1}^{k} a_i p_{k-i}(t)$$

$$= -\lambda F + \lambda \sum_{i=1}^{\infty} a_i \sum_{j=0}^{\infty} p_j(t) x^{i+j} = -\lambda F + \lambda F(t, x) \sum_{i=1}^{\infty} a_i x^i.$$

Denote

$$\Phi(x) = \sum_{i=1}^{\infty} a_i x^i,$$

then

$$\frac{\partial F}{\partial t} = \lambda[\Phi(x) - 1]F$$

or

$$\frac{\partial \ln F}{\partial t} = \lambda[\Phi(x) - 1].$$

Whence

$$F(t, x) = C(x)e^{\lambda[\Phi(x)-1]t}.$$

And, since for any x

$$F(0, x) = p_0(0) = 1,$$

we have

$$F(t, x) = e^{\lambda[\Phi(x)-1]t}. \tag{6}$$

It is known that for the generating function $F(t, x)$ for any t

$$F(t, 1) = \sum_{k=0}^{\infty} p_k(t) = 1.$$

Hence, in view of (6)

$$\Phi(1) = \sum_{k=1}^{\infty} a_k = 1.$$

This equality was derived earlier.

Equation (6) gives a general form of the generating function for an arbitrary stationary stream without aftereffects. The converse is also true and can easily be verified: if $\lambda > 0$, $a_i > 0$, $\sum_{i=1}^{\infty} a_i = 1$, there exists a stationary stream without aftereffects whose generating function is given by (6).

Assume the customer's arrival times form a simple stream with parameter λ. At each one of these times any number k of customers may arrive with probability a_k, which is independent of the number of earlier arrivals, the instant of the arrival, and the order in which the customers arrive. Thus, the given stream will clearly be stationary and without aftereffects. It is easily seen that the generating function of this stream is of the form (6).

Indeed, the probability that exactly k customers arrive in a time interval of length t is equal to

$$p_k(t) = \sum_{r=0}^{\infty} \frac{(\lambda t)^r e^{-\lambda t}}{r!} P_r(k).$$

{The probability that during time t, r arrival times occur is, by assumption, equal to $[(\lambda t)^r/r!]e^{-\lambda t}$; $P_r(k)$ denotes the probability that k customers arrive at these r instants. It is evident that $P_r(k) = 0$ for $k < r$.} Now

$$F(t, x) = \sum_{k=0}^{\infty} p_k(t)x^k = \sum_{r=0}^{\infty} \frac{(\lambda t)^r e^{-\lambda t}}{r!} \sum_{k=0}^{\infty} P_r(k)x^k.$$

But $\sum_{k=0}^{\infty} P_r(k)x^k$ is the generating function of the random variable defined to be the number of customers arriving at these r instants. Since the number(s) of arrivals at different instants are independent random variables, we have, by a well-known property of generating functions,

$$\sum_{k=0}^{\infty} P_r(k)x^k = [\Phi(x)]^r = \left[\sum_{i=1}^{\infty} a_i x^i\right]^r.$$

Thus,

$$F(t, x) = \sum_{r=0}^{\infty} \frac{[\lambda t \Phi(x)]^r}{r!} e^{-\lambda t} = e^{\lambda t[\Phi(x)-1]}.$$

2.4.7. *Concluding Remarks*

We present in this section final comments about the structure of stationary streams without aftereffects. For any stationary stream the customers' arrival times constitute a simple stream. At every instant of the customers' arrivals the probability that just k customers arrive is a_k. Thus, any stationary stream without aftereffects is fully characterized by the numbers λ—the intensity of the stream of arrival times—and a_k—the probability that exactly k customers arrive at each one of these instants:

$$\left(a_k \geqslant 0, \sum_{k=1}^{\infty} a_k = 1\right).$$

For a simple stream the intensity λ is arbitrary, $a_1 = 1$, $a_k = 0$ for $k > 1$.

Equation (6) for the generating function enables us to obtain the probability $p_k(t)$ that exactly k customers arrive in time t. The formulas for the first four values of k were derived in Section 2.4.5.

It is natural to pose the following question: What streams are characterized solely by the absence of aftereffects? In other words, if we discard not only orderliness but also stationarity, would it be possible to give a sufficiently clear description of all possible streams? This problem was posed and solved by Khinchin (1956). We shall describe his result without presenting the actual solution.

Call the mathematical expectation of the number of customers arriving during time $(0, t)$ the *leading function* of the stream; we denote it by $\Lambda(t)$. We shall call the stream *finite* if, for every t, $\Lambda(t) < +\infty$. Denote the probability that k customers arrive in the interval (t, τ) by $p_k(t, \tau)$. The time t will be called a *regular* or *singular point* of the stream, depending on whether the limit

$$\lim_{h \to 0} p_0(t - h, t + h) = p_0(t)$$

is equal to or smaller than 1. If all points of the stream are regular, the stream will be called *regular*.

The generating function of a regular stream without aftereffects is of the following form [(α, β) is a time interval]:

$$\Phi(x; \alpha, \beta) = \exp\left\{\sum_{k=0}^{\infty} [\chi_k(\beta) - \chi_k(\alpha)] x^k\right\},$$

where the functions $\chi_k(t)$ ($\chi_k(0) = 0$) have the properties:

1. they are continuous for all t;
2. they are monotone (nondecreasing for $k > 0$ and nonincreasing for $k = 0$);

3. the series $\sum k\chi_k(t)$ is convergent for all t;
4. for any $t \geqslant 0$ the equality $\sum_{k=0}^{\infty} \chi_k(t) = 0$ is valid.

A stream S is called *singular* if:

1. its events occur only at given instants of time, forming a finite or denumerable set: $t_1, t_2, \ldots$;
2. the number of events occurring at different times t_i are mutually independent random variables;
3. the probability that at time t_i k events of the stream S occur is equal to $q_k^{(i)}$; moreover, for every $t \geqslant 0$ the inequality

$$\sum_{0 \leqslant t_i \leqslant t} \sum_{k=1}^{\infty} k q_k^{(i)} < \infty$$

is satisfied.

The general theorem proved by Khinchin (1956) is as follows:

Every finite stream without aftereffects is the union (sum) of two independent streams of the same type, one regular and the other singular. The points at which the events of the singular stream occur coincide with the singular points of the initial stream, and the corresponding probabilities $q_k^{(i)}$ *are defined by the equalities*

$$q_k^{(i)} = \lim_{h \to 0} p_k(t_i - h, t_i + h).$$

2.5. The Palm-Khinchin Functions

2.5.1. *Definition of the Palm-Khinchin Functions*

In studying stationary streams, Palm (1943) introduced a function $\varphi_0(t)$, which he and a number of other authors applied very successfully to the solution of queueing theory problems. Palm himself defined the function $\varphi_0(t)$ as the conditional probability of the absence of customers in the interval $(t_0, t_0 + t)$, given that at time t_0 a customer arrives. Since the condition under which his probability is defined has in the most interesting cases probability zero, this definition has no strictly meaningful sense. Khinchin (1963) therefore refined this concept: he suggested that one consider an infinite sequence of functions similar to the Palm function, rather than a single function. For this reason the functions $\varphi_k(t)$ defined subsequently will be called the Palm-Khinchin functions.

Consider two consecutive (nonoverlapping) time intervals, the first of length τ and the second of length t; denote by $H_k(\tau, t)$ the probability of the following compound event:

at least one customer arrives in the interval τ;
at most k customers arrive in the interval t.

In general, these events may be dependent. Since in accordance with the notation introduced previously the probability of the first event is $\pi_1(\tau)$, the ratio

$$\frac{H_k(\tau, t)}{\pi_1(\tau)}$$

represents the conditional probability of the second event under the assumption that the first has occurred. In other words, this ratio expresses the probability of the arrival of at most k customers in the interval t, given that at least one customer has arrived in the interval τ.

We shall prove subsequently that for a stationary process with finite parameter λ, the limit

$$\Phi_k(t) = \lim_{\tau \to 0} \frac{H_k(\tau, t)}{\pi_1(\tau)} \tag{1}$$

exists for every k. It is natural to refer to this limit as the conditional probability that at most k customers arrive in the interval t, given that at the initial instant of this interval a customer has arrived.

Since $H_k(\tau, t)$ is a nonincreasing function of t and $\pi_1(\tau)$ does not depend on t, the functions $\Phi_k(t)$ are also nonincreasing in the region $0 < t < \infty$.

Now set for $k > 0$

$$h_k(\tau, t) = H_k(\tau, t) - H_{k-1}(\tau, t); \qquad h_0(\tau, t) = H_0(\tau, t).$$

Clearly, $h_k(\tau, t)$ is the probability that (1) at least one customer arrives in the interval τ and (2) exactly k customers arrive in the interval t.

Introduce the notations

$$\varphi_0(t) = \Phi_0(t), \qquad \varphi_k(t) = \Phi_k(t) - \Phi_{k-1}(t) \quad (k > 0). \tag{2}$$

It is clear from the above that

$$\varphi_k(t) = \lim_{\tau \to 0} \frac{h_k(\tau, t)}{\pi_1(\tau)} \quad (k = 0, 1, 2, \ldots). \tag{3}$$

We shall call the functions $\varphi_k(t)$ the Palm-Khinchin functions. By the preceding argument the function $\varphi_k(t)$ may be interpreted as the probability that exactly k customers arrive in the interval t, given that a customer arrives at the first instant of this interval.

2.5.2. *Proof of the Existence of the Palm-Khinchin Functions*

We have introduced the Palm-Khinchin functions, but have not as yet proved their existence. We shall now prove that for every stationary stream with a finite parameter, the functions $\varphi_k(t)$ exist. We shall thus verify that the probability $H_k(\tau, t)$, as a function of τ, satisfies all the conditions of the lemma proved in Section 2.3. Clearly this function is nonnegative and monotone. We must verify that it also satisfies the third condition of the lemma, i.e., that

$$H_k(\tau_1 + \tau_2, t) \leqslant H_k(\tau_1, t) + H_k(\tau_2, t).$$

We subdivide the interval τ into two parts τ_1 and τ_2, where τ_1 precedes τ_2. The compound event whose probability is $H_k(\tau, t)$ occurs if at least one of the following events takes place:

(A) In the interval τ_2 at least one customer arrives, and in the interval t at most k customers arrive [the probability of A is clearly $H_k(\tau_2, t)$].
(B) In the interval τ_1 at least one customer arrives, and in the interval $\tau_2 + t$ at most k customers arrive [the probability of the event B is $H_k(\tau_1, \tau_2 + t)$].

Since the events A and B may not be disjoint, we have

$$H_k(\tau, t) \leqslant H_k(\tau_2, t) + H_k(\tau_1, \tau_2 + t).$$

Since for fixed τ $H_k(\tau, t)$ is a decreasing function of t, we have

$$H_k(\tau_1, \tau_2 + t) \leqslant H_k(\tau_1, t).$$

Thus,

$$H_k(\tau, t) \leqslant H_k(\tau_1, t) + H_k(\tau_2, t).$$

By the previously mentioned lemma the limit

$$\lim_{\tau \to 0} \frac{H_k(\tau, t)}{\tau}$$

exists (possibly infinite). However, since

$$H_k(\tau, t) \leqslant \pi_1(\tau)$$

always, and since by assumption the ratio $\pi_1(\tau)/\tau$ approaches a finite limit λ as $\tau \to 0$, the limit

$$\lim_{\tau \to 0} \frac{H_k(\tau, t)}{\tau}$$

must be finite. Thus, since

$$\frac{H_k(\tau, t)}{\pi_1(t)} = \frac{H_k(\tau, t)/\tau}{\pi_1(\tau)/\tau},$$

the limit

$$\Phi_k(t) = \lim_{\tau \to 0} \frac{H_k(\tau, t)}{\pi_1(\tau)}$$

exists.

EXAMPLE. Let us consider a simple stream with parameter λ. In this case,

$$\pi_1(\tau) = 1 - e^{-\lambda\tau}, \qquad H_k(\tau, t) = \pi_1(\tau) \sum_{i=0}^{k} \frac{(\lambda t)^i}{i!} e^{-\lambda t},$$

so that

$$\Phi_k(t) = \sum_{i=0}^{k} \frac{(\lambda t)^i}{i!} e^{-\lambda t} \quad \text{and} \quad \varphi_k(t) = \frac{(\lambda t)^k}{k!} e^{-\lambda t}.$$

2.5.3. *The Palm-Khinchin Formulas*

The formulas presented herein were derived by Palm (1943) for $k = 0$, and by Khinchin [(1963), Section 10] for $k \geqslant 1$.

Assume that the given stream is stationary and orderly, and has a finite parameter λ. Consider a time interval of length $\tau + t$ consisting of an interval of length τ and a subsequent interval of length t. Denote by n_1 the number of customers arriving in the time interval τ and by n_2 the number of arrivals in the interval t. Clearly,

$$p_k(t + \tau) = \sum_{r=0}^{k} P\{n_1 = r, n_2 = k - r\},$$

and hence (since the stream is orderly),

$$p_k(\tau + t) = P\{n_1 = 0, n_2 = k\} + P\{n_1 = 1, n_2 = k - 1\} + o(\tau)$$

as $\tau \to 0$. However,

$$\mathsf{P}\{n_1 = 0, n_2 = k\} = \mathsf{P}\{n_2 = k\} - \mathsf{P}\{n_1 > 0, n_2 = k\} = p_k(t) - h_k(\tau, t)$$

and

$$\begin{aligned} \mathsf{P}\{n_1 = 1, n_2 = k - 1\} &= \mathsf{P}\{n_1 > 0, n_2 = k - 1\} - \mathsf{P}\{n_1 > 1, n_2 = k - 1\} \\ &= h_{k-1}(\tau, t) + o(\tau), \end{aligned}$$

we always have

$$p_k(\tau + t) = p_k(t) - h_k(\tau, t) + h_{k-1}(\tau, t) + o(\tau).$$

Hence,

$$\frac{p_k(t + \tau) - p_k(t)}{\tau} = \frac{h_{k-1}(\tau, t)}{\pi_1(\tau)} \frac{\pi_1(\tau)}{\tau} - \frac{h_k(\tau, t)}{\pi_1(\tau)} \frac{\pi_1(\tau)}{\tau} + o(1)$$

and thus in the limit ($k > 0$)

$$p_k'(t) = \lambda[\varphi_{k-1}(t) - \varphi_k(t)]. \tag{4}$$

Similarly we obtain that

$$p_0'(t) = -\lambda\varphi_0(t). \tag{5}$$

Introduce the notation

$$v_k(t) = \sum_{m=0}^{k} p_m(t).$$

Combining (4) and (5) yields

$$v_k'(t) = -\lambda\varphi_k(t) \quad (k = 0, 1, 2, \ldots). \tag{6}$$

We have thus obtained the differential equations that determine, in terms of the Palm-Khinchin functions, probabilities of the arrival of at most k customers in a given time interval t. It is expedient to write these equations in another form. Integrate (6) with respect to t from 0 to t:

$$v_k(+0) - v_k(t) = \lambda \int_0^t \varphi_k(z)\,dz \quad (k = 0, 1, 2, \ldots).$$

However, for every $k \geqslant 0$

$$v_k(+0) \geqslant v_0(+0) = p_0(+0) = 1 - \pi_1(+0).$$

Now, since the parameter of the stream $\dfrac{\pi_1(t)}{t} \to \lambda < +\infty$ as $t \to 0$, $\pi_1(+0) = 0$ and hence $v_k(+0) = 1$.

Thus, for each $t > 0$

$$1 - v_k(t) = \lambda \int_0^t \varphi_k(z)\,dz, \tag{7}$$

whence we arrive at the equations

$$p_0(t) = 1 - \lambda \int_0^t \varphi_0(z)\,dz, \tag{8}$$

$$p_k(t) = \lambda \int_0^t [\varphi_{k-1}(t) - \varphi_k(t)]\,dt, \tag{9}$$

which establish simple relations between the probabilities $p_k(t)$ and the Palm-Khinchin functions.

Examples. Consider the stream defined by

$$p_k(t) = \frac{at^k}{(a+t)^{k+1}} \quad (k = 0, 1, 2, \ldots),$$

where a is a positive constant.

In view of the above, $\lambda\varphi_k(t) = -v_k'(t)$. For the stream under consideration

$$\lambda = \frac{1}{a} \quad \text{and} \quad v_k(t) = 1 - \sum_{j=k+1}^{\infty} p_j(t) = 1 - \left(\frac{t}{a+t}\right)^{k+1},$$

and therefore the corresponding Palm-Khinchin functions are given by

$$\varphi_k(t) = (k+1)a^2 \frac{t^k}{(a+t)^{k+1}} \quad (k = 0, 1, 2, \ldots).$$

As a second example, we consider a stream for which

$$p_k(t) = p\frac{(at)^k}{k!}e^{-at} + q\frac{(bt)^k}{k!}e^{-bt} \quad (k = 0, 1, 2, \ldots),$$

where a, b, p, q are positive constants and $p + q = 1$. The parameter of this stream is $\lambda = ap + bq$.

In accordance with (6)

$$\varphi_k(t) = \frac{t^k}{k!}\frac{pa^{k+1}e^{-at} + qb^{k+1}e^{-bt}}{ap + bq}.$$

For the case $b = 0$, $a = \lambda$, we obtain

$$\varphi_k(t) = \frac{(\lambda t)^k}{k!}e^{-\lambda t}.$$

We have seen this result earlier.

2.5.4. *Intensity of a Stationary Stream*

In Chapter 1 we used the term "intensity of a stationary stream" for the mathematical expectation of the number of customers arriving in a unit time. The stream intensity was denoted by the letter μ. Since the stream is stationary, we have for every $t > 0$

$$\mu t = \sum_{k=1}^{\infty} kp_k(t).$$

We have seen in Section 1.1 that for a simple stream the relation $\mu = \lambda$ is valid and that for an arbitrary stationary stream the inequality $\mu \geqslant \lambda$ holds. What then are the conditions that will ensure the equality $\mu = \lambda$ for stationary streams? This question is very appropriate, since in both theoretical and practical problems this equality is often assumed to hold without any justification. The following theorem gives a simple necessary condition for its validity.

Theorem. *If a stationary stream has finite intensity μ, the equality $\mu = \lambda$ implies that the stream is orderly.*

PROOF.* Let $\mu = \lambda$. Since

$$\mu t = \sum_{k=1}^{\infty} kp_k(t),$$

and by the definition of the parameter λ it follows that

$$\lambda t = \sum_{k=1}^{\infty} p_k(t) + o(t)$$

we have

---

* The proof presented here is due to Zitek (1958).

$$\sum_{k=2}^{\infty} (k-1)p_k(t) = o(t).$$

Since all $p_k(t)$ are nonnegative, the inequality

$$0 \leqslant \sum_{k=2}^{\infty} p_k(t) \leqslant \sum_{k=2}^{\infty} (k-1)p_k(t)$$

is valid. Thus,

$$\sum_{k=1}^{\infty} p_k(t) = o(t).$$

This implies that the stream is orderly. □

Zitek notes that, if the stream is of infinite intensity, it does not follow from the equality $\mu = \lambda$ that the stream is orderly. Indeed, suppose that for a stationary stream $\lambda = \infty$. Since $\mu \geqslant \lambda$, we have $\mu = \infty$. Now consider a new stream in which simultaneously with each customer of the original stream two new customers arrive, while at other times no customers arrive. Clearly for the second stream $\lambda = \infty$, and the stream is not orderly.

2.5.5. *Korolyuk's Theorem*

V.S. Korolyuk observed that *if a stationary stream is orderly, the equality $\mu = \lambda$ must be valid (here the case $\mu = \lambda = \infty$ is not excluded).*

By the definition of intensity of a stream

$$\mu = \sum_{k=1}^{\infty} kp_k(1) = \sum_{k=1}^{\infty}\sum_{r=k}^{\infty} p_r(1) = \sum_{k=1}^{\infty} [1 - v_{k-1}(1)] = \sum_{k=0}^{\infty} [1 - v_k(1)]$$

and thus in view of (7) for a finite λ we have

$$\mu = \lambda \sum_{k=0}^{\infty} \int_0^1 \varphi_k(u)\,du. \tag{10}$$

However,

$$\varphi_k(u) = \lim_{\tau \to 0} \frac{h_k(\tau, u)}{\pi_1(\tau)}$$

and for any $k \geqslant 0$ and $u > 0$

$$\sum_{i=0}^{k} \frac{h_i(\tau, u)}{\pi_1(\tau)} = \frac{H_k(\tau, u)}{\pi_1(\tau)} \leqslant 1.$$

In the limit as $\tau \to 0$ we obtain that

$$\sum_{i=0}^{k} \varphi_i(u) \leqslant 1,$$

and hence for any $k \geqslant 0$,

$$\sum_{i=0}^{k} \int_0^1 \varphi_i(u)\,du \leqslant 1.$$

This implies the inequality

$$\sum_{i=0}^{\infty} \int_0^1 \varphi_i(u)\,du \leqslant 1.$$

Thus (10) yields that $\lambda \geqslant \mu$ for any finite λ. However, since $\mu \geqslant \lambda$ always, we obtain that $\mu = \lambda$. If $\lambda = \infty$, then $\mu \geqslant \lambda$ implies that $\mu = \infty$, i.e., we again arrive at $\mu = \lambda$. The orderliness of the stream was not used directly in this proof, but via the Palm-Khinchin formulas.

The results of the Sections 2.5.4 and 2.5.5 enable us to state the following theorem:

Theorem. *For any stationary stream with finite intensity μ, the necessary and sufficient condition for the stream to be orderly is that $\lambda = \mu$.*

Hence it follows that *the only stationary streams without aftereffects that satisfy the condition $\mu = \lambda$ are the simple streams.*

Indeed, $\mu = \lambda$ if and only if the stream is orderly. But every orderly stationary stream without aftereffects is by definition simple.

2.5.6. *The Case of Nonorderly Streams*

In the nonorderly case, even with a finite parameter, one can easily construct stationary streams in which the relation between the parameter and the intensity will be arbitrary. This can be accomplished by considering a stationary stream without aftereffects. Since for all such streams

$$p_0(t) = 1 - e^{-\lambda t},$$

the parameter of the stream is λ. The intensity of the stream will, however be different, provided only that the stream is not simple. In an interval of unit length there will be a random number of arrival times. Denote this number by η. At each one of these times a random number of customers arrive. Let the number of customers arriving at time t_i be ξ_i. Then the total number of customers arriving in a time interval of unit length is equal to $\xi_1 + \xi_2 + \cdots + \xi_\eta$ if $\eta \geqslant 1$, and is equal to 0 if $\eta = 0$. In view of the absence of aftereffects the variables ξ_i and η are independent, and hence

$$\mu = \mathsf{M}(\xi_1 + \xi_2 + \cdots + \xi_\eta) = \mathsf{M}\xi_1 \mathsf{M}\eta$$

(see, e.g., Gnedenko (1961), Section 28, Corollary 2 of Theorem 2.) However, in accordance with the results of Section 1.4

$$\mathsf{M}\xi_1 = \sum_{k=1}^{\infty} k a_k$$

and

$$\mathsf{M}\eta = \lambda,$$

thus,

$$\mu = \lambda \sum_{k=1}^{\infty} k a_k.$$

The sum $\sum k a_k$ can be made to be equal to any value greater than 1 by a suitable choice of the quantities a_k. In particular, if $a_k = [k(k+1)]^{-1}$, the sum is infinite, and we obtain a stationary stream with a finite parameter and infinite intensity.

2.6. Characteristics of Stationary Streams and the Lebesgue Integral

2.6.1. *A General Definition of Mathematical Expectation*

Let ξ be a discrete random variable, i.e., a variable taking on a finite or countable number of possible values x_k with probabilities p_k. Then by definition $\mathsf{M}\xi = \sum x_k p_k$ provided only $\sum |x_k| p_k < \infty$. For nonnegative discrete random variable, $\mathsf{M}\xi$ is defined in any case, even if the series diverges and the case $\mathsf{M}\xi = +\infty$ is not excluded.

If ξ is an arbitrary nonnegative random variable, then by definition

$$\mathsf{M}\xi = \lim_{n \to \infty} \mathsf{M}\xi_n, \tag{1}$$

where $\{\xi_n\}$ is a nondecreasing sequence of nonnegative random variables convergent to ξ with probability 1. Representing ξ as a functon of an elementary event ω, we will obtain an expression for $\mathsf{M}\xi$ in the form of an abstract Lebesgue integral

$$\mathsf{M}\xi = \int \xi \, d\mathsf{P} = \int \xi(\omega) \mathsf{P}(d\omega), \tag{2}$$

whose definition is in essence given by (1).

We note the following property of the Lebesgue integral for a nonnegative function $f(x)$.

If $\{A_n\}$ is a sequence of events such that $P(A_n) \to 1$, then

$$\int_{A_n} f(\omega) \mathsf{P}(d\omega) \xrightarrow[n \to \infty]{} \int_{\Omega} f(\omega) \mathsf{P}(d\omega), \tag{3}$$

where Ω is the space of elementary events.

We note also that $\int f(\omega) \mathsf{P}(d\omega)$ is a linear and monotone functional in f:

$$\int (af + bg) \, d\mathsf{P} = a \int f \, d\mathsf{P} + b \int g \, d\mathsf{P},$$

provided the r.h.s. is defined;

$$\{f \geqslant 0\} \Rightarrow \left\{ \int f \, d\mathsf{P} \geqslant 0 \right\}.$$

2.6.2. *A Refinement of the Notion of Orderliness*

In the preceding section, the following notion of an ordinary stationary stream was used. If $\pi_2(t)$ is the probability of the occurrence of at least two customers in the interval of length t, then $\pi_2(t) = o(t)$. This implies that with probability 1

$$\tau_1 < \tau_2 < \cdots < \tau_n < \cdots, \tag{4}$$

where τ_n is the time of the nth event of the stream. For the proof we introduce the following notation: $x(t)$ is the number of events of the stream in the interval $(0, t)$; $x_0(t)$ is the number of different instants of the events of the stream in the interval $(0, t)$; $x_1(t) = x(t) - x_0(t)$. Relation (1) is equivalent to relation

$$x_1(t) = 0, \qquad 0 \leqslant t < \infty, \tag{5}$$

and hence (4) is fulfilled with probability 1 as long as

$$\mathsf{P}\{x_1(t) > 0\} = 0 \tag{6}$$

for arbitrary $t > 0$. In turn

$$\begin{aligned}
\mathsf{P}\{x_1(t) > 0\} &= \mathsf{P}\left\{\bigcup_{k=1}^{n} \left\{x_1\left(\frac{k}{n}t\right) - x_1\left(\frac{k-1}{n}t\right) > 0\right\}\right\} \\
&\leqslant \sum_{k=1}^{n} \mathsf{P}\left\{x_1\left(\frac{k}{n}t\right) - x_1\left(\frac{k-1}{n}t\right) > 0\right\} \\
&\leqslant \sum_{k=1}^{n} \mathsf{P}\left\{x\left(\frac{k}{n}t\right) - x\left(\frac{k-1}{n}t\right) > 1\right\} \\
&= n\pi_2\left(\frac{t}{n}\right) = t\frac{\pi_2(t/n)}{t/n} \xrightarrow[n\to\infty]{} 0
\end{aligned}$$

and hence the equality (6) is valid.

2.6.3. *Existence of the Parameter of a Stream*

Let a stationary stream of homogeneous events $x(t)$ exist. It means that the random process $z(t) = x(t_0 + t) - x(t_0)$ for $t > 0$ has the same finite-dimensional distribution as $x(t)$. However, $z_0(t) = x_0(t_0 + t) - x_0(t_0)$ is formed from $z(t)$ in the same manner as $x_0(t)$ is formed from $x(t)$. From here it follows that $x_0(t)$ is a stationary process. Moreover, the process is clearly orderly.

Let $\Delta_{nk} = 1$ provided for $(k-1)/n \leqslant t < k/n$ at least one event of the stream $x(t)$ will occur and $\Delta_{nk} = 0$ otherwise. Set

$$x_n = \Delta_{n1} + \cdots + \Delta_{nn}.$$

Denote by A_n the event that $x_n = x_0(1)$. Then clearly $\mathsf{P}\{A_n\} \to 1$. Whence in view of (3),

$$\int_{A_n} x_n \, d\mathsf{P} = \int_{A_n} x_0(1) \, d\mathsf{P} \to \int_{\Omega} x_0(1) \, d\mathsf{P} = \mu_0, \tag{7}$$

where μ_0 is the intensity of the stream $x_0(t)$.
At the same time $x_n \leqslant x_0(1)$, whence $\int_\Omega x_n \, dP \leqslant \mu_0$; since $x_n \geqslant 0$, this together with (7) leads us to the equality

$$\int_{\Omega} x_n \, d\mathsf{P} \xrightarrow[n \to \infty]{} \mu_0. \tag{8}$$

Now note that $\int_\Omega x_n \, d\mathsf{P} = n \mathsf{M} \Delta_{nk} = n\pi_1(1/n)$. Thus, the existence of the limit

$$\lambda = \lim_{n \to \infty} \frac{\pi_1(1/n)}{1/n} \tag{9}$$

is established and at the same time the equality $\lambda = \mu_0$ is proved.

The feasibility of replacing the variable $1/n$ by a continuous variable $h \to 0$ is justified by the monotonicity of $\pi_1(1/n)$: for $1/n \leqslant h < 1/(n-1)$, we have

$$\frac{\pi_1(1/n)}{1/(n-1)} \leqslant \frac{\pi_1(h)}{h} \leqslant \frac{\pi_1(1/(n-1))}{1/n};$$

evidently both bounds tend to λ. If the initial stream is ordinary, then $x_0(t) = x(t)$, $\mu_0 = \mu$, and we arrive at Korolyuk's theorem (Section 2.5.5).

2.6.4. *Dobrushin's Theorem*

Dobrushin (1965) proved the following theorem:

Theorem. *If a stationary process is of finite intensity and satisfies property* (4), *then the process is orderly, i.e.,* $\pi_2(t) = o(t)$, $t \to 0$.

Dobrushin's theorem follows from relation (7):

$$\frac{\pi_2(1/n)}{1/n} \leqslant \mathsf{P}\{x(1) - x_n \geqslant 1\} \leqslant \mathsf{M}\{x(1) - x_n\} \leqslant \int_{\Omega \setminus A_n} x(1) \, d\mathsf{P} \xrightarrow[n \to \infty]{} 0.$$

The transition from $1/n$ to h is the same as above.

2.6.5. *The Existence of the Palm-Khinchin Function*

Let $x(t)$ be a stationary stream and τ_n be the times of its events ($\tau_1 < \tau_2 < \cdots$). Denote by $y(t|x, k)$ the number of τ_n, $0 < \tau_n < t$, such that in the interval $(\tau_n, \tau_n + x)$ exactly k events of the stream occurred. It is easy to verify that $y(t|x, k)$ is a stationary stream. We shall denote its parameter by $\lambda(x, k)$. Then, by definition of the parameter, $\varphi_k(x) = \lambda(x, k)/\lambda$ is the conditional probability that under the condition that a customer arrived in the interval $(0, dt)$, during the time interval x after him or her yet another k customers will arrive. Actually instead of the interval $(x, x+h)$ we have here the interval $(\theta, x+\theta)$, where θ is the time of arrival of a customer in the interval $(0, h)$. However, one could proceed from θ to h retaining the meaning of all the formulas if we confine ourselves only to those x for which $\lambda(x, k)$ is continuous. The probability of two or more events at different times of the interval $(0, h)$ is of order $o(h)$ in view of the orderliness of the stream $x(t)$.

2.6.6. *The k-Intensity of a Stream*

Belyaev (1969) introduced the notion of *k-intensity* of a stream $\lambda_k(t_1, \ldots, t_k)$, which means that the probability of the occurrence of events in the intervals $(t_1, t_1 + h_1), \ldots, (t_k, t_k + h)$ is equal to $\lambda_k(t_1, \ldots, t_k)h_1 \cdots h_k + o(h_1 \cdots h_k)$. This notion is successfully used to bound probabilities of events associated with streams. For example, let A be the event that in the interval $(0, t)$ at least one customer has arrived. Denote by A the event that a customer arrived in the interval

$$\left(\frac{k-1}{n}t, \frac{k}{n}t\right),$$

whence

$$\sum_{k=1}^{n} \mathsf{P}\{A_{nk}\} - \sum_{1 \leqslant k < m \leqslant n} \mathsf{P}\{A_{nk}A_{nm}\} \leqslant \mathsf{P}\{A\} \leqslant \sum_{k=1}^{n} \mathsf{P}\{A_{nk}\}.$$

Assuming that $\lambda_1(x)$ and $\lambda_2(x, y)$ are Riemann integrable, one can approach the limit as $n \to \infty$ and thus obtain

$$\int_0^t \lambda_1(x)\,dx - \iint\limits_{0<x<y<t} \lambda_2(x, y)\,dx\,dy \leqslant \mathsf{P}\{A\} \leqslant \int_0^t \lambda_1(x)\,dx.$$

2.7. Basic Renewal Theory

2.7.1. *Definition of Renewal Processes (Renewal Streams)*

Assume that a device may fail at random times. At the instant of failure the device is replaced by a new one, which is replaced by a third one if it fails, and so on. We shall plot the times of successive failures $t_1, t_2, t_3 \ldots$ on a time axis. These can naturally be regarded as the times at which the working condition of the device is renewed. Consider the sequence of random variables

$$z_1 = t_1, \quad z_2 = t_2 - t_1, \quad z_3 = t_3 - t_2, \ldots, \quad z_k = t_k - t_{k-1}, \ldots,$$

which represent for $k > 1$ the intervals between successive renewal times. The variable z_1 plays a somewhat different role since renewal does not necessarily occur at a time $t = 0$. Denote by $F_k(x)$ the distribution function of the variable z_k.

Observing the stream of calls arriving at a telephone exchange, we can arrange them in order of their arrival in the same maner as we have done for the times of equipment failures. There are many examples of events of various kinds (accidents, production of various items, landing of airplanes, etc.) following each other at random times. Much attention has been given recently to an investigation of such processes. Until quite recently, renewal processes (streams) were regarded only as sequences of times t_k, with the condition that the variables z_k are independent and identically distributed. A number of papers have now appeared discussing nonnegative variables z_k with different distributions, which are either independent or form a Markov chain.

A sequence of times t_k formed by independent (nonnegative) variables z_k with distributions $F_k(x)$ was called a *stream with limited aftereffects* by Khinchin. He referred to a stationary orderly stream with limited aftereffects as a *Palm-type stream.* The terminology suggested by Khinchin is widely used by Soviet authors. If all $F_k(x)$ with the possible exception of $F_1(x)$ coincide, i.e.,

$$F_k(x) = F(x), k \geqslant 2, \quad \text{where } F(+0) < 1,$$

we say that the sequence $\{t_k\}_{k=1}^{\infty}$ forms a *renewal process.* Thus, the notion of a renewal process is narrower than that of the stream with limited aftereffects.

It is easy to verify that for stationary *renewal streams*, the functions $F_k(x)$ must be identical for $k \geqslant 1$; for what follows we introduce the notation

$$F(x) = \mathsf{P}\{z_k < x\} = F_k(x) \quad (k = 2, 3, \ldots).$$

As we have already noted, the variable z_1, and hence also $F_1(x)$, play a special role. Later we shall obtain a connection between $F_1(x)$ and $F(x)$ for stationary renewal streams.

Of particular interest in renewal theory is the random variable N_t, defined as the number of renewals up to time t, i.e., the largest n such that

$$t_n = z_1 + z_2 + \cdots + z_n < t.$$

The mathematical expectation of the variable N_t is called the *renewal function* and is denoted by

$$H(t) = \mathsf{M}N_t.$$

This function and its asymptotic behavior as $t \to \infty$ appear in a great number of the applications of renewal theory.

Numerous investigations have been devoted to renewal theory. Good introductions are Cox (1962) and Smith (1958).

In what follows we shall assume, unless otherwise specified, that at each renewal time only one event of the stream occurs.

2.7.2. *A Property of Renewal Streams*

In any orderly, stationary renewal stream

$$\pi_{r+1}(u) \leqslant \pi_r(u)o(1),$$

for all $r > 0$ as $u \to 0$ where (as usual, in this book)

$$\pi_r(u) = \sum_{k=r}^{\infty} p_k(u),$$

i.e., the probability that at least r events occur in a time interval of length u.

PROOF. Since the event $t_{r+1} < u$ always implies the compound event $t_r < u$ and $z_{r+1} < u$, we have

$$\pi_{r+1}(u) = \mathsf{P}\{t_{r+1} < u\} \leqslant \mathsf{P}\{t_r < u, z_{r+1} < u\}.$$

Hence, since the variables t_r and z_{r+1} are independent, we have

$$\pi_{r+1}(u) \leqslant \pi_r(u)F(u).$$

Since for an orderly stream we clearly have $F(+0) = 0$, $F(u) = o(1)$, which proves the theorem. □

Observe that we do not exclude the case $\pi_r(u) = 0$ for $0 \leqslant u < u_0$: this holds in the case when $F(u_0/r) = 0$.

It should be noted that for a simple stream this property follows directly from the formulas

$$P_k(u) = \frac{(\lambda u)^k}{k!} e^{-\lambda u}.$$

2.7.3. *Relation to the Palm-Khinchin Functions*

A renewal stream (or a *recurrent stream*) is defined by assigning the distribution functions $F(x)$, $F_1(x)$. For stationary renewal streams, or, in Khinchin's terminology, Palm-type streams, it is sufficient (as we shall see) to define a single Palm-Khinchin function. This result is presented in the following theorem.

Theorem. *For stationary renewal streams the equality*

$$F_1(x) = \lambda \int_0^x \varphi_0(u)\, du \tag{1}$$

is valid, and for $k \geqslant 2$

$$F(x) = 1 - \varphi_0(x). \tag{2}$$

PROOF. By definition, $F_1(x)$ is the probability that customers arrive in the interval $(0, x)$. Thus, it is equal to

$$\pi_1(x) = 1 - p_0(x).$$

Formula (5) of Section 2.5.3. yields (1). Note that from (1) it follows that

$$\lambda \int_0^\infty \varphi_0(t)\,dt = F_1(+\infty) = 1, \tag{3}$$

which relates the stream parameter and the function $\varphi_0(t)$.

We now prove inequality (2). Consider the intervals $(0, h)$ and $(h, x + h)$. The probability that there is a customer in the first and no customer in the second is

$$\int_0^h F(x + h - \theta)\,dF_1(\theta) + o(h).$$

[The term $o(h)$ accounts for the case of more than one event in the interval of length h.] The first expression is bounded from below and above by $F(x) \cdot F_1(h)$ and $F(x + h) \cdot F_1(h)$, respectively. To arrive at (2) it is sufficient to observe that by definition of the parameter of the stream and in view of the orderliness of the stream

$$F_1(h) = \lambda h + o(h), \quad h \to 0. \tag{4}$$

Now formula (2) is obtained by dividing the expressions obtained by $F_1(h)$ as given by (4) and approaching the limit as $h \to 0$. Note that, as it follows from (1) and (2), the functions $F_1(x)$ and $F(x)$ are related by

$$F_1(x) = \lambda \int_0^x [1 - F(u)]\,du, \tag{5}$$

provided the stream is a stationary renewal stream.

It is of course natural to investigate the features of streams for which $F_1(x) = F(x)$. If this equality is satisfied, then in view of (5)

$$\lambda \int_0^x [1 - F(u)]\,du = F(x).$$

Differentiation with respect to x yields

$$\lambda[1 - F(x)] = F'(x),$$

hence

$$1 - F(x) = Ce^{-\lambda x}.$$

Thus, since $F(+\infty) = 1$, $C = 1$. Hence,

$$F(x) = 1 - e^{-\lambda x}.$$

We know that this function defines a simple stream. □

The following question arises. Is an arbitrary orderly renewal stream satisfying (5) stationary? The answer is affirmative, but the proof is postponed to subsection 5 of this section.

2.7.4. *Definition of the Palm-Khinchin Function for Stationary Renewal Streams*

Since stationary renewal streams are completely determined by the function $\varphi_0(t)$, all other Palm-Khinchin functions should be expressible in terms of this function. We shall show how one can compute any function $\varphi_k(t)$. First, we calculate the probability that in the interval $(0, t)$ exactly one customer arrives, i.e., $p_1(t)$. One customer arrives in this interval if and only if prior to time $x < t$ no customer arrives, at time x a customer arrives, and the next customer arrives only after completion of the interval $(0, t)$. Here it must be taken into account that x can assume any value between 0 and t. In view of (1) and (2) we have

$$p_1(t) = \lambda \int_0^t \varphi_0(x)\varphi_0(t-x)\,dx.$$

By (9) of Section 2.5.3.

$$p_1(t) = \lambda \int_0^t [\varphi_0(x) - \varphi_1(x)]\,dx.$$

Comparison of these two formulas yields the equality

$$\int_0^t \varphi_1(x)\,dx = \int_0^t \varphi_0(x)\,dx - \int_0^t \varphi_0(x)\varphi_0(t-x)\,dx$$

and formula

$$\varphi_1(x) = \varphi_0(x) - \frac{d}{dt}\int_0^t \varphi_0(x)\varphi_0(t-x)\,dx.$$

Similarly, we obtain

$$p_2(t) = -\lambda \int_0^t \varphi_0(x) \int_0^{t-x} \varphi_0(t-x-y)\,d\varphi_0(y)\,dx.$$

From (9) of Section 2.5.3. and the preceding expression, we have

$$\begin{aligned}\varphi_2(t) &= \frac{d}{dt}\int_0^t \left[\varphi_1(x) - \varphi_0(x)\int_0^{t-x} \varphi_0(t-x-y)\,d\varphi_0(y)\right]dx \\ &= \frac{d}{dt}\int_0^t \varphi_0(x)\left[1 - \varphi_0(t-x) + \int_0^{t-x} \varphi_0(t-x-y)\,d\varphi_0(y)\right]dx.\end{aligned}$$

Using the same method, all of the Palm-Khinchin functions can be calculated.

As an example, consider a stationary renewal stream for which

$$F'(x) = \begin{cases} 0 & \text{if } x < 0, \\ xe^{-x} & \text{if } x \geqslant 0. \end{cases}$$

An easy calculation yields

$$F(x) = 1 - e^{-x}(1 + x) \quad \text{for } x > 0.$$

In view of formula (2),

$$\varphi_0(x) = e^{-x}(1 + x).$$

Hence, from (3), $\lambda = \frac{1}{2}$, and in view of (1) we have

$$F_1(x) = 1 - e^{-x}(\tfrac{1}{2} + x).$$

We now can calculate that the Palm-Khinchin functions are of the form

$$\varphi_n(t) = e^{-t}\left[\frac{t^{2n}}{(2n)!} + \frac{t^{2n+1}}{(2n+1)!}\right].$$

2.7.5. *Basic Formulas for Renewal Processes*

Since $t_n = z_1 + z_2 + \cdots + z_n$, we have the following equality:

$$F_n(t) = \mathsf{P}\{t_n < t\} = \int_0^t F_{n-1}(t - x)\,dF(x)$$

for all $n \geqslant 2$. Since

$$\mathsf{P}\{N_t = n\} = F_n(t) - F_{n+1}(t),$$

it follows that

$$\begin{aligned} H(t) = \mathsf{M}N_t &= \sum_{n=1}^{\infty} n[F_n(t) - F_{n+1}(t)] = \sum_{n=1}^{\infty} F_n(t) \\ &= F_1(t) + \sum_{n=2}^{\infty} \int_0^t F_{n-1}(t - x)\,dF(x) \\ &= F_1(t) + \int_0^t \sum_{n=2}^{\infty} F_{n-1}(t - x)\,dF(x) = F_1(t) + \int_0^t H(t - x)\,dF(x). \end{aligned}$$

We have obtained an important integral equation

$$H(t) = F_1(t) + \int_0^t H(t - x)\,dF(x). \tag{6}$$

Equation (6) is solved by the Laplace-Stieltjes transform method. Denote

$$\psi_1(s) = \int_0^{\infty} e^{-sx}\,dF_1(x), \quad \psi(s) = \int_0^{\infty} e^{-sx}\,dF(x), \quad \varphi(s) = \int_0^{\infty} e^{-sx}\,dH(x).$$

Then from (6) we obtain

$$\begin{gathered} \varphi(s) = \psi_1(s) + \varphi(s)\psi(s), \\ \varphi(s) = \frac{\psi_1(s)}{1 - \psi(s)}. \end{gathered} \tag{7}$$

If $F_1(x) = \lambda \int_0^x [1 - F(u)]\,du$, then integration by parts yields

$$\psi_1(s) = \frac{\lambda}{s}[1 - \psi(s)],$$

so that

$$\varphi(s) = \frac{\lambda}{s}.$$

Inverting the Laplace-Stieltjes transform, we obtain $dH(x) = \lambda\, dx$. On the other hand, since $H(x)$ is a nondecreasing function, it follows from (6) that

$$H(t) \leqslant F_1(t) + H(t)F(t),$$

i.e., $H(t) \leqslant F_1(t)/[1 - F(t)]$. Hence, $H(+0) = 0$, i.e.,

$$H(x) = H(+0) + \int_{+0}^{x} dH(t) = \lambda x.$$

Thus equation (5) implies that

$$H(x) = \lambda x. \tag{8}$$

For applications, the following two interpretations of the Stieltjes integral for any renewal function are of importance.

1. At any renewal time t an impulse arises that yields an effect (action) $u(t)$ at time $t_n + t$, where $u(t)$ is a continuous function, $t \geqslant 0$. Since the total effect of impulses at time t arising before this time is equal to

$$\sum_{n=1}^{N_t} u(t - t_n) = \sum_{n=1}^{\infty} u(t - t_n),$$

where it is assumed that $u(x) = 0$ for $x < 0$. Mathematical expectation of this effect is equal to

$$\sum_{n=1}^{\infty} \int_0^t u(t - x)\, d\mathsf{P}\{t_n < x\} = \int_0^t u(t - x)\, d \sum_{n=1}^{\infty} \mathsf{P}\{t_n < x\} = \int_0^t u(t - x)\, dH(x).$$

Finally

$$\mathsf{M} \sum_n u(t - t_n) = \int_0^t u(t - x)\, dH(x). \tag{9}$$

2. Let $u(t)$ be a continuous bounded function. There corresponds to the sequence $\{t_n\}$ a random function $\xi(t)$, where $\xi(t) = 0$ for $0 \leqslant t \leqslant t_1$; $\xi(t) = u(t - t_n)$ for $t_n \leqslant t < t_{n+1}$, $n \geqslant 1$. Then

$$\mathsf{M}\xi(t) = \int_0^t u(t - x)[1 - F(t - x)]\, dH(x). \tag{10}$$

This equality follows from the formula of the total mathematical expectation

$$\mathsf{M}\xi(t) = \sum_{n=1}^{\infty} \mathsf{M}\xi(t) I_n(t),$$

where $I_n(t)$ is the indicator function of the event $\{t_n \leqslant t < t_{n+1}\}$. We have

$$\mathsf{M}\xi(t)I_n(t) = \int_0^t u(t-x)[1-F(t-x)]\,d\mathsf{P}\{t_n < x\},$$

whence the formula for $\mathsf{M}\xi(t)$ is obtained by summing over n after the order of summation and integration is interchanged. This is justified since $u(t)$ is bounded and $H(t)$ is finite. If $u(t)$ is a function of bounded variation, then (10) is valid for almost all t.

We can now solve the problem of stationarity of an orderly renewal stream, when $F_1(x)$ is given by (5).

Choose an arbitrary time $a > 0$ and denote by $t_n^{(a)} + a$ the times of the events of the stream in the interval $(a, +\infty)$. Then, as it is easy to see, stationarity means that $|t_n^{(a)}|$ has the same distribution as $\{t_n\}$—the sequence of times of events of the initial stream.

Clearly the random variables $z_n^{(a)} = t_n^{(a)} - t_{n-1}^{(a)}$, $n \geqslant 2$, are independent and have the distribution function $F(x)$. It remains to verify the equality

$$\mathsf{P}\{t_1^{(a)} < x\} = F_1(x).$$

The formula of total probability yields

$$\mathsf{P}\{t_1^{(a)} < x\} = \mathsf{P}\{a \leqslant t_1 < a + x\} + \sum_{n=1}^{\infty} [F(a+x-t) - F(a-t)]\,d\mathsf{P}\{t_n < t\}.$$

Whence in accordance with formula (10) we have

$$\begin{aligned}
\mathsf{P}\{t_1^{(a)} < x\} &= F_1(a+x) - F_1(a) + \int_0^a [F(a+x-t) - F(a-t)]\,dH(t) \\
&= F_1(a+x) - F_1(a) + \lambda \int_0^a [F(a+x-t) - F(a-t)]\,dt \\
&= \lambda \int_0^{a+x} [1-F(t)]\,dt - \lambda \int_0^a [1-F(t)]\,dt + \lambda \int_0^a [1-F(t)]\,dt \\
&\quad - \lambda \int_x^{a+x} [1-F(t)]\,dt \\
&= \lambda \int_0^x [1-F(t)]\,dt = F_1(x)
\end{aligned}$$

Q.E.D.

2.7.6. *Statements of Some Theorems on Stationary Renewal Processes*

The random variable N_t, which is equal to the number of renewals in the time interval from 0 to t, has moments of all orders. We shall prove this assertion. By assumption, $F(+0) < 1$; hence there exists an $\alpha > 0$ such that $\mathsf{P}\{z_i > \alpha\} =$

$\beta > 0$. Define now a new renewal process

$$t'_k = \sum_{n=1}^{k} z'_n$$

where

$$z'_n = \begin{cases} 0 & \text{if } z_n \leqslant \alpha, \\ \alpha & \text{if } z_n > \alpha. \end{cases}$$

Then the inequality $t'_k \leqslant t_k$ holds with probability 1, i.e.,

$$\mathsf{P}\{N_t \geqslant k\} \leqslant \mathsf{P}\{N'_t \geqslant k\},$$

where N'_t denotes the number of renewals corresponding to the process $\{t'_n\}$.

However, $\mathsf{P}\{N'_t = k\}$ is the probability that in k independent trials with probability of success β there will be exactly l successes, and then one more success occurs:

$$\mathsf{P}\{N'_t = k\} = \begin{cases} C_k^l \beta^{l+1}(1-\beta)^{k-l} & \text{for } k \geqslant l, \\ 0, & \text{for } k < l \end{cases}$$

(here $l = [t/\alpha]$).

We have

$$\mathsf{M}[N'_t]^r = \sum_{k=l}^{\infty} k^r C_k^l \beta^{l+1}(1-\beta)^{k-l} < \sum_{k=0}^{\infty} (k+l)^{r+l}(1-\beta)^k.$$

Since $\beta > 0$, this series is convergent, which proves our assertion.

In a similar manner, we can prove a stronger result:

For every distribution $F(x)$, there exists a number $\alpha > 0$ such that for any s whose real part does not exceed α, $\mathsf{M}e^{sN_t}$ exists.

One of the original general results related to the renewal function $H(t)$ is the so-called *Elementary Renewal Theorem*:

$$\frac{H(t)}{t} \to \frac{1}{a} \quad \text{as } t \to \infty,$$

where

$$a = \mathsf{M}z_2 = \int_0^{\infty} x\,dF(x).$$

This result remains true also when $a = \infty$.

If $F(x)$ is a non-lattice distribution, i.e. the points of increase of the function do not form an arithmetical progression, we have

Blackwell's Theorem:

$$H(t+\alpha) - H(t) \to \frac{\alpha}{a}, \quad t \to \infty.$$

If $F(x)$ possesses a finite second moment $\mu_2 = \mathsf{M}z_2^2$ and is non-lattice then we have the following theorem:

Smith's Theorem: *For* $t \to \infty$

$$H(t) - \frac{t}{a} \to \frac{\mu_2}{2a^2} - 1.$$

Let $Q(x)$ be an arbitrary nonnegative function defined for positive x, non-increasing and integrable in the interval $(0, \infty)$. Under these conditions Smith proved the following:

Limit Theorem: *As* $t \to \infty$

$$\int_0^t Q(t-u)\,dH(u) \to \frac{1}{a}\int_0^\infty Q(x)\,dx. \tag{11}$$

Smith called this result the *Key Renewal Theorem*. For suitable choices of the function $Q(x)$ all our previous results follow as special cases. If $\mu_2 < \infty$, then

$$\mathsf{D}N_t = \frac{\mu_2 - a^2}{a^3}t + o(t).$$

If also $\mathsf{M}|z_2|^3 < \infty$, $\mu_3 = \mathsf{M}z_2^3$, then

$$\mathsf{D}N_t = \frac{\mu_2 - a^2}{a^3}t + \left(\frac{5\mu_2^2}{4a^4} - \frac{2\mu_3}{3a^3} - \frac{\mu_2}{2a^2}\right) + o(1).$$

In many problems of renewal theory the following simple identity is useful:

$$\mathsf{P}\{t_n \leqslant t\} = \mathsf{P}\{N_t \geqslant n\}. \tag{12}$$

If $\mu_2 < \infty$ and $\sigma^2 = \mu_2 - a^2$, then

$$\mathsf{P}\left\{N_t \geqslant \frac{t}{a} - \frac{x\sigma}{a}\sqrt{\frac{t}{a}}\right\} \to \frac{1}{\sqrt{2\pi}}\int_{-\infty}^x e^{-z^2/2}\,dz.$$

A bound on the remainder in the theorem on asymptotic normality of the number of renewals is presented in Englund's (1980) paper.

We shall note another important result, which follows directly from (6).

Denote by $\gamma(t)$ the time elapsing from the instant t to the next renewal, and by $\gamma^*(t)$ the time elapsed from the previous renewal to t. Thus, if n is defined by the condition

$$t_n \leqslant t < t_{n+1},$$

then

$$\gamma(t) = t_{n+1} - t, \qquad \gamma^*(t) = t - t_n;$$

for $t_n > t$ we set $\gamma^*(t) = t$.

The variable $\gamma(t)$ is called the *amount of overjump* (over the level t) and $\gamma^*(t)$ is the *amount of underjump* (up to level t). Both variables play an important role in the theory of random walks.

Let $F_\gamma(x,t)$ and $F_{\gamma*}(x,t)$ denote the distribution functions of the random variables $\gamma(t)$ and $\gamma^*(t)$, respectively. Our goal is to find the limits of $F_\gamma(x,t)$ and $F_{\gamma*}(x,t)$ as $t \to \infty$, assuming that the interrenewal times have a finite mathematical expectation.

The event $\{\gamma(t) < x\}$ may occur in two mutually exclusive ways:

1. $t \leqslant z_1 < t + x$;
2. At some time $\tau < t$ a renewal occurs; the time elapsing until the next renewal is between $t - \tau$ and $t - \tau + x$.

The total probability formula taking (1) into account, yields

$$F_\gamma(x,t) = \mathsf{P}\{\gamma(t) < x\}$$
$$= F_1(t+x) - F_1(t) + \int_0^t [F(t-\tau+x) - F(t-\tau)]\, dH(\tau).$$

The first terms becomes infinitely small as $t \to \infty$; the conditions of Smith's theorem are not fulfilled for the second term since $F(t+x) - F(t)$ is not necessarily a nondecreasing function. However,

$$F_\gamma(x,t) = F_1(t+x) - F_1(t) + R_1(t) - R_2(t),$$

where

$$R_1(t) = \int_0^t [1 - F(t-\tau)]\, dH(\tau),$$

$$R_2(t) = \int_0^t [1 - F(t-\tau+x)]\, dH(\tau).$$

In the last two expressions the kernel $1 - F(\cdots)$ satisfies all the conditions of Smith's theorem. In that case, however,

$$R_1(t) \xrightarrow[t\to\infty]{} \frac{1}{\mathsf{M}z_2} \int_0^\infty [1 - F(\tau)]\, d\tau = 1,$$

$$R_2(t) \xrightarrow[t\to\infty]{} \frac{1}{\mathsf{M}z_2} \int_0^\infty [1 - F(\tau+x)]\, d\tau = \frac{\int_x^\infty [1 - F(\tau)]\, d\tau}{\int_0^\infty [1 - F(\tau)]\, d\tau},$$

so that

$$F_\gamma(x,t) \xrightarrow[t\to\infty]{} \frac{\int_0^x [1 - F(\tau)]\, d\tau}{\int_0^\infty [1 - F(\tau)]\, d\tau}. \tag{13}$$

The limiting distribution obtained is continuous. As is evident from the right-hand side of (13), the function

$$p_{\gamma}(x) = \frac{1 - F(x)}{\displaystyle\int_0^{\infty} [1 - F(\tau)]\, d\tau} \tag{14}$$

is the density of the limiting distribution of $\gamma(t)$ as $t \to \infty$.

We have already proved this relation earlier for stationary streams of homogeneous events. We leave it to the reader to prove, in a similar manner, the assertion that the random variable $\gamma^*(t)$ has the same limiting distribution as $\gamma(t)$ as $t \to \infty$.

2.8. Limit Theorems for Compound Streams

2.8.1. *Statement of the Problem*

We have already noted that the initial assumption in most of the literature on queueing theory is that the incoming stream is simple. However, in many cases the initial conditions defining the simple stream, studied in Chapter 1, are not appropriate to the physical situation. Indeed, streams occurring the practice are quite often different from simple streams. In view of the great diversity of conditions under which specific phenomena occur, these deviations are the rule rather than the exception. It turns out, however, that substantial discrepancies occur considerably less frequently than one would expect from *a priori* considerations.

Thus, along with the question about the reasons for the occurrence of nonsimple streams, the converse question arises: Why do simple streams so often correspond to real-world situations. This problem was dealt with by Palm (1943), Rényi (1956), Khinchin (1963), Ososkov (1956), and Grigelionis (1962a). The basic idea of these investigations was the assumption that the streams observed are sums of a large number of independent streams of small intensity, each one of which is orderly and stationary. No assumptions were made about the absence of aftereffects. Grigelionis studied this problem under somewhat more general conditions; we shall discuss his work below in more detail.

Compound streams occur very often. Indeed, the stream of calls reaching a telephone exchange is the sum of streams originating from different subscribers. The stream of ships arriving at a harbor is the sum of streams departing from various other harbors. The stream of "customers" for servicing by a repairman crew also comes from different sources: each piece of equipment represents such an elementary source of customers; the total customer stream reaching the repair crew is thus a sum of elementary streams. The reader can easily find other examples in his or her own field of activity, so there is no need for further elaboration.

We shall see that under quite general conditions on the component streams, compound streams are approximated by Poisson streams, including simple streams. Our presentation follows an idea expressed by Gnedenko and developed by Grigelionis (1962a). This idea served as a guiding principle for almost 200 years in numerous applications of probability theory, such as the theory of errors, molecular physics, ballistics, and many others. It consists of considering the observed action as a sum of elementary effects, each being a random variable independent of the others, each summand exerting a small influence in a certain sense on the total.

Pogozhev (1964) and Grigelionis (1962b) have also studied another important problem. Given that a sum of a large number of streams, each exerting only a small influence on the sum, approaches a Poisson stream, the question is how rapidly does the distribution function of the consecutive sums approach the distribution function of the limiting stream? In what follow, we shall state some of their results.

2.8.2. *Definitions and Notation*

We shall say that a random process $x(t)$ is a *step process* if the increments $x(t) - x(s)$ take on only nonnegative integer values for $t > s > 0$. We shall assume that $x(0) = 0$, i.e., the process starts at time $t = 0$. The values of the process $x(t)$ may be interpreted as the number of events of some kind occurring up to time t. These events may be calls arriving at a telephone exchange, failures of units of complex equipment, patients arriving at admission in a hospital, and so on.

Let

$$x_n(t) = \sum_{r=1}^{k_n} x_{nr}(t),$$

where $x_{nr}(t)$ are mutually independent step processes. Clearly, $x_n(t)$ is also a step process.

We shall say that a sequence of processes $x_n(t)$ *converges* to the process $x(t)$ as $n \to \infty$, if the distribution functions of the vectors

$$\{x_n(t_1), x_n(t_2), \ldots, x_n(t_k)\}$$

converge, for any k and $t_1, t_2, \ldots, t_k$, at every point of continuity to the distribution function of the vector

$$\{x(t_1), x(t_2), \ldots, x(t_k)\}.$$

In Section 2.2 we introduced the concept of a *Poisson process* $x(t)$ *with leading function* $\Lambda(t)$, as a process with independent increments such that for all $s < t$ and every nonnegative integer k

$$\mathsf{P}\{x(t) - x(s) = k\} = \frac{[\Lambda(t) - \Lambda(s)]^k}{k!} e^{-[\Lambda(t) - \Lambda(s)]}. \tag{1}$$

The function $\Lambda(t)$, which Khinchin called the *leading* function, is nonnegative, left-continuous, finite, and vanishes for $t \leqslant 0$.

Introduce the following notation:

$$p_{nr}(k, t, s) = \mathsf{P}\{x_{nr}(t) - x_{nr}(s) = k\}, \quad s < t, \quad k = 0, 1, 2, \ldots,$$

$$\Lambda_n(t, s) = \sum_{r=1}^{k_n} p_{nr}(1; t, s), \tag{2}$$

$$B_n(t, s) = \sum (1 - p_{nr}(0; t, s) - p_{nr}(1; t, s)). \tag{3}$$

The processes $x_{nr}(t)$ $(r = 1, 2, \ldots, k_n)$ are said to be infinitesimal if, for any fixed t,

$$\lim_{n \to \infty} \max_{1 \leqslant r \leqslant k_n} [1 - p_{nr}(0; t, 0)] = 0. \tag{4}$$

In other words, the processes $x_{nr}(t)$ are infinitesimal if for every $\varepsilon > 0$ and arbitrary fixed t, there exists a number n such that, uniformly in r ($r = 1, 2, \ldots, k_n$),

$$\mathsf{P}\{x_{nr}(t) > 0\} < \varepsilon.$$

2.8.3. *Statement of the Basic Result and a Proof of Necessity*

We shall now state and prove a theorem due to Grigelionis. We shall verify the necessity of its conditions using a limit theorem for sums of independent random variables that was proved almost simultaneously and independently by Gnedenko (1939) and Marcinkiewicz (1937).

Theorem. *The sums $x_n(t) = \sum_{r=1}^{k_n} x_{nr}(t)$ of independent infinitesimal processes $x_{nr}(t)$ converge to a Poisson process with the leading function $\Lambda(t)$ if and only if for any fixed s and t ($s < t$)*

$$\lim_{n \to \infty} \Lambda_n(t, s) = \Lambda(t) - \Lambda(s) \tag{5}$$

and

$$\lim_{n \to \infty} B_n(t, 0) = 0. \tag{6}$$

The proof of necessity of the theorem's conditions is based on the following assertion in the theory of sums of independent random variables due to Gnedenko (1961). If random variables $x_{n1}, x_{n2}, \ldots, x_{nk_n}$ are independent and infinitesimal, i.e., for any $\varepsilon > 0$ and $n \to \infty$

$$\sup_{1 \leqslant k \leqslant k_n} \mathsf{P}\{|x_{nk}| > \varepsilon\} \to 0,$$

then in order that the distribution function of the sum

$$s_n = x_{n1} + x_{n2} + \cdots + x_{nk_n}$$

converge to a Poisson distribution as $n \to \infty$

$$P(x) = \sum_{0 \leqslant k < x} \frac{\lambda^k}{k!} e^{-\lambda},$$

it is necessary and sufficient that the following conditions be fulfilled: for any ε $(0 < \varepsilon < 1)$ and $n \to \infty$:

1. $$\sum_{k=1}^{k_n} \int_{R_\varepsilon} dF_{nk}(x) \to 0.$$

2. $$\sum_{k=1}^{k_n} \int_{|x-1|<\varepsilon} dF_{nk}(x) \to \lambda.$$

3. $$\sum_{k=1}^{k_n} \int_{|x|<\varepsilon} x\, dF_{nk}(x) \to 0.$$

4. $$\sum_{k=1}^{k_n} \left[\int_{|x|<\varepsilon} x^2\, dF_{nk}(x) - \left(\int_{|x|<\varepsilon} x\, dF_{nk}(x)\right)^2\right] \to 0.$$

Here we introduce the notation: $F_{nk}(x) = \mathsf{P}\{x_{nk} < x\}$, R_ε is the region obtained from infinite line by omitting all the intervals $|x| < \varepsilon$ and $|x - 1| < \varepsilon$.

Note that in Grigelionis' theorem we must set

$$\lambda = \Lambda(t) - \Lambda(s), \qquad \int_{|x-1|<\varepsilon} dF_{nk}(x) = p_{nk}(1; t, s),$$

$$\int_{R_\varepsilon} dF_{nk}(x) = 1 - p_{nk}(0; t, s) - p_{nk}(1; t, s).$$

Thus the first and second conditions of the previously stated Gnedenko-Marcinkiewicz theorem completely coincide with conditions (5) and (6) of Grigelionis' theorem. The third and fourth conditions of Gnedenko-Marcinkiewicz's theorem for step processes is automatically fulfilled since in the interval $|x| < \varepsilon$ their functions possess a unique jump point, $x = 0$.

The necessity of Grigelionis' theorem thus follows from the fact that if the processes converge, their one-dimensional distributions must also converge. The convergence of the one-dimensional distributions has however been investigated in the foregoing.

2.8.4. *Proof of Sufficiency*

We must now prove that conditions (4)–(6) ensure both asymptotic independence of the increments of the process $x_n(t)$ and the convergence of the one-dimensional distributions to the corresponding Poisson distributions. The latter follows from the Gnedenko-Marcinkiewicz theorem and from the identity of its conditions with (4)–(6). Nevertheless, we shall present the arguments, since they are simple and not too tedious. Our proof will use the basic theorems of the theory of characteristic functions.

Consider vectors of the form

$$\bar{l} = (l_1, l_2, \ldots, l_m),$$

where $l_\nu \geqslant 0$ are integers:

$$\bar{0} = (\underbrace{0, 0, \ldots, 0}_{m}), \quad \bar{e}_\nu = (\underbrace{0, \ldots, 0}_{\nu-1}, 1, \underbrace{0, \ldots, 0}_{m-\nu}),$$

$$\bar{T} = (t_0, t_1, \ldots, t_m),$$

$0 < t_0 < t_1 < \cdots < t_m$ are arbitrary real numbers,

$$\bar{\alpha} = (\alpha_1, \alpha_2, \ldots, \alpha_m),$$

$$\bar{x}_{nr}(\bar{T}) = (x_{nr}(t_1) - x_{nr}(t_0), \ldots, x_{nr}(t_m) - x_{nr}(t_{m-1})),$$

$$\bar{x}_n(\bar{T}) = \sum_{r=1}^{k_n} \bar{x}_{nr}(\bar{T}).$$

In addition, denote

$$(\bar{\alpha}, \bar{\beta}) = \sum_{i=1}^{m} \alpha_i \beta_i,$$

$$p_{nr}(\bar{l}, \bar{T}) = \mathsf{P}\{\bar{x}_{nr}(\bar{T}) = \bar{l}\},$$

$$f_{nr}(\bar{\alpha}, \bar{T}) = \mathsf{M} \exp i(\bar{\alpha}, \bar{x}_{nr}(\bar{T})),$$

$$f_n(\bar{\alpha}, \bar{T}) = \mathsf{M} \exp i(\bar{\alpha}, \bar{x}_n(\bar{T})).$$

For the distributions of the vectors $\bar{x}_n(\bar{T})$ to converge to the corresponding distributions of the Poisson process, it is sufficient that their characteristic functions converge. We shall prove this assertion.

Since the processes $x_{nr}(t)$ are independent, we have

$$f_n(\bar{\alpha}, \bar{T}) = \prod_{r=1}^{k_n} f_{nr}(\bar{\alpha}, \bar{T}).$$

However,

$$f_{nr}(\bar{\alpha}, \bar{T}) = \sum_{\bar{l}} p_{nr}(\bar{l}, \bar{T}) e^{i(\bar{\alpha}, \bar{l})} = 1 + \sum_{\bar{l} \neq \bar{0}} p_{nr}(\bar{l}, \bar{T})(e^{i(\bar{\alpha}, \bar{l})} - 1),$$

where $\sum_{\bar{l}}$ denotes the summation over all the possible integer-valued vectors $\bar{l}$ with nonnegative components.

For small x

$$e^{x + O(x^2)} = 1 + x + o(x),$$

hence,

$$f_{nr}(\bar{\alpha}, \bar{T}) = \exp\left\{\sum_{\bar{l} \neq \bar{0}} p_{nr}(\bar{l}, \bar{T})(e^{i(\bar{\alpha}, \bar{l})} - 1) + O\left(\sum_{\bar{l} \neq \bar{0}} p_{nr}(\bar{l}, \bar{T})^2\right)\right\}$$

$$= \exp\left\{\sum_{\nu=1}^{m} p_{nr}(\bar{l}_\nu, \bar{T})(e^{i\alpha_\nu} - 1) + O\left(\sum_{\substack{\bar{l} \neq \bar{0}, \bar{e}_\nu \\ \nu=1,2,\ldots,m}} p_{nr}(\bar{l}, \bar{T})\right)\right.$$

$$\left. + O\left(\sum_{\bar{l} \neq \bar{0}} p_{nr}(\bar{l}, \bar{T})\right)^2\right\}.$$

Clearly

$$\sum_{\bar{l} \neq \bar{0}} p_{nr}(\bar{l}, \bar{T}) = 1 - \mathsf{P}\{x_{nr}(t_m) - x_{nr}(t_0) = 0\}$$

$$\leqslant 1 - \mathsf{P}\{x_{nr}(t_m) = 0\} = 1 - p_{nr}(0, t_m, 0),$$

$$\sum_{\substack{\bar{l} \neq \bar{0}, \bar{e}_\nu \\ \nu=1,\ldots,m}} p_{nr}(\bar{l}, \bar{T}) = \mathsf{P}\{x_{nr}(t_m) - x_{nr}(t_0) \geqslant 2\} \leqslant \mathsf{P}\{x_{nr}(t_m) \geqslant 2\},$$

$$\sum_{\bar{l} \neq \bar{0}} p_{nr}(\bar{l}, \bar{T}) \leqslant \sum_{\nu=1}^{m} p_{nr}(\bar{e}_\nu, \bar{T}) + \mathsf{P}\{x_{nr}(t_m) \geqslant 2\}. \tag{7}$$

Note that

$$p_{nr}(1; t_\nu, t_{\nu-1}) = p_{nr}(\bar{e}_\nu, \bar{T}) = \mathsf{P}\{x_{nr}(t_\nu) - x_{nr}(t_{\nu-1}) = 1,$$

$$(x_{nr}(t_{\nu-1}) - x_{nr}(t_\nu)) + (x_{nr}(t_m) - x_{nr}(t_\nu)) \neq 0\} \leqslant \mathsf{P}\{x_{nr}(t_m) \geqslant 2\}. \tag{8}$$

Using relations (7) and (8) we rewrite the expressions for the functions $f_{nr}(\bar{\alpha}, \bar{T})$ obtained:

$$f_{nr}(\bar{\alpha}, \bar{T}) = \exp\left\{\sum_{\nu=1}^{m} p_{nr}(1; t_\nu, t_{\nu-1})(e^{i\alpha_\nu} - 1) + O(x_{nr}(t_m) \geqslant 2)\right.$$

$$\left. + O\left[(1 - p_{nr}(0; t_m, 0)) \sum_{\nu=1}^{m} p_{nr}(1; t_\nu, t_{\nu-1})\right]\right\}.$$

This implies that

$$f_n(\bar{\alpha}, \bar{T}) = \exp\left\{\sum_{\nu=1}^{m} \Lambda_n(t_\nu, t_{\nu-1})(e^{i\alpha_\nu} - 1) + O(B_n(t_m, 0))\right.$$

$$\left. + O\left[\max_{1 \leqslant r \leqslant k_n} (1 - p_{nr}(0; t_m, 0))\right]\right\}.$$

Now the conditions of the theorem yield that as $n \to \infty$

$$f_n(\bar{\alpha}, \bar{T}) \to \prod_{\nu=1}^{m} \exp\{[\Lambda(t_\nu) - \Lambda(t_{\nu-1})](e^{i\alpha_\nu} - 1)\}.$$

The theorem is thus proved.

2.8.5. *The Case of Stationary and Orderly Component Streams*

Assume now, following Khinchin and Ososkov, that the component streams $x_{nr}(t)$ are stationary and orderly. Thus, we stipulate that as $t \to 0$

$$\mathsf{P}\{x_{nr}(t) \geqslant 2\} = o(t).$$

Since the streams are stationary, the limits

$$\lambda_{nr} = \lim_{t \to 0} \frac{\mathsf{P}\{|x_{nr}(t)| > 0\}}{t},$$

which we called the parameters of the streams $x_{nr}(t)$ exist; for orderly streams $\lambda_{nr} = \mathsf{M}x_{nr}(t)$, i.e., the parameter of the stream equals its intensity.

Assume that for large n the intensities are uniformly small:

$$\lim_{n \to \infty} \max_{1 \leqslant k \leqslant k_n} \lambda_{nr} = 0. \tag{9}$$

Since

$$1 - p_{nr}(0; t, 0) \leqslant \mathsf{M}x_{nr}(t) = \lambda_{nr} t,$$

it follows from (9) that the component processes $x_{nr}(t)$ are infinitesimal.

The basic tools for stating the following results are the Palm-Khinchin functions, which we introduced earlier. Recall that the Palm-Khinchin function of order k is by definition the limit

$$\varphi_{nr}(k, t) = \lim_{\tau \to 0} \frac{\mathsf{P}\{x_{nr}(t + \tau) - x_{nr}(\tau) = k; x_{nr}(\tau) > 0\}}{\mathsf{P}\{x_{nr}(\tau) > 0\}}.$$

Theorem. *A necessary and sufficient condition for a sequence of processes* $x_n(t) = \sum_{r=1}^{k_n} x_{nr}(t)$ *converge to a Poisson process with parameter* Λ *is that for every fixed* t

$$\lim_{n \to \infty} \sum_{r=1}^{k_n} \lambda_{nr} \int_0^t \varphi_{nr}(0, u)\, du = \Lambda t \tag{10}$$

and

$$\lim_{n \to \infty} \sum_{r=1}^{k_n} \lambda_{nr} \int_0^t \varphi_{nr}(1, u)\, du = 0. \tag{11}$$

PROOF. We know [Section 2.5.3., formulas (8) and (9)] that

$$p_{nr}(0; t, 0) = 1 - \lambda_{nr} \int_0^t \varphi_{nr}(0, u)\, du \tag{12}$$

and for $k \geqslant 1$

$$p_{nr}(k; t, 0) = \lambda_{nr} \int_0^t [\varphi_{nr}(k - 1, u) - \varphi_{nr}(k. u)]\, du.$$

Using the fact that the processes $x_{nr}(t)$ are stationary, the definition of the functions $\Lambda_n(t, s)$ and $B_n(t, 0)$ as well as the Palm-Khinchin formulas presented above, we obtain

$$\Lambda_n(t + s, s) = \Lambda_n(t, 0) = \sum_{r=1}^{k_n} \lambda_{nr} \int_0^t \varphi_{nr}(0, u)\,du - \sum_{r=1}^{k_n} \lambda_{nr} \int_0^t \varphi_{nr}(1, u)\,du$$

and

$$B_n(t, 0) = \sum_{r=1}^{k_n} \lambda_{nr} \int_0^t \varphi_{nr}(1, u)\,du.$$

Hence, the theorem of Section 2.8.3 yields the validity of necessity and sufficiency of conditions (10) and (11). □

If we assume that the limit

$$\lim_{n\to\infty} \sum_{r=1}^{k_n} \lambda_{nr} = \Lambda \tag{13}$$

exists (Khinchin and Ososkov in their papers assume that $\sum_{r=1}^{k_n} \lambda_{nr} = \Lambda$), the conditions of the theorem can be written in simpler form.

The Khinchin-Ososkov Theorem. *If condition* (13) *is valid, the processes* $x(t)$ *converge to a Poisson process with parameter* Λ *if and only if for every fixed* t *and* $n \to \infty$

$$\sum_{r=1}^{k_n} \lambda_{nr} \int_0^t \varphi_{nr}(0, u)\,du \to \Lambda t. \tag{14}$$

Proof. It follows from the definition of the Palm-Khinchin functions that

$$\varphi_{nr}(0, u) + \varphi_{nr}(1, u) \leqslant 1.$$

This, together with relations (13) and (14) yields

$$\lim_{n\to\infty} \sum_{r=1}^{k_n} \lambda_{nr} \int_0^t \varphi_{nr}(1, u)\,du \leqslant \lim_{n\to\infty} \sum_{r=1}^{k_n} \lambda_{nr}\left(t - \int_0^t \varphi_{nr}(0, u)\,du\right) = 0.$$

We have thus verified that (13) and (14) imply both conditions of the first theorem of this subsection. The theorem is thus proved. □

2.8.6. *Additional Remarks*

We have seen that under quite general conditions the sums of independent small streams are approximated by a Poisson stream. The problem is now to determine the rate of convergence depending on the number of component streams. This problem has been investigated by Pogozhev (1964) and Grigelionis (1963), (1964) in connection with problems of reliability theory. We shall confine ourselves to statements of a few results.

Grigelionis' Theorem. *Let a process $x_n(t)$ be a sum of infinitesimal independent stochastic processes. Then the inequality*

$$\sup_{\bar{x} \in R_m} |F_n(\bar{x}, \bar{T}) - P_n(\bar{x}, \bar{T})| \leqslant 2m \sum_{\nu=1}^{m} A_n(t_\nu, t_{\nu-1}) + 2(m+1)B_n(t_m, t_0)$$

is valid where the following notation is used: R_m is an m-space and $F_n(\bar{x}, \bar{T})$ is the distribution function of the vector $\bar{x}_n(\bar{T})$, $\bar{x} = (x_1, x_2, \dots, x_m)$,

$$P_n(\bar{x}, \bar{T}) = \prod_{\nu=1}^{m} P(x_\nu, \Lambda(t_\nu, t_{\nu-1})),$$

$$P(x, \lambda) = \sum_{0 \leqslant k < x} \frac{\lambda^k}{k!} e^{-\lambda}, \qquad A_n(t, s) = \sum_{r=1}^{k_n} p_{nr}^2(1; t, s).$$

The case of identically distributed independent renewal processes has been studied by Franken. We introduce the required notation. Consider the sequence of identically distributed independent random variables $x_1, x_2, \dots$ for which $F(x) = P\{x_i < x\}$. For every $t > 0$ we shall define a renewal process $N(t)$ as the maximal value of n for which $\sum_{i=1}^{n} x_i \leqslant t$. As we know, the renewal function $H(t) = \mathsf{M}N(t)$ and the function $F(x)$ are connected by the relation

$$H(x) = F(x) + \int_0^x H(x-u)\, dF(u).$$

Define for each n and functions $F_n(x) = F_{ni}(x)$, $H_n(x) = H_{ni}(x)$ $i = 1, 2, \dots, n$, renewal processes $N_{ni}(t)$ that are independent for all i and given n. Concerning $H_n(t)$, it is assumed that for all n and t the equality

$$H(t) = nH_n(t)$$

is fulfilled. Set

$$\zeta_n = \sum_{i=1}^{n} [N_{ni}(s+t) - N_{ni}(s)].$$

Then Franken's theorem is as follows:

Franken's Theorem. *Let $F_n(+0) < 0.5$. Then*

$$\sum_{0 \leqslant k < x} \mathsf{P}\{\zeta_n = k\} = \sum_{0 \leqslant k < x} \psi(k)\left[1 + \sum_{i=1}^{r} \frac{Q_i(x)}{n^i}\right] + O\left(\frac{1}{n^{r+1}}\right),$$

where

$$\psi(k) = \frac{[H(s+t) - H(s)]^k}{k!} e^{-[H(s+t)-H(s)]},$$

and $Q_i(k)$ are polynomials in k.

2.9. Direct Probabilistic Methods

In Section 2.8 we have proved limit theorems on the convergence of a compound stream to a Poisson stream. The basic method for proving these theorems was the method of characteristic functions. It is useful to present an example of direct probabilistic methods for solving these types of problems. These methods are more intuitive and capture the essence of the problem well.

Assume that the conditions of Grigelionis' theorem are fulfilled.

Elementary processes $x_{nr}(t)$ are represented in the form of a sum

$$x_{nr}(t) = u_{nr}(t) + v_{nr}(t),$$

where $u_{nr}(t) = \min\{x_{nr}(t), 1\}$, $v_{nr}(t) = \max\{0, x_{nr}(t) - 1\}$. Thus $u_{nr}(t)$ "marks" only the first event of the rth stream while $v_{nr}(t)$—all the remaining events. Then

$$x_n(t) = u_n(t) + v_n(t),$$

where $u_n(t)$ and $v_n(t)$ are sums of the corresponding elementary processes. We have

$$\begin{aligned} \mathsf{P}\{v_n(t) = 0, 0 \leqslant t \leqslant T\} &\geqslant 1 - \sum_r \mathsf{P}\{v_{nr}(T) \geqslant 1\} \\ &= 1 - B_n(T) \xrightarrow[n \to \infty]{} 1. \end{aligned} \tag{1}$$

Denote by t_{nk}^0 the time at which $u_n(t)$ first attains the level k, and denote by t_{nk} the time of the kth event of the compound stream $x_n(t)$. Since under the condition that $v_n(t) = 0, 0 \leqslant t \leqslant T$, all t_{nk} appearing in the time interval $[0, T]$ coincide with t_{nk}^0, we have

$$\begin{aligned} &\sup_{0 \leqslant x_1 < \cdots < x_m \leqslant T} |\mathsf{P}\{t_{n1} < x_1, \ldots, t_{nm} < x_m\} - \mathsf{P}\{t_{n1}^0 < x_1, \ldots, t_{nm}^0 < x_m\}| \\ &\leqslant B_n(T) \xrightarrow[n \to \infty]{} 0. \end{aligned} \tag{2}$$

Thus, to investigate the limiting behavior of the distribution of $(t_{n1}, t_{n2}, \ldots)$ it is sufficient to solve the corresponding problem for $(t_{n1}^0, t_{n2}^0, \ldots)$. Since, however,

$$\mathsf{P}\{t_{n1}^0 < a_1, \ldots, t_{nm}^0 < a_m\} = \prod_{k=1}^{n} \mathsf{P}\{t_{nk}^0 < a_k | t_{ni}^0 < a_i, i < k\}, \tag{3}$$

it is sufficient to consider an individual cofactor on the right-hand side of formula (3).

The following fact is well known. A and B are two events and let $B = \bigcup_y B_y$. Then

$$\inf_y \mathsf{P}\{A | B_y\} \leqslant \mathsf{P}\{A | B\} \leqslant \sup_y \mathsf{P}\{A | B_y\}.$$

We choose A to be the event $(t_{nk}^0 < a_k\}$ and B to be the event $\{t_{ni}^0 < a_i, i < k\}$. Then $B = \bigcup_y B_y$, where B_y is the event that $t_{ni}^0 = y_i$ and the numbers r_i, $i < k$, of the elementary streams that produced the events with the numbers $i < k$ of the compound stream $u_n(t)$ are known. Evidently, $0 \leqslant y_i \leqslant T$. Without loss of generality we set $r_i = i$, $i < k$. Then

$$\mathsf{P}\{A|B_y\} = 1 - \prod_{j=k}^{n} (1 - p_{nj}),$$

where $p_{nj} = \mathsf{P}\{x_{nj}(a_k) \geqslant 1 | x_{nj}(y_{k-1}) = 0\}$ and $y_0 = 0$. We have

$$p_{nj} = \frac{p_{nj}(1; a_k, y_{k-1})}{1 - p_{nj}(1; y_{k-1}, 0)}.$$

Since the expression in the numerator approaches 1 uniformly in y and r as $n \to \infty$ and the sum over j of expressions in the numerator converges to $\Lambda(a_k) - \Lambda(y_{k-1})$, then in view of the "law of rare events" the relation

$$\mathsf{P}\{A|B_y\} \xrightarrow[n \to \infty]{} 1 - \exp\{-[\Lambda(a_k) - \Lambda(y_{k-1})]\}$$

is valid. However, since $y_{k-1} = t^0_{n,k-1}$, this implies that the convergence is actually to the simplest stream.

Note that if in Grigelionis' theorem we omit condition (6), more complex streams will be obtained: after the event of the limiting stream, at random distances finite or infinite chains of the associated events will be repeated. The class of possible limiting distributions has been studied by Kovalenko (1965b).

2.10. Limit Theorem for Thinning Streams

2.10.1. *Statement of the Problem*

It quite frequently occurs in important practical problems that a customer stream passing through a number of successive servers loses a certain fraction of its elements. For example, when a mass-produced item is processed by a series of machines, defective items are rejected after each operation, and thus the initial stream is "thinned out." A similar situation occurs when proofs are read by several proofreaders in succession. After each proofreader, the number of errors missed that remain in the text decreases. The following question arises: What general statements can be made about these *thinning streams*? This question is quite appropriate, since the model may have numerous applications in physics and engineering. At a seminar on queueing theory held at the Institute of Mathematics of the Academy of Sciences of the Ukrainian SSR, streams of items on assembly lines were studied, and it was conjectured that under quite general conditions thinning streams approach Poisson streams. At approximately the same time Rényi (1956) proved the first theorem in this direction. He studied an arbitrary renewal stream in which each customer is retained in the stream with probability q and drops out with probability $1 - q = p$. By suitable changes in the time scale, the intensity of the stream remains fixed. This operation is repeated many times. Rényi has proved that a stream transformed in this manner does in fact approach a simple stream.

We thus see that simple streams are obtained not only as a result of summation of infinitesimal independent streams, but also by other limiting processes. It is of substantial interest, both from a theoretical and a practical point of view to study models that lead to simple streams.

2.10.2. *Laplace Transform of Transformed Streams*

Assume that a renewal stream for which $F_1(x) = F(x)$ is subjected to the following thinning operation: each customer is retained in the stream with probability q and drops out with probability $p = 1 - q$. Simultaneously, another process takes place; the time scale is altered; the new time unit is an interval of length q^{-1}. This double transformation will be called a T_q transformation.

Denote by $t_1, t_2, \ldots$ the arrival times of customers in the stream. By assumption, the random variables

$$z_1 = t_1, \quad z_2 = t_2 - t_1, \quad z_3 = t_3 - t_2, \ldots$$

are independent and identically distributed, their distribution function is $F(x)$. We introduce the notation

$$F_1(x) = F(x), \qquad F_n(x) = \int_0^x F_{n-1}(x - z)\, dF_1(z) \quad \text{for } n = 2, 3, \ldots.$$

It is known that $F_n(x)$ is the distribution function of the sum of n independent random variables each with distribution function $F(x)$. Thus, $F_2(x)$ may be viewed as the distribution of the time interval between the arrival of the kth and the $(k + 2)$th customers. In general, $F_n(x)$ represents the distribution function of the time interval between the arrival of the kth and the $(k + n)$th customers.

Denote by $T_qF(x)$ the distribution function of the interarrival times in a stream obtained by applying the transformation T_q to the initial stream. We shall prove that for any $x > 0$

$$T_qF(x) = \sum_{k=1}^{\infty} qp^{n-1} F_n\left(\frac{x}{q}\right). \tag{1}$$

Indeed, after transforming T_q adjacent in time of arrival customers remain in the stream with probability q, every other customer is retained with probability $pq, \ldots$, and $n - 1$ successive customers will be omitted (drop out) from the stream with probability $p^{n-1}q$. But since the distributions of the interarrival times of these customers are $F_1(x)$, $F_2(x)$, $\ldots$, $F_n(x)$, respectively, the total probability formula yields (1).

Let $\varphi(s)$ denote the Laplace-Stieltjes transform for the function $F(x)$, i.e., set

$$\varphi(s) = \int_0^\infty e^{-xs}\, dF(x).$$

We shall prove that the Laplace-Stieltjes transform of the transformed stream is

$$T_q\varphi(s) = \frac{q\varphi(qs)}{1 - p\varphi(qs)}. \tag{2}$$

Indeed, by the theory of the Laplace-Stieltjes transforms, the transform of the sum of independent terms is equal to the product of the corresponding transforms of the summands. Thus, the Laplace transform of the function $T_qF(x)$ in accordance with formula (1) is equal to

$$T_q\varphi(s) = q \sum_{n=1}^{\infty} p^{n-1}\varphi^n(qs).$$

From here simple algebraic transformations yield (2).

2.10.3. *Some Properties of the T-Operation*

We shall prove that *successive application of transformations T_{q_1} and T_{q_2} to a stream is equivalent to the single transformation $T_{q_1 q_2}$.*

In accordance with (2)

$$T_{q_2}(T_{q_1}\varphi(s)) = \frac{q_2 T_{q_1}\varphi(q_2 s)}{1 - p_2 T_{q_1}\varphi(q_2 s)}$$

and hence

$$T_{q_2}(T_{q_1}\varphi(s)) = \frac{\dfrac{q_2 q_1 \varphi(q_1 q_2 s)}{1 - p_1\varphi(q_1 q_2 s)}}{1 - p_2 \dfrac{q_1\varphi(q_1 q_2 s)}{1 - p_1\varphi(q_1 q_2 s)}} = \frac{q_1 q_2 \varphi(q_1 q_2 s)}{1 - (1 - q_1 q_2)\varphi(q_1 q_2 s)}$$

Q.E.D.

Since the correspondence between distribution functions and their Laplace-Stieltjes transforms is one-to-one, we conclude that

$$T_{q_2}T_{q_1}F(x) = T_{q_1 q_2}F(x).$$

A T_q-transformation does not alter the mean value of the interarrival time of a stream. This property is proved using the well-known equality

$$\int_0^\infty x\,dF(x) = -\varphi'(0).$$

Simple calculations yield

$$\left[\frac{d}{ds}(T_q\varphi(s))\right]_{s=0} = \left[\frac{q^2\varphi'(qs)}{(1 - p\varphi(qs))^2}\right]_{s=0} = \frac{q^2\varphi'(0)}{(1 - p)^2} = \varphi'(0).$$

Q.E.D.

2.10.4. *The T_q-Transformation for a Simple Stream*

Let the initial stream be simple and with intensity λ. For such a stream

$$F(x) = 1 - e^{-\lambda x}, \qquad \int_0^\infty x\,dF(x) = \frac{1}{\lambda}.$$

It is easy to evaluate the Laplace transform $F(x)$ to be

$$\varphi(s) = \frac{\lambda}{\lambda + s}.$$

We shall show that simple streams are invariant under T_q-transformations. Indeed,

$$T_q\varphi(s) = \frac{q\dfrac{\lambda}{\lambda + qs}}{1 - p\dfrac{\lambda}{\lambda + qs}} = \frac{\lambda q}{q(\lambda + s)} = \frac{\lambda}{\lambda + s} = \varphi(s).$$

2.10.5. *Rényi's Limit Theorem*

Rényi's Limit Theorem. *Let A be a renewal stream with finite intensity λ to which the transformations $T_{q_1}, T_{q_2}, T_{q_3}, \ldots$ are successively applied; also let*

$$Q_n = q_1 q_2 \cdots q_n \to 0$$

as $n \to \infty$; then as $n \to \infty$ the stream $T_{q_n} T_{q_{n-1}} \cdots T_{q_1}$ approaches a simple stream with the same intensity λ.

Proof. As it was previously shown

$$T_{q_n} T_{q_{n-1}} \cdots T_{q_1} = T_{Q_n}.$$

Now

$$T_{Q_n}\varphi(s) = \frac{Q_n\varphi(Q_n s)}{1 - (1 - Q_n)\varphi(Q_n(s))}.$$

However, since

$$T_{Q_n}\varphi(s) = \frac{\varphi(Q_n s)}{\dfrac{1 - \varphi(Q_n s)}{Q_n} + \varphi(Q_n(s))} = \frac{\varphi(Q_n s)}{\dfrac{\varphi(0) - \varphi(Q_n s)}{Q_n} + \varphi(Q_n s)}$$

and since the stream has finite intensity the derivative of $\varphi(s)$ exists,

$$\lim_{Q_n \to 0} \frac{1 - \varphi(Q_n s)}{Q_n} = -s\varphi'(0) = \frac{s}{\lambda}.$$

Thus, as $n \to \infty$

$$T_{Q_n}\varphi(s) \to \frac{\lambda}{\lambda + s}.$$

This relation proves the theorem. □

2.11. Additional Limit Theorems for Thinning Streams

2.11.1. *Belyaev's Theorem and its Generalizations*

Belyaev (1963) proved a rather simple but quite important theorem on convergence of a thinning stream to a simple one. It often happens that proofs of a mathematical theorem in particular cases are more complicated than a proof of a general theorem that encompasses all of these particular cases. This observation also applies to Belyaev's theorem.

We shall state a somewhat more general theorem (which allows for the case of a nonstationary limiting stream) retaining the ideas of Belyaev's proof.

Theorem. *Let a sequence of streams be given, $x_n(t)$ be the number of events of the* nth *stream in the interval* $(0, t)$. *The events of the* nth *stream are independently thinned,* p_{nk} *be the probability of omitting the* kth *event of the* nth *stream;* $q_{nk} = 1 - p_{nk}$. *If*

$$\sup_k q_{nk} \xrightarrow[n\to\infty]{} 0 \tag{1}$$

and for some sequence $N = N(n) \to \infty$ for any fixed $t > 0$

$$\frac{x_n(t)}{N(n)} \xrightarrow[n\to\infty]{} A(t), \tag{2}$$

$$\sum_{k \leqslant tN(n)} q_{nk} \xrightarrow[n\to\infty]{} B(t), \tag{3}$$

where $A(t)$ and $B(t)$ are nondecreasing functions, $B(t)$ being a continuous function, then the nth *thinned stream converges as $n \to \infty$ to a Poisson stream with the leading function*

$$\Lambda(t) = B(A(t)). \tag{4}$$

Corollary. *If $q_{nk} = q_n \to 0$,*

$$\frac{x_n(t)}{N(n)} \to at, \qquad q_n N(n) \to b,$$

then the limiting stream is a simple stream with parameter $\lambda = ab$.

PROOF. Introduce a time scale θ, corresponding to each $k = 1, 2, \ldots$ the value of $\theta = \theta_k$ such that $\sum_{i=1}^{k} q_{ni} = \theta$. If the kth event of the initial stream is retained, we say that the event of the θ-stream at time θ_k has occurred. The numbers of stream events in disjoint time intervals are independent; by the theorem on rare events in the interval of length θ asymptotically, a Poisson number of events with parameter 1 takes place. Hence, the θ-stream in the limit becomes a simple stream with parameter 1.

A thinning of the initial n-stream will be called a t-stream. It remains to obtain the correspondence between the t-stream and θ-stream.

Denote by $\Delta(t_1, t_2)$ and $\delta(\theta_1, \theta_2)$ the number of events of t- and θ-streams in the intervals (t_1, t_2) and (θ_1, θ_2), respectively.

Let $\varepsilon > 0$ be an arbitrary fixed number. Select a $\delta > 0$ such that $|z| < \delta$ implies

$$|B(A(t_i) + z) - B(A(t_i))| < \varepsilon, \quad i = 1, 2.$$

Since with probability greater than $1 - \varepsilon$

$$\left| \frac{x_n(t_i)}{N(n)} - A(t_i) \right| < \delta, \quad i = 1, 2, \quad \text{for } n > n_0,$$

to the interval $[t_1, t_2]$ there corresponds an interval $[k_1, k_2]$ of values of k such that $[A(t_i) - \delta]N \leqslant k_i \leqslant [A(t_i) + \delta]N$, $i = 1, 2$. For definiteness, let us consider the upper bound. Then

$$\sum_{k \leqslant k_i} q_{nk} \leqslant \sum_{k \leqslant [A(t_i) + \delta]N} q_{nk} \xrightarrow[n \to \infty]{} B(A(t_i) + \delta) < B(A(t_i)) + \varepsilon.$$

Hence, for n sufficiently large,

$$\sum_{k \leqslant k_i} q_{nk} < B(A(t_i)) + 2\varepsilon,$$

i.e., $\theta_2 < B(A(t_i)) + 2\varepsilon$. The lower bound is obtained in a similar manner. Thus, an interval $[\theta_1, \theta_2]$ will correspond to the interval $[t_1, t_2]$ with probability as close to 1 as desired. Here $|\theta_i - B(A(t_i))| < 2\varepsilon$.

Whence

$$\begin{aligned} &\mathsf{P}\{\delta(B(A(t_1)) - 2\varepsilon, B(A(t_2)) + 2\varepsilon) \leqslant k\} - \varepsilon \\ &\quad \leqslant \mathsf{P}\{\Delta(t_1, t_2) \leqslant k\} \leqslant \mathsf{P}\{\delta(B(A(t_1)) - 2\varepsilon, B(A(t_2)) + 2\varepsilon) \leqslant k\} + \varepsilon. \end{aligned}$$

Since the probabilities in the upper and lower bounds are distribution functions of Poisson laws with parameters $B(A(t_2)) - B(A(t_1)) \pm 4\varepsilon$ and since $\varepsilon > 0$ is arbitrary, we obtain that $\Delta(t_1, t_2)$ is distributed in the limit according to the Poisson distribution with parameters $B(A(t_2)) - B(A(t_1))$.

The limiting independence of $\Delta(t_1, t_2)$ for disjoint intervals follows from the fact that the "inner" intervals of the type $[B(A(t_1)) - 2\varepsilon, B(A(t_2)) + 2\varepsilon]$ are disjoint and the number of events in the complement to these intervals with respect to the intervals of the type $[\theta_1, \theta_2]$ differs from zero only with the probability of order ε. The theorem is thus proved. □

2.11.2. *Rare Events in the Scheme of a Regenerative Process*

Solov'yev (1971) studied possible limit distributions of the time of a rare event for a thinning renewal stream in the "scheme of series." Of special interest is the limit theorem on convergence of the distribution of the time of occurrence of a rare event to an exponential distribution.

Let $F_n(x)$ be the distribution function of the random variable $\xi_n \geqslant 0$ and $T_n = \int_0^\infty \bar{F}_n(x)\,dx$. Following Solov'yev we say that ξ_n converges to zero in the Khinchin sense if for any $x > 0$

$$\frac{1}{T_n}\int_x^\infty \bar{F}_n(x)\,dx \xrightarrow[n\to\infty]{} 0.$$

Let there exist for any n a renewal process such that the intervals between renewals are distributed as random variables ξ_n. A "rare" event may occur in any interrenewal intervals with probability q. Moreover, the time of observing the event may depend on the time elapsed since the last renewal; the event itself may depend on the length of the corresponding renewal interval.

Solov'yev's theorem. *If $q_n\xi_n/T_n$ tends to zero in the Khinchin sense, then $q_n\zeta_n/T_n$, where ζ_n is the first moment of a rare event, is an asymptotically exponential function with parameter* 1.

NOTES

Zakusilo (1972a, 1972b) studied necessary and sufficient conditions for the convergence of thinning semi-Markov processes. It was noted by Gnedenko that these problems are closely related to the study of distributions of sums of a random number of random summands. In the later works of Zakusilo (1973a, 1973b, 1973c) necessary and sufficient conditions for such sums for random variables defined on certain random processes were considered.

Necessary and sufficient conditions on convergence of superpositions of independent streams to a simple one were studied in Zakusilo and Meleshchuk's paper (1976).

Along with the above mentioned we note the following important papers on streams of homogeneous events and their applicaton to queueing theory: Brémaud (1981); Belyaev (1969), Çinlar (1972), Franken and Streller (1980).

In the papers of Grigelionis (1975) and Kabanov, Liptser and Shiryaev (1975) a martingale approach to the construction of the theory of streams of homogeneous events was developed.

3
Some Classes of Stochastic Processes

Markov stochastic processes are of special importance in queueing theory. The reader has seen in Chapter 1 that queueing processes in a very wide range of systems can be described by Markov processes with finitely or denumerably many states. However, Markov processes involve maximal analytical assumptions: both arrivals of new customers and completion of servicing customers that are in the system should not depend on previous history. There is no need to emphasize that in the majority of practical situations this type of condition is far from being satisfied. Thus, it is necessary to utilize stochastic processes of a more complex character. In this chapter we shall discuss classes of processes that are most fruitfully applied in queueing theory.

3.1. Kendall's Method: Semi-Markov Processes

3.1.1. *Semi-Markov Processes and Embedded Markov Chains*

The general tendency in queueing theory is to determine a stochastic process associated with the servicing process that can be viewed as a Markov process.

Denote by $v(t)$ a process describing the state of the queueing system at an arbitrary time; it is assumed that by a realization of the random function $v(t)$ all changes occurring in the system can be tracked, such as arrival times of customers or times of service completion. The number of customers in the system at an arbitrary time t is an example of such a process $v(t)$. Sometimes it is expedient to consider the process $v(t)$ as a set of several parameters having certain physical meanings. Thus, when dealing with a system involving failures of servers, it is natural to consider the process $v(t)$ as a two-dimensional one:

$$v(t) = \{v_1(t), v_2(t)\},$$

where $v_1(t)$ is the number of customers in the system at time t and $v_2(t)$ is the number of failed servers at the same time. In this and the succeeding chapters we shall present numerous examples in which the functioning of a queueing system is described by stochastic processes.

As soon as the probability laws governing the incoming customer stream are given, and the distribution of the service time and the service discipline are known, $\nu(t)$ becomes a well-defined stochastic process. Kendall, one of the leading experts in queueing theory, has proposed the method of *embedded Markov chains*. Kendall's basic paper (1953) is devoted to this method. The idea is as follows: times $\{t_n\}$ $(t_n < t_{n+1})$ are selected so that the values of the process $\{\nu(t_n)\}$ form a Markov chain. The distribution of the random variables $\nu(t_n)$ is then studied by the standard methods for Markov chains. Finally, inference is made based on this distribution about the properties of the original process $\nu(t)$. This last step is omitted in many cases, since the variables $\nu(t_n)$ themselves provide exhaustive information about the operation of the queueing system.

Very often one considers the case when the set of possible values of the process $\nu(t)$, hence also of the embedded Markov chain, is either finite or denumerable. Many problems can be reduced to such a situation. However, this restriction is not necessary; a continuous set of states may also be considered. From this point of view the method of embedded Markov chains (*Kendall's method*) includes the theory of random walks whose applications in queueing problems will frequently be used in what follows.

Thus the embedded Markov chain is a sequence of values of the process at specially chosen times t_n; these values form a Markov chain. It must be emphasized that the times t_n are, as a rule, random and depend on the behavior of the process $\nu(t)$ itself. An embedded Markov chain can also be determined in the case when the times t are not described by a Markov chain.

We now present a definition of a semi-Markov process with a finite or countable set of states.

Let X be a finite or quantable set; the elements of X will be denoted by letters $i, j, \ldots$. We shall assume that a homogeneous Markov chain $\{\nu_n, n \geqslant 1\}$ with values in X and transition matrix $\|P_{ij}\|$ is given.

A semi-Markov process (*SMP*) is defined as a step linear process $\nu(t)$, $t \geqslant 0$, with the following properties. In the half interval $[0, t_1)$, $\nu(t) = \nu_1$; in the half interval $[t_1, t_2)$, $\nu(t) = \nu_2$; and so on. For a fixed realization of the Markov chain $\nu_n = i_n$, $n \geqslant 1$, of duration t_1, $t_2 - t_1$, $t_3 - t_2, \ldots$ the sojourn of $\nu(t)$ in the states $i_1, i_2, i_3, \ldots$ are independent; moreover, each one of these variables depends only on the state in which the process is at present and on the next state; also distribution functions

$$\mathsf{P}\{t_n - t_{n-1} < x \mid \nu_n = i, \nu_{n+1} = j\} = F_{ij}(x), \quad n \geqslant 1,$$

are defined where to achieve generality it is assumed that $t_0 = 0$. In place of $\|p_{ij}\|$ and $\{F_{ij}(x)\}$ one can assign only the functions

$$P_{ij}(x) = p_{ij} F_{ij}(x),$$

which possess the following interpretation. If at a given instant the process arrives at the state i, then with probability $P_{ij}(x)$ the next transition of the process will occur during a period shorter than x and will proceed into the state j. The function $P_i(x) = \sum_j P_{ij}(x)$ is the distribution function of the time until the next transition of the process. Note that the states ν_n and ν_{n+1} of a semi-Markov process are not necessarily different; in this case a "transition" is a return to the initial state. Moreover, an initial distribution $p_i^{(0)} = \mathsf{P}\{\nu_i = i\}$ should be defined.

Now let $\nu(t)$ be a SMP, (ν_0, t_0) be a two-dimensional random variable ($t_0 > 0$), which does not depend on the trajectory of $\nu(t)$ given that ν_1 is known. Then the random process

$$\nu_1(t) = \begin{cases} \nu_0 & \text{for } 0 \leqslant t < t_0, \\ \nu(t - t_0) & \text{for } t \geqslant t_0 \end{cases}$$

is called a *semi-Markov process with delay*. To characterize this process statistically in addition to $P_{ij}(x)$, one should define the distribution of the random variable (ν_0, ν_1, t_0). The transition times of a SMP with delay is more convenient to denote as $t_0, t_1, t_2, \ldots$ rather than $t_0, t_0 + t_1, t_0 + t_2, \ldots$.

One can define SMP in a somewhat different manner. Consider a system that is at the initial $t = 0$ at a random state ν_0 and that changes its states by jumping at times $t_1, t_2, \ldots$.

Let the system reach the state i at time t_n. A number of factors act on the system, among them the jth one, which transfers this system into the state j provided this state exhibits itself before the other states. The time η_j at which the jth factor manifests itself counted from the time t_n possesses the distribution function $\Phi_{ij}(x)$; the random variables η_j are independent.

Evidently $t_{n+1} - t_n = \min_j \eta_j$; if η_j possesses densities $f_{ij}(x)$, then

$$P_{ij}(x) = \int_0^x \left(\prod_k \bar{\Phi}_{ik}(t)\right)(f_{ij}(t)/\bar{\Phi}_{ij}(t))\,dt, \quad x \geqslant 0,$$

where $\bar{\Phi}_{ij}(t) = 1 - \Phi_{ij}(t)$.

The Markov chain $\{\nu_n\}$ is called an *embedded Markov chain* of the given semi-Markov process. The intervals (t_{n-1}, t_n) are called *cycles*.

Note that a Markov process with transition intensities λ_{ij} and output intensities $\lambda_i = \sum_{j \neq i} \lambda_{ij}$ is a SMP; for $x \geqslant 0$

$$P'_{ij}(x) = \begin{cases} \lambda_{ij} e^{-\lambda_i x}, & j \neq i, \\ 0, & j = i, \end{cases}$$

$$\Phi'_{ij}(x) = \begin{cases} \lambda_{ij} e^{-\lambda_{ij} x}, & j \neq i, \\ 0, & j = i. \end{cases}$$

We make the following remarks:

1. In the definition of SMP an infinite value for the duration in the state i is admissible. Thus, the probability $P_i(\infty)$ may be less than 1; the probability that the process after arriving into state i remains there forever is equal to $1 - P_i(\infty)$.
2. In a general case a SMP is defined by the characteristics of SMP introduced above only for $t \leqslant t^* = \lim_{n\to\infty} t_n$. However, if the embedded Markov chain

$\{v_n\}$ possesses an ergodic distribution, then $t^* = \infty$ always, and, hence, the process is defined for all $t \geqslant 0$.

3. If we extend the set of states of the process, namely, introduce the variable $\bar{v}(t) = (v_n, v_{n+1})$, where $v_n = v(t)$ and v_{n+1} is the state of the processes after the next transition, then in the new variables $F_{ij}(x)$ will depend only on i but not on j.

3.1.2. *Some Results from the Theory of Markov Chains*

The possibility of solving queueing problems by the method of embedded Markov chains discovered by Kendall stimulated the development of methods in the theory of Markov chains reflecting the special features of an applied nature. First, ergodic theorems are of interest for queueing theory. Using these theorems inference can be made about the existence of a steady-state mode of the system that is independent of the initial state.

Consider a homogeneous Markov chain. The states of this system will be denoted by $v_0, v_1, \ldots, v_n, \ldots$ Thus, v_0 is the initial state and v_n is the state after the nth step or at the nth instant of time. The set of states is finite or denumerable; we shall denote this set by X.

Let p_{ij} be the probability of transition from state i to state j in one step; $p_{ij}^{(n)}$ is the probability of a similar transition in n steps. (In particular p_{ij} equals $p_{ij}^{(1)}$.)

The condition for ergodicity of the chain $\{v_n\}$ is of special interest to us; it will be understood here in the following sense.

As $n \to \infty$, the transition probabilities possess limits which do not depend on the initial state:

$$p_{ij}^{(n)} \xrightarrow[n \to \infty]{} \pi_j, \quad i, j \in X, \tag{1}$$

where

$$\sum_{j \in X} \pi_j = 1$$

(the case of departure to infinity is excluded if i is viewed as a numerical index).

Before giving the condition for ergodicity of a Markov chain, we state two essential requirements for the validity of this property.

1. *Irreducibility*: transition from any state i to any other state j is possible (in a number of steps). More precisely, for arbitrary states i and j there exist states $i_1, i_2, \ldots, i_n$ (here n may depend on i and j) such that the probabilities of transition in one step from i to i_1, i_1 to $i_2, \ldots, i_{n-1}$ to i_n, and finally from i_n to j are positive.

2. *Aperiodicity*: the greatest common divisor of the positive integers n such that $p_{ij}^{(n)} > 0$ is 1 (for all i and j).

Markov chains with properties (1) and (2) are called *irreducible aperiodic chains*.

Conditions for the ergodicity of irreducible aperiodic chains have been derived by many authors in somewhat different forms. We refer the reader to

Feller (1950) (Chapter 15), Sarymsakov (1954), and Foster's paper (1953). We shall not discuss the general ergodic theorem here, which may present difficulties for an engineering-oriented reader. Instead, we shall state a theorem that is useful for queueing theory.

Let a nonnegative functon $f(i)$, $i \in X$, exist with the following properties:

1. For some $\varepsilon > 0$ and all $i \in X$ except possibly a finite number of them

$$\mathsf{M}\{f(v_{n+1})|v_n = i\} \leqslant f(i) - \varepsilon. \tag{2}$$

2.
$$\mathsf{M}\{f(v_{n+1})|v_n = i\} < \infty, \quad i \in X. \tag{3}$$

Theorem. *If a Markov chain $\{v_n\}$ is irreducible and aperiodic and satisfies the conditions* (2)–(3), *then the chain is ergodic.*

The proof is presented in Klimov's paper (1964).

We cite yet another useful test for ergodicity of a Markov chain (Tweedie (1975)).

Let $\{v_n\}$ be an irreducible, aperiodic Markov chain with the states 0, 1, 2, ...:

$$p_{ij}^{(n)} = \mathsf{P}\{v_n = j|v_0 = i\};$$

$$\mu(i) = \mathsf{M}\{v_{n+1} - v_n|v_n = i\}.$$

Then for the ergodicity of $\{\xi_n\}$ it is sufficient that i and $N > 0$ exist such that

$$\varlimsup_{n\to\infty} \sum_{j\geqslant N} p_{ij}^{(n)}\mu(j) < 0.$$

In practical problems it is important to know, in addition to the ergodic distribution, how fast the transient distribution approaches the limiting distribution.

Of special interest is the case when an exponential bound can be obtained of the form

$$|p_{ij}^{(n)} - \pi_j| < M_{ij}\lambda_{ij}^n,$$

where the constants λ_{ij} are smaller than 1.

If (2) is satisfied following Kendall, we speak of *geometric ergodicity*. The appropriate conditions are discussed in Kendall's paper (1960).

Vêre-Jones (1962) has proved a theorem on the existence of a uniform bound

$$|p_{ij}^{(n)} - \pi_j| \leqslant M_{ij}\lambda^n, \quad 0 < \lambda < 1.$$

If in a given state i the bound

$$|p_{ii}^{(n)} - \pi_i| < M_{ii}\lambda_{ii}^n$$

holds, the state is called *geometrically ergodic*. Vêre-Jones' result may be stated as follows:

If all states of an irreducible aperiodic Markov chain are geometrically ergodic, then the uniform bound holds for some $\lambda < 1$.

Vêre-Jones (1964) applied this result to the study of the time required for a queueing process to reach stationary conditions.

The corresponding theorem for the case when the number of possible states of the Markov chain is finite is useful. In this situation geometric ergodicity does not need to be specially proved.

Theorem. *If a Markov chain is irreducible and aperiodic and the set N of the states is finite, then the uniform bound* (4) *holds for some* $\lambda < 1$.

For the proof, see Feller (1950).

The classical Bernstein inequality provides a bound for the values of λ: if $p_{ii_0} \geqslant 1 - \lambda$, then

$$\sum_j |p_{ij}^{(n)} - \pi_j| \leqslant 2\lambda^n. \tag{4}$$

3.1.3. *Basic Relations for Semi-Markov Processes*

Denote by $B_{ij}(t)$ the probability of the event $\{v(t) = j\}$ under the condition $v(0) = i$, by $B_j(t)$ the unconditional probability of the same event.

By the formula of total probability

$$B_j(t) = \sum_i p_i^{(0)} B_{ij}(t), \tag{5}$$

where $p_i^{(0)} = \mathsf{P}\{v_0 = i\}$. Thus $B_{ij}(t) = B_j(t)$ when the initial probabilities are of the form $p_k^{(0)} = \delta_{ik}$.

We have the stochastic relation

$$v(t) = \begin{cases} v_1, & t_1 \geqslant t, \\ v^*(t - t_1), & t_1 < t, \end{cases}$$

where $v^*(t)$ is a SMP with the same characteristics as $v(t)$ and starts from the state to which the initial process arrives at time t. Proceeding from stochastic relations to probabilities we obtain the equation

$$B_{ij}(t) = \bar{P}_i(t)\delta_{ij} + \sum_k \int_0^t B_{kj}(t - x)\, dP_{ik}(x), \quad t \geqslant 0. \tag{6}$$

One can use the method of Laplace transforms introducing the functions

$$B_{ij}^*(s) = \int_0^\infty e^{-st} B_{ij}(t)\, dt, \tag{7}$$

$$\pi_{ij}(s) = \int_0^\infty e^{-st}\, dP_{ij}(t), \tag{8}$$

$$\pi_i(s) = \sum_j \pi_{ij}(s) = \int_0^\infty e^{-st}\, dP_i(t). \tag{9}$$

(The first is meaningful for $\operatorname{Re} s > 0$, the others for $\operatorname{Re} s \geqslant 0$.)

Applying Laplace transforms to (6) we arrive at

$$B_{ij}^*(s) = \frac{1}{s}[1 - \pi_i(s)]\delta_{ij} + \sum_k B_{kj}^*(s)\pi_{ik}(s). \tag{10}$$

For fixed s, relation (10) is a system of linear algebraic equations.

Assume that s is positive and let it tend to infinity. Then for all i, j, $\pi_{ij}(s)$ tends to zero. Consequently, the determinant, as is easily seen, is an analytical function for $\operatorname{Re} s > 0$, it can vanish only at isolated points (otherwise it would be identically zero, which, as we have seen, is impossible). Thus, Eq. (10) has a unique solution provided there is a finite number of states in the system.

An attentive reader may note a lacuna in our arguments. Actually, our reasoning is valid only when finitely many renewals occur in a finite period with probability 1. Otherwise one cannot speak of the value of the process $v(t)$ at the instant t, since it may be undefined.

To exclude this possibility it is sufficient to require that there exist a positive ε such that for every j

$$F_j(\varepsilon) \leqslant 1 - \varepsilon.$$

In practical problems this condition is always satisfied. In this connection we remind the reader of Feller's theorem discussed in Chapter 1 on incoming streams.

Equations (10) are the inverse equations of a SMP or, to use Feller's terminology (1950), backward equations. Now we shall set up the forward equations.

Denote by $H_j(t)$ the average number of transitions of the process $v(t)$ in the interval $(0, t)$ after each one of which the process arrives at the state j. The set of times at which such transitions occur forms a renewal process; thus, $H_j(t)$ is a renewal function. For $F_{ij}(+0) = 0$ the differential of this function $dH_j(t)$ can be interpreted as the probability that in the interval $(t, t + dt)$ a transition into the state j occurs. Whence

$$dH_j(t) = \sum_i p_i^{(0)}\, dP_{ij}(t) + \sum_k \int_0^t dH_k(t - x)\, dP_{kj}(x), \quad t \geqslant 0. \tag{11}$$

We can apply the Laplace-Stieltjes transform to the equations (11). Denoting

$$h_j(s) = \int_0^\infty e^{-st}\, dH_j(t),$$

we obtain

$$h_j(s) = \sum_i p_i^{(0)}\pi_{ij}(s) + \sum_k h_k(s)\pi_{kj}(s), \quad \operatorname{Re} s > 0. \tag{12}$$

We now return to the problem at hand. The event $\{v(t) = j\}$ may occur in two ways: either $v(0) = j$ and $t_1 > t$ or at some time $x < t$ the transition into the state j occurred and after that during time $t - x$ no transitions took place. Whence

$$B_j(t) = p_j^{(0)}\bar{P}_j(t) + \int_0^t \bar{P}_j(t-x)\,dH_j(x). \tag{13}$$

Hence,

$$sB_j^*(s) = p_j^{(0)}[1 - \pi_j(s)] + [1 - \pi_j(s)]h_j(s). \tag{14}$$

We note a generalization of formulas (13) and (14). Assume that to each trajectory of a semi-Markov process there corresponds a random process $\eta(t)$ with the following property: If it is known that at time τ the transition $\nu(t)$ into the state i occurred and the next transition occurs after time $\tau + x$, then $\eta(\tau + x)$ does not depend on the trajectory of $\nu(t)$ up until time τ and possesses mathematical expectation $f_i(x)$. Then

$$\mathsf{M}\eta(t) = \sum_i p_i^{(0)}\bar{P}_i(t)f_i(t) + \sum_j \bar{P}_j(t-x)f_j(t-x)\,dH_j(x). \tag{15}$$

Taking Laplace transforms we obtain

$$\int_0^\infty e^{-st}\mathsf{M}\eta(t)\,dt = \sum_i p_i^{(0)}(\bar{P}_i f_i)^*(s) + \sum_j (\bar{P}_j f_j)^* h_j(s), \quad \mathrm{Re}\, s > 0,$$

where $(\bar{P}f_i)^*(s)$ is the Laplace transform of the function $\bar{P}_i(t)f_i(t)$.

3.1.4. *Ergodic Properties of a Semi-Markov Process*

Let a SMP with delay $\nu(t)$ with a finite or countable set of states be given. The delay time as well as the sojourn time in any state is assumed to be finite with probability 1. Denote

$$\tau_i = \int_0^\infty x\,dP_i(x)$$

and assume that $\tau_i < \infty$, $i \in X$. Finally, assume that a Markov chain $\{\nu_n\}$ possesses an ergodic distribution $\{\pi_j\}$.

Theorem 1. *Let T_j be the instant of time in the interval $(0, T)$ during which the process $\nu(t)$ is in the state j. Then with probability 1 for $\{\sum \tau_i\pi_i < \infty\} \vee \{\tau_j < \infty\}$*

$$\lim_{T\to\infty} (T_j/T) = \tau_j\pi_j \Big/ \sum_i \tau_i\pi_i. \tag{16}$$

Theorem 2. *Let the conditions of Theorem 1 be fulfilled and let, at least for one i, such that $\pi_i > 0$, $P_i(x)$ be a nonlattice distribution.* Then*

$$B_j(t) \underset{t\to\infty}{\longrightarrow} \tau_j\pi_j \Big/ \sum_i \tau_i\pi_i \tag{17}$$

for any initial distribution $\{p_i^{(0)}\}$.

---

* For this to be valid it is sufficient to require the existence of a positive derivative of function $P_i(x)$ for at least one $x > 0$.

Theorem 3. *Let the conditions of Theorem 2 be fulfilled and let $f_j(t)$ be integrable functions such that $|f_j(t)| \leqslant C$. Then under the conditions of formula* (15)

$$\lim_{t \to \infty} \mathsf{M}\eta(t) = \frac{1}{\sum_j \tau_j \pi_j} \sum_j \pi_j \int_0^\infty \bar{P}_j(t) f_j(t)\, dt. \tag{18}$$

We suggest that the reader compare Theorem 3 with the key renewal theorem (Section 2.6).

3.1.5. *Method of "Catastrophes"*

In queueing theory a probabilistic interpretation of various integral transformations is used, which allows us to derive equations directly, for example, for the Laplace transform of the transition function of the queue length. This method was developed by Klimov (1964) and is most efficiently used in the study of priority queueing systems; cf. Gnedenko et al. (1973).

To explain the idea behind the method we shall consider the simplest example. Let there be a renewal process with the distribution function $F(t)$ of the interrenewal times and the renewal function $H(t)$:

$$\psi(s) = \int_0^\infty e^{-st}\, dF(t); \qquad \varphi(s) = \int_0^\infty e^{-st}\, dH(t).$$

It is required to derive an equation for $\varphi(s)$. Assume that $s = \lambda > 0$. Define the random variable ξ independent of the renewal process and exponentially distributed with parameter λ. Integration by parts of the expression defining $\varphi(\lambda)$ yields

$$\varphi(\lambda) = \int_0^\infty H(t)\lambda e^{-\lambda t}\, dt = \mathsf{M}H(\xi).$$

We shall call ξ the time of a "catastrophe"; we then see that $\varphi(\lambda)$ is the mathematical expectation of the number of renewals up to the time of a catastrophe.

We have

$$\gamma = I(1 + \gamma_1), \qquad \mathsf{M}\gamma = \mathsf{M}I + \mathsf{M}I\gamma_1,$$

where I is the indicator of the event {a catastrophe did not occur up to the first renewal}, γ_1 is the number of renewals up to the occurrence of a catastrophe without counting the first one of them.

Clearly,

$$\mathsf{M}I = \int_0^\infty e^{-\lambda t}\, dF(t) = \psi(\lambda).$$

In view of the properties of an exponential distribution, if a catastrophe did not occur up to the first renewal, the duration until its occurrence does not depend on the past. Thus, under the condition $\{I = 1\}$, γ_1 is a random variable having the same distribution as γ. Thus,

$$\mathsf{M}I\gamma_1 = \mathsf{M}I \cdot \mathsf{M}\{\gamma_1 | I = 1\} = \psi(\lambda)\varphi(\lambda).$$

Finally, we obtain

$$\varphi(\lambda) = \psi(\lambda) + \psi(\lambda)\varphi(\lambda) = \psi(\lambda)/[1 - \psi(\lambda)];$$

this equation was obtained in Chapter 2 using another method. Actually, from $\lambda > 0$ one should proceed to complex variable s, but this does not involve any difficulties using the principle of analytic continuation: it is sufficient to note that $\psi(x)/[1 - \psi(s)]$ is an analytic function in the half-plane $\operatorname{Re} s > 0$. Since it coincides with $\varphi(s)$ for $s \in \mathbb{R}^+$, it follows that it will be equal to $\varphi(s)$ in the whole half-plane.

3.2. Linear-Type Markov Processes

3.2.1. *Definition*

We shall describe a class of Markov processes proposed by Belyaev (1962). Let a system be given such that no more than one operation can be performed in it simultaneously. The state of the system is described by a Markov process $\xi(t)$, the set of states of which X consists of two subsets: X_0 and $X_1 \times \mathbb{R}^+$. The sets X_0 and X_1 are finite or countable; if $\xi(t) \in X_0$, then operations are *not* carried out at time t in the system. The set $X_1 \times \mathbb{R}^+$ consists of elements of the form (l, j, z) where l is the index that determines the type of operation being performed, j is an additional discrete parameter, z is the time from the beginning of the operation. The distribution function of the time of an operation of the lth type is $F_l(x)$. Thus, for $\xi(t) = (l, j, z)$ the probability of completion of the operation during time dt is $[F_l(z + dt) - F_l(z)]/\bar{F}_l(z)$.

If at time t the operation is completed and if immediately before that the state of the process was (l, i, z), then with probability $p_{li}(m, j)$ the process moves into the state $(m, j, 0)$ (here an operation of the type m commences) and with probability $p_{li}(j)$ into the state $j \in X_0$. Besides transitions of the process associated with terminations of operations, spontaneous transitions are also possible. Namely, for $\xi(t) = i \in X_0$ during time dt with probability $\lambda_i(j)\,dt$ a transition of the process into state $j \in X_0$ may occur and with probability $\lambda_i(l, j)\,dt$ into the state $(l, j, 0)$; if $\xi(t) = (l, i, z)$, then with probability $\lambda_{li}(j)\,dt$ the process passes into the state $(l, j, z + dt)$. It is required that the random process $\xi(t)$ be Markovian. The process defined above will be called a *lined Markov process* (or an *L-process*).

3.2.2. *Basic Equations*

Assume that the set of states of discrete components of a process is finite, $F_l(x)$ are absolutely continuous functions, $p_l(x)$ are the corresponding probability densities. Belyaev has shown that under an absolutely continuous initial probability distribution of states of the process there exist continuous functions

$$p_i(t) = \mathsf{P}\{\xi(t) = i\}, \quad i \in X_0$$

and

$$q_{li}(t, x) = [\bar{F}_l(x)]^{-1} \frac{d}{dx} \mathsf{P}\{\xi(t) \in \{(l, i, y), y < x\}\},$$

satisfying the system of differential equations

$$p_j'(t) + \lambda_j p_j(t) = \sum_{j \neq i \in X_0} \lambda_i(j) p_i(t) + \sum_{(l,i) \in X_1} p_{li}(j) \int_0^\infty q_{li}(t, z)\, dF_l(z), \quad i \in X_0, \quad (1)$$

$$\frac{\partial q_{lj}(t, x)}{\partial t} + \frac{\partial q_{lj}(t, x)}{\partial z} + \lambda_{lj} q_{lj}(t, x) = \sum_{i \neq j} \lambda_{li}(j) q_{li}(t, x) \quad (2)$$

with the boundary conditions

$$q_{lj}(t, 0) = \sum_{i \in X_0} \lambda_i(l, j) p_i(t) + \sum_m \sum_{i \neq j} p_{mi}(l, j) \int_0^\infty q_{mi}(t, z)\, dF_m(z), \quad (3)$$

where

$$\lambda_i = \sum_{i \neq j \in X_0} \lambda_i(j) + \sum_{(l,j) \in X_1} \lambda_i(l, j), \qquad \lambda_{li} = \sum_{j \neq i} \lambda_{li}(j).$$

If the set of states of discrete components is infinite, to justify (1), (2), and (3) additional regularity conditions are required. The following simple condition

$$\lambda_i \leqslant c, \qquad \lambda_{li} \leqslant c, \quad dF_l(z) \leqslant c\, dz \quad (4)$$

is sufficient. One can associate with a lined Markov process $\xi(t)$ an "embedded" Markov chain $\xi_n = \xi_0(t_n + 0)$ where t_n is the nth—in increasing order—time when either a certain operation is terminated or the process proceeds from a state in the set X_0 into some other state (belonging to X_0 or to $X_1 \times \mathbb{R}^+$), $\xi_0(t) = i$ for $\xi(t) = i \in X_0$, $\xi_0(t) = (l, i)$ for $\xi(t) = (l, i, z)$. The transition probabilities $\{\xi_n\}$ are of the form:

$$p_{ij} = \lambda_i(j)/\lambda_i, \quad (5)$$

$$p_{i,(l,j)} = \lambda_i(l, j)/\lambda_i, \quad (6)$$

$$p_{(l,i),j} = \sum_k g_{li}(k) p_{lk}(j), \quad (7)$$

$$p_{(l,i),(m,j)} = \sum_k g_{li}(k) p_{lk}(m, j), \quad (8)$$

where

$$g_{li}(k) = \int_0^\infty u_{ik}^{(l)}(t)\, dF_l(t),$$

$u_{ik}^{(l)}(t)$ is the transition probability function of a homogeneous Markov process

$\gamma(t)$ with the transition intensities

$$\frac{1}{dt}\mathsf{P}\{\gamma(t+dt)=k|\gamma(t)=i\}=\lambda_{li}(k).$$

Let $H_j(t)$ be the mathematical expectation of the number of n such that $\xi_n = j$, $0 < t_n < t$, $H_{lj}(t)$ is the analogous mathematical expectation for (l, j) in place of j. Introduce the notation

$$B_j(t)=\mathsf{P}\{\xi(t)=j\},\quad B_{lj}(t)=\mathsf{P}\{\xi_0(t)=(l,j)\}.$$

We have

$$B_j(t)=p_j^{(0)}e^{-\lambda_j t}+\int_0^t e^{-\lambda_j(t-x)}\,dH_j(x), \tag{9}$$

$$B_{lj}(t)=\sum_i p_{li}^{(0)}\bar{F}_l(t)u_{ij}^{(l)}(t)+\sum_i\int_0^t \bar{F}_l(t-x)u_{ij}^{(l)}(t-x)\,dH_{li}(x). \tag{10}$$

Let

$$h_j(s)=\int_0^\infty e^{-st}\,dH_j(t),\qquad h_{ij}(s)=\int_0^\infty e^{-st}\,dH_{lj}(t).$$

Then for $h_j(s)$ the system of equations of the form (12) presented in Section 3.1, namely:

$$\begin{aligned} h_j(s)&=\sum_i p_i^{(0)}\lambda_i(j)/(s+\lambda_i)+\sum_{l,i}p_{li}^{(0)}\pi_{li}(j,s)\\ &\quad+\sum_i h_i(s)\lambda_i(j)/(s+\lambda_i)+\sum_{l,i}h_{li}(s)\pi_{li}(j,s), \end{aligned} \tag{11}$$

$$\begin{aligned} h_{lj}(s)&=\sum_i p_i^{(0)}\lambda_i(l,j)/(s+\lambda_i)+\sum_{m,i}p_{m,i}^{(0)}\pi_{mi}(l,j,s)\\ &\quad+\sum_i h_i(s)\lambda_i(l,j)/(s+\lambda_i)+\sum_{m,i}h_{mi}(s)\pi_{mi}(l,j,s), \end{aligned} \tag{12}$$

is fulfilled, where

$$\pi_{li}(j,s)=\sum_k\int_0^\infty e^{-st}u_{ik}^{(l)}(t)\,dF_l(t)p_{lk}(j), \tag{13}$$

$$\pi_{mi}(l,j,s)=\sum_k\int_0^\infty e^{-st}u_{ik}^{(m)}(t)\,dF_m(t)p_{mk}(l,j). \tag{14}$$

Note that it is often possible to express expressions of a similar type in an elementary manner in terms of $\psi_l(s)=\int_0^\infty e^{-st}\,dF_l(t)$. This is because $u_{ik}^{(m)}(t)$ is a solution of a system of linear differential equations with constant coefficients. For example, in the case where a given function is represented in the form of a linear combination of exponents $u(t)=e^{-\rho t}$ it is sufficient to observe that

$$\int_0^\infty e^{-st}u(t)\,dF_l(t)=\psi_l(s+\rho).$$

3.2.3. *The Ergodic Theorem for Lined Processes*

We define a SMP $\nu(t)$ as follows. If $\xi(t) = i$, then $\nu(t) = i$; if $\xi(t + 0) = (l, i, 0)$, then $\nu(t) = (l, i)$ starting from time t up until the time of completion of an operation that started at time t. At discontinuity times we shall define $\nu(t)$ by the continuity from the right. Thus, $\nu(t)$ in a sense traces the changes in the states of $\xi(t)$ but loses this property at those intervals where operations take place. It is easy to see that $H_j(t)$, $H_{lj}(t)$ defined by means of the process $\xi(t)$ also has an analogous meaning for the new process: $H_j(t)$ ($H_{lj}(t)$) is the mean number of arrivals of $\nu(t)$ into the state j ((l, j)) in the interval $(0, t)$. The mean sojourn time of $\nu(t)$ is in the state j

$$\tau_j = \lambda_j^{-1},$$

and in the state $(1, j)$

$$\tau_{lj} = \int_0^\infty x\, dF_l(x).$$

Theorem. *Let the Markov chain with the set of states $X_0 \cup X_1$ and transition probabilities* (5)–(8) *possess an ergodic distribution $\{\pi_i; \pi_{li}\}$ and*

$$T_0 = \sum_i \tau_i \pi_i + \sum_{l,i} \tau_{li} \pi_{li} < \infty.$$

Then

$$B_j(t) \underset{t \to \infty}{\longrightarrow} \pi_j \tau_j / T_0, \tag{15}$$

$$B_{lj}(t) \underset{t \to \infty}{\longrightarrow} \frac{1}{T_0} \sum_i \pi_{li} \int_0^\infty u_{ij}^{(l)}(t) \bar{F}_l(t)\, dt. \tag{16}$$

The *proof* is completely trivial: it is sufficient to use Smith's theorem (Chapter 2).

3.2.4. *The Method of Integrodifferential Equations*

When $\xi(t) = i$, set $\xi_0(t) = i$; if, however, $\xi_0(t) = (l, i, z)$, set $\xi_0(t) = (l, i)$, $\xi_1(t) = z$. In this case $\xi_1(t)$ is an "additional" component of the random process $\xi(t)$. Processes with additional components are commonly employed in queueing theory; these components are chosen in a manner that the resulting process becomes Markovian.

Denote

$$F_{li}(t, x) = \mathsf{P}\{\xi_0(t) = (l, i), \xi_1(t) < x\}.$$

We shall clarify the basic principle behind the derivation of Eqs. (1)–(3). Assume that

$$F_{li}(t,x) = \int_0^x p_{li}(t,y)\,dy,$$

where $p_{li}(t,x)$ are continuously differentiable functions. Then $p_{li}(t,x)\,dx$ is the probability of the event $\{\xi_0(t) = (l,i), x < \xi_1(t) < x + dx\}$. Thus, for $\Delta > 0$

$$p_{lj}(t+\Delta, x+\Delta)\,dx = \mathsf{P}\{\xi_0(t+\Delta) = (l,j), x+\Delta < \xi_1(t+\Delta) < x+\Delta+dx\}.$$

The event written in braces can occur in the following mutually exclusive ways:

1. $x < \xi(t) < x + dx$; during time Δ not a single spontaneous change in the state of the lined process occurred and the operation did not terminate.
2. $x < \xi(t) < x + dx$; $\xi_0(t) = (l,i)$, $i \neq j$, and in the interval $(t, t+\Delta)$ a spontaneous change from (l,i) to (l,j) occurred; the operation did not terminate.
3. $x < \xi(t) < x + dx$; $\xi_0(t) = (l,i)$; during the time from t to $t + \Delta$ at least two spontaneous changes occurred.

Using the probabilities of the above stated events we obtain

$$\begin{aligned} p_{lj}(t+\Delta, x+\Delta)\,dx &= [1 - \lambda_{lj}\Delta + o(\Delta)]p_{lj}(t,x)[\bar{F}_l(x+\Delta)/\bar{F}_l(x)]\,dx \\ &\quad + \sum_{i\neq j} [\lambda_{li}(j)\Delta + o_i(\Delta)]p_{li}(t,x)\frac{\bar{F}_l(x+\Delta)}{\bar{F}_l(x)}dx \\ &\quad + \sum_i o_i'(\Delta)p_{li}(t,x)\,dx, \end{aligned} \tag{17}$$

where $o_i(\Delta)$ and $o_i'(\Delta)$ denote infinitesimal random variables in comparison with Δ.

Equality (17) can be rewritten as follows:

$$\begin{aligned} q_{ij}(t+\Delta, x+\Delta)\,dx &= [1 - \lambda_{lj}\Delta + o(\Delta)]q_{lj}(t,x)\,dx \\ &\quad + \Delta \sum_{i\neq j} \lambda_{li}(j)q_{li}(t,x)\,dx + o(\Delta)\,dx, \end{aligned} \tag{18}$$

if one assumes that the bounds $o_i(\Delta)$ and $o_i'(\Delta)$ are uniform in i as $\Delta \to 0$. Dividing both sides of (18) by $\Delta\,dx$ and rearranging the summands we obtain

$$\frac{1}{\Delta}[q_{lj}(t+\Delta, x+\Delta) - q_{lj}(t,x)] = -\lambda_{lj}q_{lj}(t,x) + \sum_{i\neq j} \lambda_{li}(j)q_{li}(t,x) + o(1). \tag{19}$$

Equation (19) immediately implies (2), since for the continuously differential function $f(t,x)$ the difference ratio

$$\frac{1}{\Delta}(f(t+\Delta, x+\Delta) - f(t,x))$$

converges as $\to 0$, to,

$$\frac{\partial f}{\partial t} + \frac{\partial f}{\partial x}.$$

The derivation of Eq. (1) is analogous to that of (2); only one point requires an explanation. If $\xi(t+\Delta) = j$, then it may happen that $\xi(t) = (l, i, z)$. Moreover, the operation should be completed during time Δ and the process should move on into the state i. The probability of this event is

$$p_{li}(j) \int_0^\infty p_{li}(t, z)([F_l(z+\Delta) - F_l(z)]/\bar{F}_l(z))\, dz$$

$$= p_{li}(j) \int_0^\infty q_{li}(t, z)[F_l(z+\Delta) - F_l(z)]\, dz.$$

The integral on the right-hand side of this equality actually extends only over the interval $(0, t)$ since $\xi_1(t) \leqslant t$. Assume that $F_l(z) \geqslant \beta(z) > 0$ for any $z > 0$. Then

$$\int_a^\infty q_{li}(t, z)(\cdots)\, dz \leqslant \frac{1}{\beta(a)} \int_a^t p_{li}(t, z)(\cdots)\, dz;$$

since the series $\sum_{l,i} \int p_{li}(t, z)\, dz$ converges as a sum of probabilities of disjoint events, the required limiting transition follows from the continuity of $F_l'(z)$. In view of monotonicity in a, we can then set $a = 0$.

We now turn to Eq. (3). We have

$$q_{lj}(t, 0)\Delta + o(\Delta) = \mathsf{P}\{\xi_0(t) = (l, j), \xi_1(t) < \Delta\}. \tag{20}$$

Since $\xi_1(t) < \Delta$, the operation that takes places at time t starts in the interval $(t-\Delta, t)$. This is possible only under the following conditions: 1) in the given interval the preceding operation terminated, 2) in the given interval an operation commenced due to a spontaneous change in the state of the process.

Subsequent arguments are analogous to those described above.

3.2.5. *Lined Processes with a Fixed Remainder*

Let $\xi(t)$ be a lined random process. If $\xi(t) = (l, i, z)$, then based on the realization $\xi(s)$, $s > t$, we can determine the time $t + \gamma$ when an operation will be completed that lasted at time t. Set $\gamma(t) = \gamma$ and call $\gamma(t)$ the size of a jump at time t. A random process $\zeta(t)$ that is equal to $\xi(t)$ for $\xi(t) \in X_0$ and is equal to $(\xi_0(t), \gamma(t))$ otherwise is a Markov process.

Considering the random process $\zeta(t)$ instead of $\xi(t)$ allows us to somewhat simplify definitions and notation. Indeed, for the definition of a lined Markov process the index l was essential: this index determines the probability with which an operation can be completed during time Δ provided it has already lasted time z. In the case of the process $\zeta(t)$, the remaining time is fixed, i.e., l is inessential here.

We shall now present a new definition of a lined Markov process (with a fixed remainder). The set of states of a process is $X_0 \cup (X_1 \times \mathbb{R}^+)$, where X_0, X_1 are finite or countable sets, $X_1 \times \mathbb{R}^+$ is the set of pairs (i, z), z is a variable admitting any positive values. The values i for which $i \in X_1$ do not appear in X_0.

For $\xi(t) = i \in X_0$ with probability $1 - \lambda_i\Delta + o(\Delta)$, the equality $\xi(t + \Delta) = i$ is satisfied; with probability $\lambda_{ij}\Delta + o(\Delta)$ the equality $\xi(t + \Delta) = j \neq i, j \in X_0$; with probability $\lambda_{ij}\Delta B_j(x) + o(\Delta)$, the equality $\xi(t + \Delta) = (j, z)$, $z < x$, where $j \notin X_0$. Moreover, $\lambda_i = \sum_j \lambda_{ij}$. For $\xi(t) = (i, x)$ with probability $1 - \lambda_i\Delta + o(\Delta)$ we have $\xi(t + \Delta) = (i, x - \alpha_i\Delta)$, where $\alpha_i \geqslant 0$ is a constant, with probability $\lambda_{ij}\Delta + o(\Delta)$ at time $t + \theta$ in the interval $(t, t + \Delta)$ a transition of $\xi(t)$ into the new state such that

$$\xi(t + \Delta) = (j, x - \alpha_i\theta - \alpha_j(\Delta - \theta)).$$

occurs.

Now let $\xi(t - 0) = (i, 0)$. Then with probability $p_{ij}^{(1)}$ we have $\xi(t) = j \in X_0$, with probability $p_{ij}^{(1)}B_j(x)$ we have $\xi(t) = (j, z), j \in X_1$, where $z < x$. Finally, we complete the definition of $\xi(t)$ at the instants of changes of states by the continuity from the right.

Note that $\xi(t)$ is the same as $\zeta(t)$ in the preceding section; the second component of $\xi(t)$ is $\gamma(t)$, however, for uniformity we shall denote it by $\xi_1(t)$ retaining for the first "discrete" component the notation $\xi_0(t)$.

The process $\xi(t)$ defined above will be called a *lined Markov process with fixed remainder.*

Note that the introduction of constants α_j allows us to proceed from a "temporal" to an "energy" interpretation of operations: α_j is the rate of performance of an operation at state j of a discrete component, $B_j(x)$ is the distribution of the amount of work associated with an operation that begins in state j. If in all the cases $\alpha_j = 1$, then the amount of work becomes the time of an operation execution.

Note that by extending the set of states of the process one can make $B_j(x)$ dependent on the preceding or succeeding state of the process (as in the case of SMP). When solving specific problems, readers may carry out this modification—if necessary—on their own.

3.2.6. *Differential Equations*

Denote

$$p_j(t) = \mathsf{P}\{\xi_0(t) = j\}, \quad j \in X_0;$$
$$F_j(t, x) = \mathsf{P}\{\xi_0(t) = j, \, \xi_1(t) < x\}, \quad j \in X_1.$$

For given functions a system of integro-differential equations can be derived analogously to the corresponding characteristics of a lined Markov process. The reader is requested to construct this system as an exercise. We shall consider the most important case of a stationary distribution: $p_j(t) = p_j$, $F_j(t, x) = F_j(x)$. Let $j \in X_1$. Assume that $\lambda_j \leqslant c$, $\alpha_j \leqslant c$;

$$\lim_{x \downarrow 0} \sup_j B_j(x) < 1. \tag{21}$$

We have $p_j = \mathsf{P}\{\xi_0(\Delta) = j\}$.

The following cases are possible:

1. $\xi_0(0) = j$; there are no transitions in the interval $(0, \Delta)$.
2. $\xi_0(0) = i \neq j$; in the interval $(0, \Delta)$ there is a transition from i into j.
3. $\xi(0) = (i, z)$, $z < \Delta$; at time z there is a transition into j.
4. In the interval $(0, \Delta)$ at least two events occur.

The probabilities of occurrences of these cases are equal to $p_j(1 - \lambda_j\Delta + o(\Delta))$, $p_i\lambda_{ij}\Delta + o(\Delta)$, $F_i(\alpha_i\Delta)p_{ij}^{(1)}$, $o(\Delta)$, respectively. Only the last assertion requires explanation.

Let t be the time of spontaneous change of the state of $\xi_0(t)$, and t' be the next such time. Then with any information on the trajectory of the process up to time t the random variable $t' - t$ is stochastically larger than the exponential variable with parameter c. Whence, if $s_0(T)$ is a number of spontaneous changes in the interval $(0, T)$, then $\mathsf{M}s_0(T) \leqslant cT$. If τ is now the time of an operation completion and τ' is the next such time, then $\tau' - \tau$ under any additional assumptions on the trajectory of the process up to time τ is stochastically larger than the random variable with the distribution function $\Phi(t) = \sup_j B_j(ct)$, which in accordance with condition (21) possesses a positive (not necessarily finite) mathematical expectation. Whence, we have $\mathsf{M}s(T) \leqslant c_1 T$ for the number $s(T)$ of completions of operations in the interval $(0, T)$ by a property of a renewal function. Thus, the intensity of the stream of events is finite ($\leqslant c + c_1$).

Equating the probabilities associated with times Δ and 0 we obtain

$$\begin{aligned} p_j = p_j(1 - \lambda_j\Delta + o(\Delta)) &+ \sum_{j \neq i \in X_0} p_i(\lambda_{ij}\Delta + o(\Delta)) \\ &+ \sum_{i \notin X_0} F_i(\alpha_i\Delta)p_{ij}^{(1)} + o(\Delta). \end{aligned} \tag{22}$$

Whence,

$$\lambda_j p_j = \sum_{j \neq i \in X_0} \lambda_{ij} p_i + \sum_{i \in X_1} \alpha_i F_i'(0) p_{ij}^{(1)}. \tag{23}$$

The limiting transition is justified since $\{\xi_i(0) < \Delta\}$ implies the completion of an operation in the interval $(0, \Delta)$, whence

$$\sum_{i \in X_1} F_i(\Delta) \leqslant c_1 \Delta. \tag{24}$$

On the other hand, $F_i'(0)$ is by definition the value of the parameter of a stationary stream of homogeneous events, namely, the times of completion of operations when the discrete component is in state i.

Now consider the event $\{\xi_0(\Delta) = j, \xi_1(\Delta) < x\}$. At time 0 we have the following:

No.	Event	Probability
1	$\xi_0(0) = j$, $\alpha_j\Delta < \xi_1(0) < x + \alpha_j\Delta$; in the interval $(0, \Delta)$ there are no spontaneous transitions.	$(F_j(x + \alpha_j\Delta) - F_j(\alpha_j\Delta)) \times (1 - \lambda_j\Delta + o(\Delta))$
2	$\xi_0(0) = i \in X_i$; at time $\Delta - \theta$ a spontaneous transition into j occurred; $\alpha_i(\Delta - \theta) + \alpha_j\theta < \xi_1(0) < x + \alpha_i(\Delta - \theta) + \alpha_j\theta$	$(F_i(x + v') - F_i(v')) \times (\lambda_{ij}\Delta + o(\Delta))$, $v' = O(\Delta)$
3	$\xi_0(0) = i$, $\xi_1(0) < \alpha_i\Delta$; at time $\Delta - \theta$ the transition into the state (j, z), occurred $\alpha_j\theta < z < x + \alpha_j\theta$.	$F_i(\alpha_i\Delta)p_{ij}^{(1)}B_j(x + \alpha_j v)$, $0 < v < \Delta$
4	$\xi_0(0) = i \in X_0$; at time $\Delta - \theta$ the transition into the state (j, z), occurred $\alpha_j\theta < z < x + \alpha_j\theta$	$p_i(B_j(x + v'') - B_j(v'')) \times (\lambda_{ij}\Delta + o(\Delta))$, $v'' = O(\Delta)$
5	At least two events in the interval $(0, \Delta)$ occurred.	$o(\Delta)$

We arrive at the equality

$$\begin{aligned} F_j(x) = {} & [F_j(x + \alpha_j\Delta) - F_j(\alpha_j\Delta)](1 - \lambda_j\Delta) \\ & + \sum_i F_i(x + v')\lambda_{ij}\Delta + \sum_i F_i(\alpha_i\Delta)p_{ij}^{(1)} \\ & + \sum_{i \equiv X_0} p_i\lambda_{ij}[B_j(x + v'') - B_j(v'')]\Delta + o(\Delta). \end{aligned} \tag{25}$$

Analogously to the preceding we have the differential equation

$$\alpha_j F_j'(x) - \alpha_j F_j'(0) - \lambda_j F_j(x) + \sum_i \lambda_{ij}F_i(x) + \sum_i \alpha_i F_i'(0)p_{ij}^{(1)} + \sum_{i \equiv X_0} p_i\lambda_{ij}B_j(x) = 0 \tag{26}$$

at the points where all the $B_i(x)$ are continuous, i.e., everywhere in the interval $(0, \infty)$.

Thus, the stationary distribution of a lined Markov process with a fixed remainder satisfies the system of equations (23), (26). Evidently the normalizing condition

$$\sum_{i \in X_0} p_i + \sum_{i \in X_1} F_i(\infty) = 1 \tag{27}$$

is also satisfied. For any $x \geqslant 0$ it follows from (25) that

$$\frac{1}{\Delta}[F_j(x + \alpha_j\Delta) - F_j(x)] \leqslant \frac{1}{\Delta}F_j(\alpha_j\Delta) + \lambda_j + O(\Delta) \sim c_j F_j'(0) + \lambda_j;$$

thus $F_j(x)$ are absolutely continuous functions.

Therefore, if one introduces the Laplace-Stieltjes transform $b_j(s) = \int_0^\infty e^{-st}\,dF_j(t)$, then in this integral $dF_j(t) = F_j'(t)\,dt$. From (26) we obtain the system of equations

$$s\alpha_j b_j(s) - \alpha_j F_j'(0) - \lambda_j \psi_j(s) + \sum_i \lambda_{ij} b_i(s) + \sum_{i \in X_1} \alpha_i F_i'(0) p_{ij}^{(1)} + \sum_{i \in X_0} p_i \lambda_{ij} \psi_j(s) = 0, \tag{28}$$

where $\psi_j(s)$ is the Laplace-Stieltjes transform of $B_j(x)$. Formally, the system of equations (28), (23), (27) contains "more" unknowns than equations. The missing equations can be obtained from analytic considerations. Thus, when there is a finite number of discrete states of the process, by solving the system (28) with respect to $\{b_j(s)\}$ we arrive at

$$b_j(s) = \Delta^{-1}(s) L_j(s), \tag{29}$$

where $\Delta(s)$—as the determinant of the system—is a polynomial in s; $L_j(s)$ is an expression linear in p_i, $i \in X_0$; $c_i F_i'(0)$, $i \in X_1$. Since $b_j(s)$ are analytic functions for $\operatorname{Re} s > 0$, we have

$$L_j(s) = 0, \qquad L_j'(s) = 0, \ldots, L_j^{r-1}(s) = 0 \tag{30}$$

for each root of $\Delta(s)$ in the right-half-plane, where r is the multiplicity of this root.

We now present an alternative method for constructing equations directly from the unknown constants based on an embedded Markov chain.

Let t_n be the time of completion of the nth operation, $v_n = \xi_0(t_n - 0)$. Then $\{v_n\}$ is a homogeneous Markov chain. Denote

$u_{ij}(t)$ to be the transition function of a homogeneous Markov process with transition intensities λ_{ij};

$\bar{u}_{ij}(t)$ to the transition function of the same process with a prohibition to return to the set of the states in X_0;

$$a_{ij} = \int_0^\infty u_{ij}(t)\, dB_i(t); \tag{31}$$

v_{ij} is the probability that the Markov chain with transition probabilities λ_{ij}/λ_i under the initial state i will leave the set of states X_0 for the first time passing over to the state $j \in X_1$.

Let t_n^0 be the time of the beginning of the nth operation, $v_n^0 = \xi_0(t_n^0 + 0)$. Denote

$$x_j = \lim_{n \to \infty} \mathsf{P}\{v_n = j\}, \quad j \in X_1;$$

$$y_j = \lim_{n \to \infty} \mathsf{P}\{v_n^0 = j\}, \quad j \in X_1.$$

We have the system of equations

$$x_j = \sum_{i \in X_1} y_i a_{ij}, \tag{32}$$

$$y_j = \sum_{i \in X_1} x_i \left(p_{ij}^{(1)} + \sum_{k \in X_0} p_{ik}^{(1)} v_{kj} \right). \tag{33}$$

Under the condition that the random process $\xi(t)$ is ergodic, the solution of the system (32)–(33) is unique up to a constant factor provided that $\sum |x_j| < \infty$ (or $\sum |y_j| < \infty$). The ergodic probabilities $p_j = \lim \mathsf{P}\{\xi_0(t) = j\}$ are expressed in terms of this solution: $t \to \infty$:

$$p_j = \mu \sum_i x_i \sum_{k \in X_0} p_{ik}^{(1)} \int_0^\infty \bar{u}_{kj}(t)\, dt, \quad j \in X_0; \tag{34}$$

$$p_j = \mu \sum_i y_i \int_0^\infty u_{ij}(t) \bar{B}_i(t)\, dt, \qquad j \in X_1. \tag{35}$$

The constant μ is uniquely determined by the normalizing condition

$$\sum_j p_j = 1. \tag{36}$$

The constants v_{ij}, $i \in X_0, j \in X_1$ can be determined as either the solution of the system of equations

$$\lambda_i v_{ij} = \lambda_{ij} + \sum_{k \in X_0} \lambda_{ik} v_{kj}, \quad i \in X_0, \tag{37}$$

or directly by means of the formula

$$v_{ij} = \sum_{k \in X_0} \lambda_{kj} \int_0^\infty \bar{u}_{ik}(t)\, dt. \tag{38}$$

Functions $u_{ij}(t)$ are determined as a solution of the system of forward Kolmogorov's differential equations

$$u_{ij}'(t) + \lambda_j u_j(t) = \sum_{k \in X_1} \lambda_{kj} u_{ik}(t), \quad i \in X_1, \quad j \in X_1, \tag{39}$$

under the initial condition

$$u_{ij}(0) = \delta_{ij}. \tag{40}$$

The functions $\bar{u}_{ij}(t)$ are determined by the solution of the system of equations

$$\bar{u}_{ij}'(t) + \lambda_j \bar{u}_{ij}(t) = \sum_{k \in X_0} \lambda_{kj} \bar{u}_{ik}(t), \quad i \in X_0, \quad j \in X_0. \tag{41}$$

Note that $\alpha_j F_j'(+0)$ are proportional to the constants x_j.

3.3. Piecewise-Linear Markov Processes

3.3.1. *Method of Additional Variables*

In this section we shall present a generalized mathematical framework for random processes that will enable us to describe a wide class of queueing systems. Development of such a scheme is necessary for the following reasons. First, owing to the ever-increasing complexity of specific queueing systems, the existing concepts (processes with waiting, losses, etc.) are no longer sufficient to encompass all the problems of practical interest; it is necessary to develop analytical methods capable of taking into account the variety of diverse factors present. Second, it is necessary to ascertain what properties of queueing processes of importance in practice can be defined analytically (in the same manner as there exist theories for integrating irrational functions, solution algebraic equations by means of radicals, etc.); such a theory is impossible if the appropriate class of objects is not clearly specified. Third, we require a general framework for algorithmization of approximating calculations and for simulation of queueing systems by means of the Monte Carlo method. Finally, there are many problems of synthesis of optimal queueing systems that cannot be incorporated into a systematic theory without having a general framework.

We are convinced that at the present time the algorithmic approach is basic for queueing theory, in view of the development of various computer software systems. Advances in the solutions of specific problems must be interpreted with an eye to construction of algorithms that will aid in the solution of classes of problems.

The previously outlined problem has been studied by many authors; numerous generalized schemes for describing queueing processes are available in the literature.

In our opinion, Cox's (1955) model is the most fruitful. This model is described in general terms as follows. Consider a random process $v(t)$ with a finite or countable set of states X_0. This process except for certain special cases is not Markovian; however, jointly with additional components $\xi_j(t)$ becomes one. A particular number of additional components correspond to each possible value i. The latter increase in a unit rate and have the meaning of time from the beginning of some event; the duration of the existence of an additional component is random and does not depend on other analogous quantities. We shall describe a scheme of piecewise-linear Markov processes, which generalizes Cox's model. In various degrees of generality, piecewise-linear Markov processes were, proposed by Kovalenko (1964a, 1964b, 1965a), Kalashnikov (1978) and Tien (1977).

3.3.2. *Piecewise-Linear Markov Process*

Consider a Markov random process $\zeta(t) = (v(t), \bar{\xi}(t))$ defined as follows. The space of states of the process X is the set of pairs $(i, \bar{\xi}_i)$, where i is the element of a finite or countable set. $\bar{\xi}_i$ is the vector $(\xi_1, \ldots, \xi_{|i|})$, $|i| \geqslant 0$ is the "rank" of the state i, $\xi_j \geqslant 0$. Let $\zeta(t) = (i, \bar{y})$, $\bar{y} = (y_1, \ldots, y_{|i|})$. Then with probability $\lambda_{ij}\,dt$ during time dt a spontaneous transition $v(t)$ to the state j takes place. After the transition a new value of $\bar{\xi}(t)$ is random and possesses a measurable in $\bar{y}$ distribution function

$$B_{ij}^{(0)}(\bar{x}\,|\,\bar{y}) = \mathsf{P}\{\bar{\xi}(t + dt) < \bar{x}\,|\,\bar{\xi}(t) = \bar{y}, v(t + dt) = j\}.$$

The probability is $o(h)$ that two or more spontaneous transitions will occur during a short period of time h. In the absence of spontaneous transitions in the interval $(t, t+dt)$, we have $v(t+dt) = v(t)$, $\bar{\xi}(t+dt) = \bar{\xi}(t) - \bar{\alpha}_i\,dt$, where $\bar{\alpha}_i = (\alpha_{i1}, \ldots, \alpha_{i|i|})$ is a vector with nonnegative components. Note that, if $\xi_j(t)$ is viewed as the time until the completion of an operation actually performed in a given state, then $\alpha_{ij} = 1$; if $\xi_j(t)$ is the remaining amount of work, then α_{ij} is the rate of its execution (the capacity of the server). Assume that $\lambda_i = \sum_j \lambda_{ij} < \infty$. The possibility that $\lambda_{ii} > 0$ is not excluded: for example, during a spontaneous transition $v(t)$ is not changed, but $\bar{\xi}(t)$ undergoes a jump.

At time t let a component of the vector $\bar{\xi}(t)$ vanish, i.e., $v(t-0) = i$, $\bar{\xi}(t-0) = \bar{y}$, $y_j = 0$. Then at time t the transition $\zeta(t)$ into a new random state $(k, \bar{\xi}(t+0))$ takes place;

$$H(k, \bar{x} | i, \bar{y}) = \mathsf{P}\{v(t+0) = k, \bar{\xi}(t+0) < \bar{x} | v(t-0) = i, \bar{\xi}(t-0) = \bar{y}\}.$$

If the probability that both components of $\bar{\xi}(t)$ vanish simultaneously is zero, then, instead of $H(\cdots)$, it is sufficient to assign probability $p_{ik}^{(j)}$ of the transition of $v(t)$ from i into k when $\xi_j(t-0) = 0$ and the conditional distribution $B_{ij}(\bar{x}|\bar{y})$ of the random vector $\bar{\xi}(t+0)$ under these conditions.

$\zeta(t)$ is defined at the jump points by continuity from the right.

The random process $\zeta(t)$ defined previously is called a *piecewise-linear Markov process* (PLMP). The component $v(t)$ is called the discrete ("qualitative") component (variable), $\xi_j(t)$ are the additional components (variables), $\bar{\xi}(t)$ is the vector of additional components (variables). As any Markov process, PLMP is determined—in addition to the transition characteristics—by the initial distribution $P^{(0)}(A)$ or by the initial state $(i_0, \bar{x}_0)$.

3.3.3. *Regularity Conditions*

A PLMP is called *regular in an interval* $(0, t)$ if it has in it, with probability 1, a finite number of jumps for any initial state $(i_0, \bar{x}_0)$. A PLMP is called *regular* if it is regular in the interval $(0, T)$ for any T.

We state two versions of sufficiency conditions for regularity of PLMP.

Condition A. The set X_0 is finite; if $n_\varepsilon(\bar{x})$ is the number of components of vectors x smaller than ε, then for some $\varepsilon > 0$

$$\mathsf{P}\{n_\varepsilon(\bar{\xi}_{\bar{y}}) < n_\delta(\bar{y})\} \xrightarrow[\delta \to 0]{} 1, \tag{1}$$

where $\bar{\xi}_{\bar{y}} = \bar{\xi}_{\bar{y}}(i, k)$ is a random vector with the distribution function $H(k, \bar{x} | i, \bar{y}) / H(k, \overline{\infty} | i, \bar{y})$, uniformly in $\bar{y}$ such that $n_\delta(\bar{y}) \geqslant 1$.

Condition B. The quantities $|i|$, λ_i, α_{ij} are bounded from above and (1) is fulfilled uniformly in i, k, $\bar{y}$ $[n_\delta(\bar{y}) \geqslant 1]$.

3.3.4. *Two Reductions*

When solving theoretical problems, it is sometimes more convenient to consider the PLMP model without spontaneous transitions ($\lambda_i = 0$). An arbitrary

PLMP can be reduced to a process without spontaneous transitions by adjoining $\xi_0(t)$ as an additional component, which decreases with unit rate and for any transition $v(t)$ from state i to state k taking on a value that is equal to an exponentially distributed random variable with parameter λ_k. The vanishing of $\xi_0(t)$ replaces the possible spontaneous transitions in a natural manner.

We also note a reduction of PLMP with a finite $s = \max_i |i|$ to a service system with a constant number s of objects. Such a system is useful for statistical simulation of queueing systems as well as in many theoretical problems. It has been developed in the book by Franken et al. (1984). In this model there is a constant number s of objects, each one of which can be in a finite set of states "active" and "passive." In the first, operations with rates depending on the state of the system are performed. A transition from one state to another occurs at the time of completion of an operation.

We shall describe a reduction algorithm. (Clearly, in specific cases, taking the special features of a process into account, one can construct more economical methods of reduction.)

The variable $\xi_j(t)$ will correspond to an operation that takes place at the jth object. Objects with numbers $j > |v(t)|$ are in passive states. Under such reduction in the general case at any given transition time all the components $\xi_j(t)$ may change simultaneously. However, in real-world examples the majority of the components do not undergo jumps. This fact may be reflected in the algorithm retaining the old numeration for the variables that are not subject to jumps.

3.3.5. *Embedded Markov Chain*

For definiteness consider PLMP $\zeta(t)$ without spontaneous transitions. Denote by t_n, $n \geqslant 1$, the time of the nth jump of the process and introduce two sequences of random variables $\zeta_n^- = \zeta(t_n - 0) = (v_n^-, \bar{\xi}_n^-)$ and $\zeta_n^+ = \zeta(t_n + 0) = (v_n^+, \xi_n^+)$. Each one of these is a homogeneous Markov chain—the embedded Markov chain of the process $\zeta(t)$. The sequences (ζ_n^-, t_n) and (ζ_n^+, t) possess the same property.

Denote

$$F_n^{\pm}(j, \bar{x} | i, \bar{y}) = \mathsf{P}\{v_n^{\pm} = j, \bar{\xi}^{\pm} < \bar{x} | \zeta_0^+ = (i, \bar{y})\}.$$

The functions introduced, directly by the definition, satisfy the recurrence relations

$$F_n^+(j, \bar{x} | i, \bar{y}) = \sum_k \int_{\Gamma_k} H(j, \bar{x} | k, \bar{z})\, dF_n^-(k, \bar{z} | i, \bar{y}), \tag{2}$$

$$F_n^-(j, \bar{x} | i, \bar{y}) = \int_{\substack{\bar{z} = \bar{u} + \bar{\alpha}_j t,\, t > 0, \\ \bar{u} < \bar{x},\, \bar{u} \in \Gamma_j}} dF_{n-1}^+(j, \bar{z} | i, \bar{y}), \quad \bar{x} \in \Gamma_j, \tag{3}$$

where Γ_j is the set of vectors of dimension $|j|$ all of whose components are nonnegative and at least one of them equals zero.

In the same manner one can also obtain relations for transition probabilities of Markov chains $(\zeta_n^{\pm}, t_n)$. These relations are written in the simplest manner for densities; however, in the formulas a delta-type component appears. Therefore, it is better to introduce the density $\psi_i(\bar{y})$ under the condition $v_0^+ = i$ in place of the condition $\bar{\xi}_0^+ = \bar{y}$.

Let

$$B_{ij}(\bar{x}|\bar{y}) = \int_0^{\bar{x}} b_{ij}(\bar{u}|\bar{y})\, du_1 \cdots du_{|j|}, \quad \text{where } \int_0^{\bar{x}} = \int_0^{x_1} \cdots \int_0^{x_{|j|}};$$

$$f_{ij}^{(n)\pm}(t, \bar{x}|\psi)\, dt\, d\bar{x} = \mathsf{P}\{t < t_n < t + dt, v_n^{\pm} = j, \bar{x} < \xi_n^{\pm} < \bar{x} + d\bar{x}|i, \psi\},$$

where i, ψ are abbreviated expressions of the condition $\{v_0^+ = i; \bar{\xi}_0^+$ possesses the densty $\psi(\bar{y})\}$. Then

$$f_{ij}^{(n)+}(t, \bar{x}|\psi) = \sum_{k,l} p_{kj}^{(l)} \int_{\{\bar{z}: z_l = 0\}} b_{kj}(\bar{x}|\bar{z}) f_{ik}^{(n)-}(t, \bar{z}|\psi)\, dz_1 \cdots dz_{|k|}; \tag{4}$$

$$f_{ij}^{(n)-}(t, \bar{x}|\psi) = \alpha_{jl} \int_0^t f_{ij}^{(n-1)+}(t - \tau, \bar{x} + \bar{\alpha}_j \tau|\psi)\, d\tau, \quad \min_k x_k = x_l; \tag{5}$$

$$f_{ij}^{(0)+}(t, \bar{x}|\psi) = \delta(t)\delta_{ij}\psi(\bar{x}). \tag{6}$$

In terms of characteristics of an embedded Markov chain the distribution of the process $\zeta(t)$ is expressed as follows (provided that the process is regular in a given interval):

$$\mathsf{P}\{v(t) = j, \bar{\xi}(t) < \bar{x}|i, \psi\} = \int_0^{\bar{x}} du_1 \cdots du_{|j|} \int_0^t \left(\sum_{n=0}^{\infty} f_{ij}^{(n)+}(t - \tau; \bar{u} + \bar{\alpha}_j \tau|\psi) \right) d\tau. \tag{7}$$

In this expression the role of $dH(t - \tau)$ is played by $\sum_{n=0}^{\infty} f_{ij}^{(n)+}(\cdots)\, d\tau$, where $H(t)$ is a renewal function (cf. Chapter 2).

3.4. Other Important Classes of Random Processes

An interesting generalization of homogeneous walks, which serves as a model for a single-server queueing system, is the Markov processes that are homogeneous in the second component; these processes were introduced and studied by Ezhov and Skorokhod (1969). Such a process is of the form $(x(t), y(t))$, where $x(t)$ is a Markov process in an arbitrary measurable space of states; $y(t)$ takes on values in a multidimensional Euclidean space. Let $x(t) = x$, $y(t) = y$. Then for $s > t$ the probability that $x(s)$ falls into the set A and $y(s)$ falls into the set B remains unchanged when this set, simultaneously with vector y, is shifted by an arbitrary vector Δy. In the paper cited various functionals of the process that generalize characteristics of known single-server systems are studied.

Borozdin and Ezhov (1976) introduced strongly regenerative random processes. A process from this class is a regenerative process whose behavior in a regeneration cycle consists of an exponentially distributed sojourn phase in a marked state and an arbitrary distributed sojourn phase in other states. For such processes it turned out to be possible to study analytically many interesting characteristics and in particular distributions of the times when the additive functional of the process attains a given level.

Nonhomogeneous-in-space random walks are also widely used in queueing theory. In particular, they serve as models of dam theory, inventory theory, and other theories. We introduce a model of a dam due to Moran (1969). Let $z(t)$ be the level of a dam at time t. Owing to a continuous discharge, this level decreases with the rate $R(z(t))$. At the same time replenishment of the dam takes place governed by a random process $x(t)$. This results in a stochastic differential equation

$$dz(t) = -R(z(t))\,dt + dx(t), \tag{1}$$

where in Moran's model $x(t)$ is a homogeneous process with independent increments.

Moran constructed a solution to Eq. (1) when $x(t)$ is a generalized Poisson process, $R(u)$ is a continuous nondecreasing locally Lipschitz function such that $R(0) = 0$, $R(z) > 0$ for $z > 0$. Moran investigated local properties of a solution of (1) and developed a numerical method. In Çinlar and Pinsky's papers (1971, 1972), the existence and uniqueness of the solution of Eq. (1) was studied under more general conditions, and conditions for ergodicity of the Markov process $z(t)$ were obtained.

Harrison and Resnick (1976) obtained the generating operator of this process (under certain restrictions of the type $\int_0^x [1/R(y)]\,dy < \infty$) and expressed the stationary distribution in the form of a series.

Brockwell (1977) obtained certain conditions for the existence of a stationary distribution. Thus, if $R(z)$ is a function that satisfies only conditions of monotonic decrease and $R(+0) > 0$, then it is necessary and sufficient for the existence of a stationary distribution that the inequality $R(\infty) > \mathsf{M}x(1)$ be fulfilled.

Smith and Yeo (1981) studied the case when $x(t)$ is a step process such that the times of its jump form a renewal process. Brockwell and Resnick (1982) and Tweedie (1975) considered a model in which $x(t)$ is a general process with independent increments, and $R(u)$ is positive for $u > 0$, left-continuous, and posseses a limit from the right; these authors obtained necessary and sufficient conditions in terms of $R(z)$ and the spectral measure of the process $x(t)$ for the existence of a stationary distribution for the solution of (1). These are given under the condition of existence of a nonnegative integrable solution of an integral equation. Simpler sufficient conditions are also available.

Notes

We cite the following important papers dealing with ergodicity and continuity of countable Markov chains: Nagayev (1965), Popov (1977), and Malyshev and Men'shikov (1979).

Limit theorems for semi-Markov processes and renewal theory for Markov chains are discussed in the paper by Athreya et al. (1979). Books by Cohen (1976) and Krein and Lemuan (1982) are devoted to the method of regenerative processes in queueing theory.

We mention the books by Çinlar (1975), Brodi and Pogosyan (1973), and Nollau (1980) among those dealing with various generalizations of semi-Markov processes. Piecewise-linear Markov processes are generalized in Tien's paper (1977). Various classes of constructively defined processes, which are of interest in particular in queueing theory, are collected in the handbook by Kovalenko et al. (1960).

4
Semi-Markov Models of Queueing Systems

4.1. Classification of Queueing Systems

Queueing systems can be classified according to the logical structure of the service process (number of servers, order priorities, possibility of waiting, etc.) as well as the analytical assumptions concerning the incoming stream of customers and the distribution of service time. The classification proposed by Kendall is the one commonly used. Many authors extended this classification [see, e.g., Borovkov's book (1972)]. For our purposes Kendall's classification is sufficient.

A queueing system is expressed by a set of symbols of the form $A|B|m$ or of the form $A|B|m|r$. The interpretation is as follows: A is the symbol of the incoming stream, B is the symbol of the distribution of the service time, m is the symbol of the number of servers, r denotes the number of locations available for waiting. The variables A, B can take the following values:

$$A = G, GI, E_k, D, M; \qquad B = G, E_k, D, M.$$

If $A = G$ (general), then the incoming stream is more general than a renewal stream; if $A = GI$ (general independent), the stream is a renewal one. If $A = E_k$, we have the Erlang stream of the kth order, i.e., a stream formed by every kth customer of a simple stream; D is a regular (deterministic) stream, M denotes the simple (Markovian) stream. As to service time, it is assumed that in all cases the durations of servicing different customers are jointly independent and identically distributed, and do not depend on the incoming stream.

The equation $B = G$ signifies that the distribution of service time is general; the meaning of E_k, D and M is the same as in the case of the incoming stream. For example M means that the service is exponential.

The symbol m denotes the number of servers. If a Latin letter such as m, c etc. is used it means that the number of servers is arbitrary; sometimes a specific number of servers is indicated (1, 2 and so on). The same applies to the symbol r. A system with waiting $A|B|m|\infty$ is coded in a simpler manner as: $A|B|m$. Systems with losses (holding time) can now be coded as $A|B|m|0$.

It is usually assumed that the system $A|B|m|r$ is governed by a general queueing discipline; the customers are served in order of their arrivals. If a system has special features these are added verbally to the symbols introduced above. We say for example: "a system $M|G|2$ with an inverse order of service", "a system $GI|G|1$ with an unreliable server". Verbal description abolishes the corresponding usual assumptions implicit in the preceding symbols.

4.2. $M|G|1$ System

4.2.1. *Statement of the Problem, Notation*

The theory discussed in Section 1.5 is based on very narrow conditions: the customer stream is simple and the service time is exponentially distributed. Erlang himself attempted to dispense with the second condition and considered the case of service time governed by a distribution that is uniform in a certain interval (a, b) or it is the Erlang distribution for some integer k. It has long been evident that the problem of service with waiting should be treated under more general assumptions on the customer's stream and the distribution of service time. This is important for theoretical development as well as for practical applications. For a long time the opinion prevailed that any stream may be considered a close approximation to a simple stream, at least in problems of telephone traffic and that it is of substantial importance to study problems of formation and servicing of queues with an arbitrarily distributed service time. As we have already pointed out, the study of the distribution of waiting time for the beginning of service is especially important for serving systems with waiting (namely, queues). Considerable progress has been made, and there is substantial literature on the subject, especially in the case of a single server. We shall restrict ourselves to this case and study the problem of determining the distribution of the waiting time for the beginning of service under the following conditions:

1. the customers are served on a "first-come, first-served" basis;
2. the customer stream is simple, of intensity λ;
3. the distribution of the service time, denoted by $B(x)$ is arbitrary:

$$B(x) = \mathsf{P}\{\eta < x\}, \qquad \bar{B}(x) = 1 - B(x),$$

   where η is the service time; $\mathsf{M}\eta = \tau < \infty$;
4. there is one server who does not fail and is able, on finishing his or her service of one customer, to proceed immediately to the next one.

The first solution of the problem under these conditions is due to Khinchin (1932). Later he published a somewhat modified version in his monograph (1955) to which we have referred frequently.

The problem was studied later under more general conditions, using other methods, by Lindley (1952), Smith (1958), Pollaczek (1961), Takács (1967) and others.

4.2.2. *Embedded Markov Chain*

Denote by t_n the time at which the service of the nth customer is completed ($n = 1, 2, 3, \ldots$). Let $v_n = v(t_n + 0)$, i.e., v_n is the number of customers remaining in the system after the nth customer has left the system. It is easy to see that the sequence $\{v_n, n \geqslant 1\}$ is a Markov chain.

This Markov chain is irreducible. Indeed, from any state t one can with positive probability reach the state 0: it is sufficient that during t consecutive durations of servicing no new customers will arrive. This chain is also aperiodic since one can return from state 0 to the same state in one step (the condition is that no customer arrives during the time of servicing of one customer).

We verify the ergodicity conditions stipulated in Section 3.1.2.

If $v_{n-1} \geqslant 1$, then $v_n = v_{n-1} + \Delta_n - 1$, where Δ_n is the number of customers arriving during the nth customer's service time η_n. If $v_{n-1} = 0$, then $v_n = \Delta_n$.

We shall find the distribution of the random variable Δ_n. If $\eta_n = x$, the probability that k customers arrive is

$$e^{-\lambda x}(\lambda x)^k/k!;$$

averaging with respect to the distribution of the random variable η_n we obtain

$$\mathsf{P}\{\Delta_n = k\} = \frac{1}{k!}\int^{\infty} e^{-\lambda x}(\lambda x)^k \, dB(x) = f_k, \quad k \geqslant 0. \tag{1}$$

Hence $\mathsf{M}\Delta_n = \sum_{k=0}^{\infty} k f_k = \lambda\tau$. The transition matrix $\{v_n\}$ is of the form

$$P = \begin{pmatrix} f_0 & f_1 & f_2 & f_3 & \cdots \\ f_0 & f_1 & f_2 & f_3 & \cdots \\ & f_0 & f_1 & f_2 & \cdots \\ & & f_0 & f_1 & \cdots \\ & & & f_0 & \cdots \\ & & & & \cdots \end{pmatrix}. \tag{2}$$

For $i \geqslant 1$

$$\mathsf{M}\{v_{n+1} | v_n = i\} = \mathsf{M}\Delta_n - 1 = \lambda\tau - 1;$$

$$\mathsf{M}\{v_{n+1} | v_n = 0\} = \mathsf{M}\Delta_n = \lambda\tau.$$

We introduce the parameter $\rho = \lambda\tau$, which is called the *load* of a single-server queueing system and plays a most important role in the study of behavior of queues (cf. Section 1.2).

Let $\rho < 1$. Then denoting $\varepsilon = 1 - \rho > 0$, we obtain

$$\mathsf{M}\{v_{n+1} | v_n = i\} = i - \varepsilon, \quad i \geqslant 1;$$

$$\mathsf{M}\{v_{n+1} | v_n = 0\} < \infty.$$

Thus the conditions of the theorem in Section 3.1.2 are fulfilled for $f(v) = v$.

Hence, the Markov chain $\{v_n\}$ possesses the ergodic distribution

$$\pi_k = \lim_{n\to\infty} \mathsf{P}\{v_n = k\} \quad (k = 0, 1, 2, \ldots). \tag{3}$$

As is well known from the theory of Markov chains, $\{\pi_k\}$ is determined by the system of equations

$$\pi_k = \sum_{j=0}^{\infty} \pi_j p_{jk} \quad (k = 0, 1, 2, \ldots), \tag{4}$$

where p_{jk} are the elements of the matrix P, with the additional condition

$$\sum_{k=0}^{\infty} \pi_k = 1. \tag{5}$$

In view of (2), Eq. (4) have the following form:

$$\pi_k = \sum_{i=0}^{k} \pi_{k-i+1} f_i + \pi_0 f_k, \quad k \geqslant 0. \tag{6}$$

The system (5)–(6) is conveniently studied by the method of generating functions. Denote

$$f(z) = \sum_{k=0}^{\infty} f_k z^k, \qquad \pi(z) = \sum_{k=0}^{\infty} \pi_k z^k,$$

assuming that $|z| \leqslant 1$ and

$$\psi(s) = \int_0^{\infty} e^{-sx}\, dB(x), \quad \operatorname{Re} s \geqslant 0.$$

Multiply both sides of (6) by z^k and sum over k:

$$\begin{aligned}\pi(z) &= \sum_{k=0}^{\infty} z^k \sum_{i=0}^{k} \pi_{k-i+1} f_i + \pi_0 f(z) = \sum_{i=0}^{\infty} f_i z^i \sum_{k=i}^{\infty} z^{k-i} \pi_{k-i+1} + \pi_0 f(z) \\ &= z^{-1} f(z)(\pi(z) - (1-z)\pi_0);\end{aligned}$$

hence,

$$\pi(z) = \pi_0 (1-z) f(z)/[f(z) - z]. \tag{7}$$

When $z \to 1$, $f(z) - z \sim (1-z)[1 - f'(1)]$, so that

$$1 = \pi(1) = \pi_0/[1 - f'(1)],$$

whence $\pi_0 = 1 - f'(1)$ and

$$\pi(z) = [1 - f'(1)](1-z) f(z) \Big/ [f(z) - z].$$

Since $f(z) = \sum_{k=0}^{\infty} f_k z^k$ we have

$$\begin{aligned} f(z) &= \sum_{k=0}^{\infty} \frac{1}{k!} \int_0^{\infty} e^{-\lambda x} (\lambda x z)^k\, dB(x) = \int_0^{\infty} e^{-\lambda x} \sum_{k=0}^{\infty} \frac{1}{k!} (\lambda x z)^k\, dB(x) \\ &= \int_0^{\infty} e^{-\lambda x(1-z)}\, dB(x) = \psi(\lambda(1-z)). \end{aligned} \tag{8}$$

The interchange of summation and integration signs is justified since

$$\int_A^\infty e^{-\lambda x} \sum_{k=0}^\infty \frac{1}{k!}(\lambda xz)^k \, dB(x) \leqslant \int_A^\infty dB(x) \xrightarrow[A\to\infty]{} 0$$

and for sufficiently large n

$$\sum_{k=n+1}^\infty \int_0^A \frac{1}{k!}(\lambda xz)^k e^{-\lambda x} \, dB(x) < \frac{1}{n!}(\lambda Az)^n e^{-\lambda A}.$$

Differentiating (8) we obtain

$$f'(z) = -\lambda\psi'(\lambda(1-z)),$$

hence,

$$1 - f'(1) = 1 - \lambda\tau = 1 - \rho,$$

and finally

$$\pi(z) = \frac{(1-\rho)(1-z)\psi(\lambda(1-z))}{\psi(\lambda(1-z)) - z}. \tag{9}$$

The variable v_n studied previously is the number of customers who remain in the system after the service of the nth customer has been completed. Now let v'_n be the number of customers in the system at the time of arrival of the nth customer. Thus, if t'_n is the time of arrival of the nth customer, then $v'_n = v(t'_n - 0)$.

We shall study the limiting distribution of the random variables v'_n. These variables do not form a Markov chain. To construct a Markov chain, we shall extend the phase space.

Let $l = l(n)$ be the largest value $i \leqslant n$ such that $v'_i = 0$, $l' = l'(n) = n - l(n)$. Then consider the random sequence

$$\bar{v}_n = (l'; v'_{l'+1}, v'_{l'+2}, \ldots, v'_n).$$

This sequence is a homogeneous Markov chain.

If $v'_n = k$, then, if no incoming customers arrive during the k intervals of service, we have $\bar{v}_{n+k} = (0; 0)$. Thus, $\{\bar{v}_n\}$ is an irreducible Markov chain. The chain is also nonperiodic, since it may return to the null state in one step. For $n \geqslant 2$ the equality $\bar{v}_n(0; 0)$ is fulfilled each time that $v_{n-1} = 0$. Whence

$$\mathsf{P}\{\bar{v}_n = (0; 0)\} \xrightarrow[n\to\infty]{} \lim_{n\to\infty} \mathsf{P}\{v_{n-1} = 0\} = \pi_0 = 1 - \rho > 0.$$

Hence the Markov chain $\{\bar{v}_n\}$ possesses an ergodic distribution.

Denote by $N_n(k \to k+1)$ the number of i's $(1 \leqslant i \leqslant n)$ such that $v'_i = k$ and by $N_n(k+1 \to k)$ the number of i's $(1 \leqslant i \leqslant n)$ such that $v_i = k$. The notation is justified because for $v'_i = k$ at time t'_i the function $v(t)$ passes from state k into the state $k+1$ and for $v_i = k$ at time t_i the transition is reversed. Until time t_n inclusive, n customers left the system and $n + v_n$ arrived. At the same time in any interval the number of transitions $k \to k+1$ may differ from the number of reverse transitions by at most 1. Whence,

$$|N_{n+v_n}(k \to k+1) - N_n(k+1 \to k)| \leqslant 1.$$

However,

$$|N_{n+v_n}(k \to k+1) - N_n(k \to k+1)| \leqslant v_n.$$

Hence,

$$|N_n(k \to k+1) - N_n(k+1 \to k)| \leqslant v_n + 1.$$

Since $(1/n)N_n(k+1 \to k) \to \pi_k$ and $v_n/n \to 0$ in probability, it follows that $(1/n)N_n(k \to k+1) \to \pi_k$ in probability. The latter means that

$$\lim_{n\to\infty} \mathsf{P}\{v_n' = k\} = \pi_k. \tag{10}$$

4.2.3. *Pollaczek-Khinchin Formula*

Introduce a random process $v^*(t)$ equal to v_n when n is the number of customers served until time t; $v^*(t) = 0$ for $n = 0$. This process is a semi-Markov one; $\{v_n\}$ is its embedded Markov chain. From Theorem 3 of Section 3.1.4. we conclude that

$$\pi_k^* = \lim_{t\to\infty} \mathsf{P}\{v(t) = k\} = c \sum_j \pi_j \int_0^\infty P_j(t) f_{jk}(t)\,dt, \tag{11}$$

where c is a constant,

$$P_j(t) f_{jk}(t) = \mathsf{P}\{v(t_n + t) = k,\ t_{n+1} - t_n > t \mid v_n = j\}.$$

Denoting by $\Delta(t)$ the Poisson random variable with parameter λt we obtain

$$P_j(t) f_{jk}(t) = \mathsf{P}\{\eta > t,\ \Delta(t) = k - j\} = e^{-\lambda t}(\lambda t)^{k-j}\bar{B}_j(t)/(k-j)!, \quad j \geqslant 1,$$

whence

$$\int_0^\infty P_j(t) f_{jk}(t)\,dt = \frac{1}{(k-j)!}\int_0^\infty (\lambda t)^{k-j} e^{-\lambda t}\bar{B}(t)\,dt.$$

In the same manner

$$P_0(t) f_{00}(t) = e^{-\lambda t}$$

$$\begin{aligned} P_0(t) f_{0k}(t) &= \lambda \int_0^t e^{-\lambda u}\bar{B}(t-u) e^{-\lambda(t-u)} \frac{\lambda^{k-1}(t-u)^{k-1}}{(k-1)!}\,du \\ &= e^{-\lambda t}\frac{\lambda^k}{(k-1)!}\int_0^t u^{k-1}\bar{B}(u)\,du, \quad k \geqslant 1, \end{aligned}$$

whence

$$\int_0^\infty P_0(t) f_{00}(t)\,dt = 1/\lambda,$$

$$\int_0^\infty P_0(t) f_{0k}(t)\,dt = \frac{\lambda^{k-1}}{(k-1)!}\int_0^\infty u^{k-1} e^{-\lambda u}\bar{B}(u)\,du, \quad k \geqslant 1.$$

Introduce the generating function

$$a(z) = \sum_{n=0}^{\infty} z^n \frac{1}{n!} \int_0^{\infty} (\lambda t)^n e^{-\lambda t} \bar{B}(t)\, dt$$
$$= \int_0^{\infty} e^{-\lambda t(1-z)} \bar{B}(t)\, dt = \frac{1}{\lambda(1-z)}[1 - \psi(\lambda(1-z))].$$

It follows from (11) and subsequent expressions for the integrals appearing on the right-hand side of this equality that the generating function of the sequence $\{\pi_k^*/c\}$ is of the form

$$[\pi(z) - \pi_0]a(z) + \pi_0(1/\lambda + za(z))$$
$$= \pi_0/\lambda + [\pi(z) - \pi_0(1-z)]a(z) = (1-\rho)/\lambda$$
$$+ [\pi(z) - (1-\rho)(1-z)]a(z) = \frac{(1-\rho)(1-z)}{\lambda} \frac{\psi(\lambda(1-z))}{\psi(\lambda(1-z)) - z}.$$

This expression differs only by a constant factor from $\pi(z)$. Using the normalization condition we obtain

$$\sum_{k=0}^{\infty} \pi_k^* z^k = \pi(z) = \frac{(1-\rho)(1-z)\psi(\lambda(1-z))}{\psi(\lambda(1-z)) - z}. \tag{12}$$

Formula (12) is called the *Pollaczek-Khinchin formula.*

4.2.4. *Mathematical Law of a Stationary Queue*

We have just established a remarkable fact: under stationary conditions the distribution of the process $\nu(t)$ at an arbitrary time coincides with the distribution at the time of a customer's arrival. This phenomenon was called by Khinchin the *mathematical law of a stationary queue.* It was discovered later that this coincidence is valid for a great variety of systems with the simple incoming stream. We shall refer the reader interested in the current state of the problem to books by Franken et al. (1984) and Shurenkov (1981) and present a theorem that is easily interpreted in various applications.

Theorem. *Let the behavior of a queue be described by a random process* $\xi(t)$, f *be a bounded function, the incoming stream be simple with parameter* λ, t_n *be the arrival time of the* nth *customer. Assume that* $\{\xi(t), t \leqslant s\}$ *does not depend on the times of arrivals of customers for* $t > s$. *Then if*

$$\mathsf{M}f(\xi(t_n - 0)) \xrightarrow[n\to\infty]{} a, \tag{13}$$

then

$$\frac{1}{T}\int_0^T Mf(\xi(t))\, dt \xrightarrow[T\to\infty]{} a. \tag{14}$$

PROOF. It is easy to see that only the case $\lambda = 1$ should be considered. Set $n = [T]$ and consider the expression

$$b_n = \mathsf{M} \int_0^{t_n} f(\xi(t))\,dt = \sum_{k=1}^{n} \mathsf{M} \int_{t_{k-1}}^{t_k} f(\xi(t))\,dt,$$

where we set $t_0 = 0$. Since the distribution of $t_k - t_{k-1}$ is exponential with parameter 1, we have

$$\begin{aligned}\mathsf{M} \int_{t_{k-1}}^{t_k} f(\xi(t))\,dt &= \mathsf{M} \int_0^{\infty} e^{-x}\,dx \int_0^x f(\xi(t_{k-1} + t))\,dt \\ &= \mathsf{M} \int_0^{\infty} f(\xi(t_{k-1} + t))\,dt \int_t^{\infty} e^{-x}\,dx \\ &= \mathsf{M} \int_0^{\infty} e^{-t} \mathsf{M} f(\xi(t_{k-1} + t))\,dt.\end{aligned}$$

However

$$\mathsf{M} f(\xi(t_k - 0)) = \mathsf{M} f(\xi(t_{k-1} + (t_k - t_{k-1}) - 0)) = \mathsf{M} \int_0^{\infty} e^{-t} f(\xi(t_{k-1} + t))\,dt,$$

which coincides with the preceding expression.

Thus,

$$b_n = \sum_{k=1}^{n} \mathsf{M} f(\xi(t_k - 0)) \sim na, \quad n \to \infty. \tag{15}$$

At the same time

$$b_n = \mathsf{M} \int_0^T f(\xi(t))\,dt + \mathsf{M} \int_T^{t_n} f(\xi(t))\,dt,$$

whence,

$$\left| b_n - \mathsf{M} \int_0^T f(\xi(t))\,dt \right| \leqslant \bar{f} \mathsf{M} |t_n - T|, \tag{16}$$

where $\bar{f} = \sup_x |f(x)|$.

We have the bound

$$\mathsf{M}|t_n - t| \leqslant T - n + \mathsf{M}|t_n - n| < 1 + \sigma\{t_n\} = 1 + \sqrt{n}$$

(σ is the mean square deviation). The required assertion thus follows from this bound in conjuction with (15) and (16). □

4.2.5. *Virtual Waiting Time*

Denote by $\gamma(t)$ a stochastic process defined at each instant t as the time interval elapsing from t until the server completes serving customers entering the queue before t. If at the instant t the server is free, $\gamma(t) = 0$.

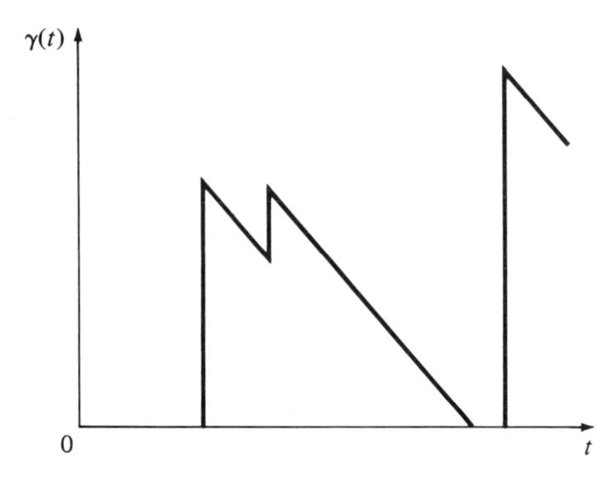

FIGURE 4

Denote the arrival times of the customers by t_1, t_2, t_3, ... Then for $t_n < t < t_{n-1}$ the process $\gamma(t)$ is defined by

$$\gamma(t) = \begin{cases} 0 & \text{if } \gamma(t_n) \leqslant t - t_n, \\ \gamma(t_n) - (t - t_n) & \text{if } \gamma(t_n) \geqslant t - t_n. \end{cases}$$

For $t = t_n$ we have the equality $\gamma(t_n + 0) = \gamma(t_n - 0) + \eta_n$, where η_n denotes the service time for customers arriving at time t_n.

The general form of the process $\gamma(t)$ is presented in Fig. 4. The process jumps at the customers' arrival times; it remains unchanged until the arrival of the next customer when it reaches zero value and decreases with the unit rate at those intervals where it is positive.

Under our assumptions $\gamma(t)$ is a Markov process. Indeed, the number of customers arriving between t and $t + s$ does not depend on the number of customers arriving prior to t. The value of $\gamma(t + s)$ is determined, first by $\gamma(t)$, second by the number of customers arriving in the time interval $(t, t + s)$, and third by the service time for these newly arrived customers. These three quantities are not affected by the values of $\gamma(t)$ at times τ preceding t.

We assume that $\gamma(0) = 0$ in order that the process be completely defined. This condition can be treated as a convention that the servicing process commences at the time $t = 0$.

The process $\gamma(t)$ was introduced by Takács (1967) and is called the *virtual* (*possible*) *waiting time*: if at time t a customer arrives, it will have to wait until the beginning of the service the amount of time equal to $\gamma(t)$.

We note an interesting analogy when the same equation serves as a suitable description of two completely different physical processes.

At times $t_1, t_2, \ldots, t_n, \ldots$, let amounts of water of volumes $\eta_1, \eta_2, \ldots, \eta_n, \ldots$, respectively, flow into a water supply tank. The water discharge is uniform; a unit volume of water flows out. Then $\gamma(t)$ is equal to the volume per unit time of water in the tank at time t.

4.2.6. *The Limiting Distribution of the Waiting Time*

Denote by w_n the waiting time of the nth customer. We shall assume that $\rho < 1$. From the formula of the total probability

$$\mathsf{P}\{w_n \geqslant x\} = \sum_{i=1}^{n} \mathsf{P}\{v_i' = 0\}\,\mathsf{P}\{v_{i+1}' > 0, \ldots, v_n' > 0, \quad w_n \geqslant x \,|\, v_i' = 0\}. \quad (17)$$

Since the second factor of the summand depends only on x and $n - i$, and $\mathsf{P}\{v_n' = 0\} \to 1 - \rho$, there exists the limit

$$\lim_{n \to \infty} \mathsf{P}\{w_n < x\} = F(x).$$

It is directly evident that $F(x)$ is a nondecreasing left-continuous function. We shall prove that $F(\infty) = 1$.

A Markov chain $\{v_n'\}$ is ergodic; therefore, the average return time into the zero state is finite, and can be represented as

$$\sum_{m=1}^{\infty} \mathsf{P}\{v_{i+1}' > 0, \ldots, v_{i+m}' > 0\}.$$

Consequently, on the right-hand side of (17) the sum with respect to i from 1 to $n - N$ for an appropriately fixed N is less than ε, and the sum from $n - N + 1$ up to n can be made arbitrarily small by an appropriate choice of x. Whence $\lim_{n \to \infty} F_n(x) \geqslant 1 - 2\varepsilon$, $x > x_0$, i.e., $F(\infty) = 1$.

Based on the Theorem in Section 4.2.4 we conclude that

$$F(x) = \lim_{T \to \infty} \frac{1}{T} \int_0^T \mathsf{P}\{\gamma(t) < x\}\, dt \quad (18)$$

and the limit on the right-hand side exists.

We set $x = +0$ in (18) [this is justified since in formula (17) in place of "$\geqslant x$" one can write "$\in A$"]. We then obtain (18)—after taking into account relation

$$\mathsf{P}\{w_n = 0\} = P\{v_n' = 0\} \xrightarrow[n \to \infty]{} 1 - \rho$$

and that

$$\lim_{T \to \infty} \frac{1}{T} \int_0^T \mathsf{P}\{\gamma(t) = 0\}\, dt = 1 - \rho, \quad (19)$$

i.e., the average fraction of time for which $\gamma(t) = 0$ is equal to $1 - \rho$ the "underload" of the system.

Denote $F(t, x) = \mathsf{P}\{\gamma(t) < x\}$. Choose two instants of time t and $t + h$ $(h > 0)$. If in the interval $(t, t + h)$ no customer arrived, then the events $\{\gamma(t + h) < x\}$ and $\{\gamma(t) < x + h\}$ are equivalent. If one customer arrived with service time η, then

$$\{\gamma(t) + \eta < x\} \subset \{\gamma(t + h) < x\} \subset \{\gamma(t) + \eta < x + h\}.$$

Noting that two or more customers can arrive during time h only with probability $o(h)$, we obtain

$$(1 - \lambda h)F(t, x + h) + \lambda h \int_0^x B(x - y)\,dF(t, y) \leqslant F(t + h, x)$$

$$\leqslant (1 - \lambda h)F(t, x + h) + \lambda h \int_0^{x+h} B(x + h - y)\,dF(t, y) + o(h).$$

Averaging with respect to t in the limits from 0 to T and letting T approach ∞ we obtain, in view of (18),

$$(1 - \lambda h)F(x + h) + \lambda h \int_0^x B(x - y)\,dF(y) \leqslant F(x)$$

$$\leqslant (1 - \lambda h)F(x + h) + \lambda h \int_0^{x+h} B(x + h - y)\,dF(y) + o(h).$$

Whence

$$\lambda F(x + h) - \lambda \int_0^{x+h} B(x + h - y)\,dF(y) + o(1)$$

$$\leqslant \frac{1}{h}[F(x + h) - F(x)] \leqslant \lambda F(x + h) - \lambda \int_0^x B(x - y)\,dF(y). \quad (20)$$

We conclude from (20) that the difference ratio on the right-hand side is bounded and hence $F(x)$ is absolutely continuous for $x > 0$. Since the integral in (20) is a distribution function, it follows that for almost all $x > 0$ the lower bound as $h \to 0$ converges to the upper one. Thus, for almost all $x > 0$

$$F'(x) = \lambda F(x) - \lambda \int_0^x B(x - y)\,dF(y),$$

or, equivalently,

$$F'(x) = \lambda \int_0^x \bar{B}(x - y)\,dF(y). \quad (21)$$

It will be established subsequently that $F(+0) = 1 - \rho$. Therefore, (21) can be represented as

$$F'(x) = \lambda \int_{+0}^x \bar{B}(x - y)F'(y)\,dy + \lambda(1 - \rho)\bar{B}(x). \quad (22)$$

We shall solve Eq. (22) using the Laplace transform denoting

$$\Phi(s) = \int_0^\infty e^{-sx}\,dF(x) = 1 - \rho + \int_0^\infty e^{-sx}F'(x)\,dx.$$

We have

$$\Phi(s) = 1 - \rho + \frac{\lambda}{s}[1 - \psi(s)]\Phi(s),$$

whence we obtain

$$\Phi(s) = \frac{1-\rho}{1-\dfrac{\lambda}{s}[1-\psi(s)]}. \tag{23}$$

Formula (23) is called *Khinchin's formula.* Using the method of renewal processes one easily proves that $\gamma(t)$ possesses for $t \to \infty$ a limiting distribution; this implies that the Laplace-Stieltjes transform of this distribution is of the form (23).

Khinchin's formula allows us to calculate the moments of the distribution of the random variable w, which is the stochastic limit of w_n as $n \to \infty$. To evaluate these moments we write (23) in the form

$$[s + \lambda\psi(s) - \lambda]\Phi(s) = s(1-\rho).$$

Successively differentiating this relation and setting $s = 0$ we obtain the moments $\mathsf{M}w^r = (-1)^r\Phi^{(r)}(0)$ in terms of the moment $\mathsf{M}\eta^j = (-1)^j\psi^{(j)}(0)$ of the service time. Thus, differentiating twice we obtain

$$\lambda\psi''(s)\Phi(s) + 2[1 + \lambda\psi(s)]\Phi'(s) + [s + \lambda\psi(s) - \lambda]\Phi''(s) = 0,$$

whence

$$\mathsf{M}w = \frac{\lambda\mathsf{M}\eta^2}{2(1-\rho)} = \frac{\rho\tau}{2(1-\rho)}\left(1 + \left(\frac{\sigma}{\tau}\right)^2\right), \tag{24}$$

where σ^2 is the variance of the random variable η.

Formula (24) shows that 1) as $\rho \to 1$ the mean waiting time tends to infinity and 2) the mean waiting time increases as the coefficient of variation of the service time increases; the mean waiting time is minimal for $\sigma = 0$, i.e., in the case of the system $M|D|1$. Formula (24)—which is the most important corollary of Khinchin's formula—has been used for solving a variety of problems of efficient organization of service. Formulas for the moments of w of arbitrary order are given, for example, in Riordan's book (1962).

4.2.7. *The Case $\rho \geqslant 1$*

We shall prove that the relation

$$\lim_{t\to\infty} F(x,t) = 0$$

is valid when $\rho \geqslant 1$.

For this purpose we shall establish the following two lemmas.

Lemma 1. *If $\rho < 1$, then the inequality*

$$F(x) \leqslant (1-\rho)e^{\lambda x}$$

is valid.

Indeed in view of (21) $F'(x) \leqslant \lambda F(x)$, whence

$$F(x) \leqslant F(+0)e^{\lambda x};$$

this proves the required inequality since $F(+0) = 1 - \rho$.

Lemma 2. *Let two processes $\gamma(t)$ and $\gamma'(t)$ be given corresponding to the distribution functions of service times $B(x)$ and $B'(x)$ and the same initial conditions. Then if $B(x) \leqslant B'(x)$ for all x, it follows that for all $t > 0$ and for all x*

$$\mathsf{P}\{\gamma(t) < x\} = F(x,t) \leqslant F'(x,t) = \mathsf{P}\{\gamma'(t) < x\}.$$

Proof. Each one of the processes $\gamma(t)$ and $\gamma'(t)$ is determined by defining sequences of arrival times of customers and sequences of their service times. We shall establish a one-to-one correspondence between realizations of random processes $\gamma(t)$ and $\gamma'(t)$.

If realization of the process $\gamma(t)$ is characterized by the sequence $\{t_n\}$ of the times of customer's arrivals and a sequence $\{\gamma_n\}$ of the service times, we shall agree to correspond to this realization a realization of the process $\gamma'(t)$ with the same times of the customer's arrivals and with the service times γ'_n that are determined by the equation

$$B'(\gamma'_n) = \alpha_\gamma,$$

where α_γ is a random variable uniformly distributed on the interval $[B(\gamma), B(\gamma + 0)]$ (in particular, if $B(x)$ is continuous at point γ, then this interval degenerates into a point). Under this correspondence the service time of each customer in the first case will be larger or equal to the service time in the second case. Evidently, we would then have for any $t > 0$ the inequality

$$\gamma(t) \geqslant \gamma'(t).$$

Since this inequality is valid for any realizations of $\gamma(t)$ and $\gamma'(t)$ in view of the correspondence established above, it follows that $F(t,x) \leqslant F'(t,x)$.

We now return to the proof of the theorem. Let $\lambda\tau \geqslant 1$. If $B(x)$ is the distribution function of service time for the given system, denote

$$B'(x) = \begin{cases} B(x), & x \leqslant x_0, \\ 1, & x > x_0, \end{cases}$$

where x_0 is selected in such a manner that

$$1 > \lambda\tau' = \lambda \int_0^\infty [1 - B'(x)]\,dx \geqslant 1 - \varepsilon$$

and ε is an arbitrarily small positive number.

Then in accordance with Lemma 1 we have the inequality $F'(x) \leqslant \varepsilon e^{\lambda x}$. Hence for t sufficiently large $F'(x,t) \leqslant 2\varepsilon e^{\lambda x}$. In view of Lemma 2 we have all the more that $F(x,t) \leqslant 2\varepsilon e^{\lambda x}$. Since x and ε are arbitrary positive numbers, it follows that

$$\lim_{t\to\infty} F(x,t) = 0,$$

which proves the assertion of the theorem. □

4.3. Nonstationary Characteristics of an $M|G|1$ System

4.3.1. *The Busy Period*

For practical purposes, one of the most important characteristics of queueing systems is the busy period of the server. Consider the sequence of instants t_k at which the server passes from the busy state to the free state. Let t_k' be the instants at which the server passes from the free state to the busy state. Then the difference between t_k and the greatest t_k' that does not exceed it is called the busy period of a server. An investigation of busy periods is of interest in cases where technical features of the server and its capacity for continuous operation must be taken into account.

Denoting the successive in time busy periods of the server by ζ_n, we easily see that $\{\zeta_k\}$ is a sequence of independent uniformly distributed random variables.

We shall determine the distribution function of the busy period

$$G(x) = \mathsf{P}\{\zeta_k < x\}.$$

4.3.2. *An Integral Equation*

We shall study the structure of the busy period. Assume that it starts at the instant $t = 0$. This means that at $t = 0$ one customer arrives and finds the server free and service commences. Assume that the service time is y. Two cases are possible: 1) no customer arrives; 2) $n > 0$ customers arrive. In the first case the busy period ends at the time y; in the second case the situation is more complicated. In order to clarify this it is convenient to assume that the customers are served according to the so-called "last-come–first-served" discipline.

It is clear that with this queue discipline the distribution of the busy period is identical with that of the service on a "first-come–first-served" basis, i.e., in the order of the queue.

Thus, we assume that the customers are served in the reverse order to that of their arrival, and that at time y n customers arrive. Then the first customer is served when the busy period is completed (in accordance with the queue discipline we have assumed). If we denote by $l_1, l_2, \ldots, l_n$ the distribution of the times at which service of each of the n customers commences (in ascending order), we see that $l_{i+1} - l_i$ is the period in which the $(n - i + 1)$th customer and all customers arriving after him or her are served. In other words, the structure of the interval $l_{i+1} - l_i$ is the same as that of the busy period. By construction the intervals $l_{i+1} - l_i$ are independent. Thus, under the condition that n customers arrive during the period y, the busy period is equal to y plus the sum of n independent random variables and the distribution of each one of them is the same as that of the busy period. Since the probability of this condition is given by $[(\lambda y)^n/n!]e^{-\lambda y}$ (recall that the stream is simple) we obtain the integral equation

$$G(x) = \int_0^x \sum_{n=0}^{\infty} e^{-\lambda y} \frac{(\lambda y)^n}{n!} G_n(x-y)\, dH(y), \tag{1}$$

where

$$G_n(x) = P\{\zeta_1 + \zeta_2 + \cdots + \zeta_n < x\}.$$

4.3.3. *Functional Equation*

Introduce the Laplace-Stieltjes transform of the distribution of the busy period, i.e.,

$$g(s) = \int_0^{\infty} e^{-sx}\, dG(x).$$

Applying the Laplace-Stieltjes transform to both sides of (1) we obtain

$$g(s) = \sum_{n=0}^{\infty} g^n(s) \int_0^{\infty} \exp\{-(s+\lambda)x\}(\lambda x)^n\, dB(x)$$

$$= \sum_{n=0}^{\infty} \frac{1}{n!}(-\lambda g(s))^n \psi^{(n)}(s+\lambda) = \psi(s+\lambda-\lambda g(s)).$$

Thus

$$g(s) = \psi(s+\lambda-\lambda g(s)) \tag{2}$$

is valid. The question arises: Does this formula actually yield the value of $g(s)$, i.e., whether (2) has a unique solution? The answer is given by the following theorem.

Theorem. *The function $g(s)$ is the unique analytical solution of the functional equation* (2) *for* $\operatorname{Re} s > 0$ *subject to the condition* $|g(s)| \leqslant 1$, *and which is real for all real* $s > 0$. *Denote by* p^* *the smallest positive number for which the condition*

$$\psi(\lambda(1-p^*)) = p^* \tag{3}$$

is fulfilled. Then

$$G(\infty) = p^*. \tag{4}$$

If $\rho \leqslant 1$, *then* $p^* = 1$ *and* $G(x)$ *is a proper distribution function; if, however,* $\rho > 1$, *then* $p^* < 1$ *and* $G(x)$ *is a singular distribution function, i.e., the busy period may be infinite with probability* $1 - p^*$.

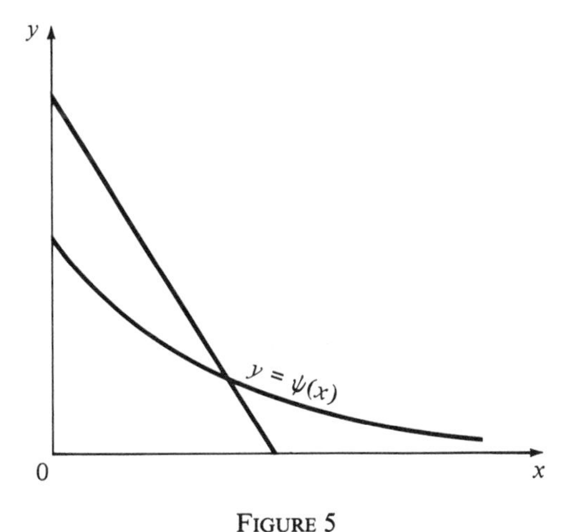

FIGURE 5

PROOF. Assume that (2) has a unique solution under the conditions

$$s > 0, \qquad |g(s)| \leqslant 1.$$

Denote

$$s + \lambda - \lambda g(s) = x$$

and consider Eq. (2) for real positive s. In our new notation, we have to prove that the equation

$$(s + \lambda - x)/\lambda = \psi(x) \tag{5}$$

has a unique solution in x for $s > 0$ such that

$$s \leqslant x \leqslant s + \lambda.$$

The right-hand and left-hand sides of (5) for real x are presented in Fig. 5. When $x = s$, the left-hand side exceeds the right-hand side. Since the straight line representing the left-hand side of (5) has a negative slope and $\psi(x)$ is continuous and positive, it will intersect the curve at a point; this point will correspond to a root of equation (5). Since $h(x)$ is convex, no other root exists. Consequently, $g(s)$ is uniquely determined on the positive real semiaxis. By analytic continuation it can be extended uniquely over the whole right-half-plane.

As $s \to 0$ the slope of the line tends to $-1/\lambda$; at the same time $\psi'(0) = -\tau$. Since the function $\psi(x)$ is convex for $\rho > 1$, the limiting line will intersect the graph of $\psi(x)$ for some $x = x^* > 0$. Hence, $\lim_{s \to 0} x = x^*$. Then $p^* = g(0) = (\lambda - x^*)/\lambda < 1$. For $\rho \leqslant 1$ the limiting straight line intersects the graph of $\psi(x)$ at the unique point $x = 0$ and then $p^* = 1$. It remains to note that $G(\infty) = g(0)$. The theorem is thus proved. □

Differentiating (2) and substituting $s = 0$ we obtain

$$g'(0) = \psi'(0)(1 - \lambda g'(0));$$

hence, the mean duration of the busy period is

$$-g'(0) = \tau/(1 - \rho). \tag{6}$$

Formula (6) can easily be interpreted from ergodic considerations. Let T_0 and T_1 be the mean durations of idle and busy periods, respectively. Then $T_1/(T_0 + T_1) = \rho$ is the average fraction of the time during which the server is occupied. Noting that $T_0 = 1/\lambda$ we have $T_1 = \tau/(1 - \rho)$.

4.3.4. *Distribution of the Number of Customers Served During the Busy Period*

Let g_k be the probability that during a busy period k customers will be serviced, $\{g_k^{(n)}\}$ be the n-fold convolution of the distribution of $\{g_k\}$,

$$g_k = \sum_{n=0}^{\infty} f_n g_{k-1}^{(n)},$$

and hence the generating function $g(z)$ of the sequence $\{g_k\}$ satisfies the functional equation

$$g(z) = zf(g(z)),$$

or

$$g(z) = z\psi(\lambda - \lambda g(z)). \tag{7}$$

For $\rho > 1$ the number of customers serviced during the busy time is infinite with some positive probability; for $\rho \leqslant 1$ it is finite with probability 1. The proof is analogous to one in the case of the busy period. If $\rho < 1$, then this number has a finite mathematical expectation a. From (7) we obtain

$$g'(z) = \psi(\lambda - \lambda g(z)) - \lambda z\psi'(\lambda - \lambda g(z))g'(z).$$

Substituting $z = 1$ into this equation yields $a = 1 + \rho a$, i.e.,

$$a = 1/(1 - \rho). \tag{8}$$

Formula (8) has the following intuitive meaning: during time T on the average $\lambda T(1 - \rho)$ busy periods occurred—the same as the number of customers during the free state of the server; thus, per one busy period there are on the average $\lambda T/[\lambda T(1 - \rho)] = 1/(1 - \rho)$ serviced customers.

4.3.5. *Distribution of Time until the First Disengagement of the Server*

Let $\gamma(0) = x \geqslant 0$. Denote by T_x the first time at which $\gamma(t)$ vanishes. Note that if one averages the distribution of T_x over the distribution $B(x)$, the distribution of the busy period is obtained. Let σ_x be the number of customers arriving into the system during the interval $(0, T_x)$ and

$$G_n(x, t) = \mathsf{P}\{T_n < t, \sigma_x = n\}.$$

Prabhu (1965) has obtained an elegant and very useful formula for this distribution, namely,

$$G_n(x, t) = \int_x^t e^{-\lambda u}\frac{(\lambda u)^{n-1}}{n!}\,dB_n(u - x), \qquad t > x, \quad n \geqslant 0, \tag{9}$$

where $B(x)$ is the n-fold convolution of the distribution function $B(x)$.

We shall prove (9) following Prabhu's argument. Clearly formula (9) is valid for $n = 0$, since

$$B_0(x) = \begin{cases} 0 & \text{for } x \leqslant 0, \\ 1 & \text{for } x > 0. \end{cases}$$

For $b \geqslant 1$ in order that $T_x < t$ and $\sigma_x = n$ be valid, it is necessary that a customer arrives at the system at time $u < x$. If η is the service time of this customer, then $\gamma(u + 0) = x - u + \eta$ and hence the event $\{T_x < t, \sigma_x = n\}$ occurs if and only if $T_{x-u+\eta} < t - u$. Thus,

$$G_n(x, t) = \lambda \int_0^x e^{-\lambda u}\,du \int_0^{t-x} G_{n-1}(x - u + v, t - u)\,dB(v), \quad t > x. \tag{10}$$

Formula (9) is equivalent to a less cumbersome formula in differentials

$$dG_n(x, t) = e^{-\lambda t}\frac{(\lambda t)^{n-1}}{n!}\,dB(t - x).$$

Assume that this formula is valid for $n - 1$. From (10) we obtain (assuming without loss of generality that $\lambda = 1$):

$$\begin{aligned} dG_n(x, t) &= \int_0^x e^{-u}\,du \int_0^{t-u} dG_{n-1}(x - u + v, t - u)\,dB(v) \\ &= \int_0^x e^{-u}\,du \int_0^{t-u} e^{-t+u}\frac{(t-u)^{n-2}}{(n-1)!}(x - u + v)\,dB_{n-1}(t - x - v)\,dB(v). \end{aligned}$$

Whence the required formula is obtained after interchanging the order of integration.

4.3.6. *Nonstationary Distribution of the Virtual Waiting Time*

In Section 4.2 the random process $\gamma(t)$ defined to be 0 if at time t the server is free and if the server is busy the process is defined as the time from instant t until the server is disengaged from all the customers who have arrived prior to t. The Laplace-Stieltjes transform $\Phi(s, t)$ of the distribution of the random variable $\gamma(t)$ satisfies

$$\Phi(s,t) = e^{st-(1-\psi(s))\Lambda(t)}\left[\Phi(s,0) - s\int_0^\infty e^{-su+(1-\psi(s))\Lambda(u)} F(u_1 + 0)\,du\right], \quad (11)$$

where $F(t,x) = \mathsf{P}\{\gamma(t) < x\}$. In order to use this formula $F(t, +0)$ must be defined, i.e., the probability that the service system at arbitrary time u will be free. This is the result that we shall consider now.

The following assumptions are imposed :

1. The servicing time possesses density, i.e.,

$$B(x) = \int_0^x b(t)\,dt,$$

where $b(x)$ has the property that for some $c \geqslant 0$ the function $e^{-cx}b(x)$ is square-integrable on the interval $(0, \infty)$.
2. The parameter of the incoming stream $\lambda(t)$ is square-integrable on the interval $(0, \infty)$.

Theorem. *Under the preceding assumptions the function $F(t, +0)$ is determined as the unique continuous solution of the Volterra equation of the second kind*

$$g'(t) = F(t, +0) + \int_0^t k(t,u)F(u, +0)\,du,$$

where

$$k(t,u) = \frac{1}{2\pi i}\frac{d}{dt}\lim_{M\to\infty}\int_{x-iM}^{x+iM} e^{(t-u)s-[\Lambda(t)-\Lambda(u)][1-\psi(s)]}\frac{ds}{s},$$

$$g'(t) = \frac{1}{2\pi i}\frac{d}{dt}\int_{x-i\infty}^{x+i\infty} \Phi(s,0)e^{ts-[1-\psi(s)]\Lambda(t)}\frac{ds}{s^2}.$$

In the last integral necessarily $x > c$.

We shall not give the proof here; the reader is referred to Reich (1958).

4.3.7. *Nonstationary Conditions of the Queueing System for a Simple Incoming Stream*

Assume it is known that at $t = 0$ the server is free. Denote by t_n the final instant of the server's nth busy period ($n \geqslant 1$). $\{t_n\}$ is a renewal process. The server is free at time t if and only if one of the following disjoint events occurs:

1. no customer arrives in the time interval from 0 to t;
2. a busy period ends at some time $\tau < t$; no customer arrives during the following $t - \tau$ units of time.

Denote the renewal function of the process $\{t_n\}$ by $H(t)$. Then the probability that the server is free is

$$F(t, +0) = e^{-\lambda t} + \int_0^t e^{-\lambda(t-\tau)}\, dH(\tau). \tag{12}$$

We shall derive the renewal function of the process $\{t_n\}$. The random variable $z_n = t_n - t_{n-1}$ (putting $t_0 = 0$) is evidently the sum of two independent random variables, the interval ξ_n from t_{n-1} to the arrival time of the next customer, and the busy interval ζ_n. Thus, if $g(s)$ is the unique analytic solution of the functional equation

$$g(s) = \psi(s + \lambda - \lambda g(s))$$

for which $|g(s)| \leqslant 1$ in the right half-plane and which is real for all real $s \geqslant 0$, then

$$\mathsf{M}e^{-s\zeta_n} = \lambda g(s)/(s + \lambda). \tag{13}$$

According to Section 2.6 the Laplace-Stieltjes transform of $H(t)$ is

$$h(s) = \int_0^\infty e^{-sx}\, dH(x) = \frac{\mathsf{M}e^{-s\zeta_n}}{1 - \mathsf{M}e^{-s\zeta_n}} = \frac{\lambda g(s)}{s + \lambda - \lambda g(s)}. \tag{14}$$

Taking the Laplace-Stieltjes transform of both sides of (12) and using (14) we obtain

$$\int_0^\infty e^{-st} F(t, +0)\, dt = \frac{1}{s + \lambda}\left(1 + \frac{\lambda g(s)}{s + \lambda - \lambda g(s)}\right),$$

or

$$\int_0^\infty e^{-st} F(t, +0)\, dt = \frac{1}{s + \lambda - \lambda g(s)}. \tag{15}$$

4.4. A System of the $GI|M|m$ Type

4.4.1. *Construction of an Embedded Markov Chain*

Queueing systems of a very general structure with incoming stream with a limited aftereffect and exponentially distributed service time can be studied using the method of embedded Markov chains (Kendall's method).

We shall stipulate the following assumptions.

A queueing system is described by a random process $\nu(t)$. The process takes integral nonnegative values 0, 1, 2, ... At times $\{t_n\}$—which form a random renewal stream—customers arrive at the system. If at time $t_n = 0$ the state of the process was i, then at time t_n the process moves into the state j with probability p_{ij} where

$$\sum_{j=0}^{\infty} p_{ij} = 1$$

independently of the random times associated with the behavior of the process until time t. In the intervals between $\{t_n\}$, the process $v(t)$ is a homogeneous Markov process with transition intensities λ_{ij}. To complete the definition of the random process $v(t)$ we specify either its initial distribution

$$p_k(0) = \mathsf{P}\{v(0) = k\} \quad (k = 0, 1, 2, \ldots),$$

or the distribution at the time preceding the first renewal:

$$p_k^*(1) = \mathsf{P}\{v(t_1 - 0) = k\} \quad (k = 0, 1, 2, \ldots).$$

Since $\{t_n\}$ is a renewal process, it follows that

$$\begin{aligned} F_n(x) &= \mathsf{P}\{t_n - t_{n-1}(1 - \delta_{n1}) < x\} \\ &= F_1(x)\delta_{n1} + F(x)(1 - \delta_{n1}), \quad n \geqslant 1. \end{aligned} \tag{1}$$

Introduce the notation

$$v_n = v(t_n - 0) \quad (n = 1, 2, \ldots).$$

The following characteristics are of interest:

$$p_k^*(n) = \mathsf{P}\{v_n = k\}, \qquad p_k^* = \lim_{n\to\infty} p_k^*(n),$$

$$p_k(t) = \mathsf{P}\{v(t) = k\}, \qquad p_k = \lim_{t\to\infty} p_k(t).$$

If $F(x)$ is a nonlattice distribution* that possesses a finite first moment τ, then the existence of a nonsingular distribution $\{p_k^*\}$ implies immediately (in view of Smith's theorem in the theory of regenerative random processes) the existence of a nonsingular distribution $\{p_k\}$, where

$$p_k = \frac{1}{\tau}\int_0^\infty \sum_{i=0}^{\infty} p_i^* \sum_{j=0}^{\infty} p_{ij} f_{jk}(t)[1 - F(t)]\, dt, \tag{2}$$

and $f_{jk}(t)$ is the probability that a homogeneous Markov process with transition intensities λ_{ij} proceeds during time t from the state j into the state k. [It is assumed that $\sum_{k=0}^{\infty} f_{jk}(t) = 1$ for all t.] These arguments imply that to study a stationary condition of the process $v(t)$ it is sufficient to study the Markov chain $\{v_k\}$. We shall derive a formula for transition probabilities of this chain $\pi_{ij} = \mathsf{P}\{v_{k+1} = j | v_k = i\}$.

Since under the condition $t_{n-1} - t_n = t$ the transition probability coincides with the function $\sum_k p_{ik} f_{kj}(t)$ and the random variable $z_{n+1} = t_{n+1} - t_n$ is

---

* In the case of a lattice distribution we consider the limit

$$p_k^{(h)} = \lim_{n\to\infty} P_k(nh + b),$$

where h is the maximal step of the distribution, $0 \leqslant b < h$.

distributed with the distribution $F(t)$, we have

$$\pi_{ij} = \sum_k p_{ik} \int_0^\infty f_{kj}(t)\, dF(t). \tag{3}$$

The stationary distribution of the Markov chain $\{v_n\}$ (the final probabilities) are derived from the relations

$$p_k^* = \sum_{i=0}^{\infty} \pi_{ik} p_i^* \quad (k = 0, 1, 2, \ldots),$$

$$\sum_{k=0}^{\infty} p_k^* = 1.$$

Thus the probabilities of the states of the queueing system can be obtained either as characteristics of a lined random process or an embedded Markov chain.

4.4.2. *Example*

Consider a system with waiting consisting of m identical servers. Let the stream of customers form a renewal process $\{t_n\}$ in accordance with formula (1). The service time is assumed to be an exponential random variable with parameter μ.

We shall interpret $v(t)$ as the number of customers in the system at time t. We now obtain the characteristics of this process.

Clearly

$$p_{ij} = \delta_{i+1,j},$$

i.e., at time t_n the random function $v(t)$ possesses a jump equal to 1. Furthermore, for $j \geqslant m$

$$f_{ij}(t) = \begin{cases} 0 & \text{if } j > i, \\ \dfrac{(m\mu t)^{i-j}}{(i-j)!} e^{-m\mu t} & \text{if } j \leqslant i \end{cases} \tag{4}$$

(since in the interval between t_n and t_{n-1} the number of customers in the system can only decrease following a pure death process with parameter $m\mu$ as long as the queue exists).

Consider the case when $0 \leqslant j < m$. There are two possibilities: (1) $i \leqslant m$ or (2) $i > m$. We shall investigate each one of them separately.

1. $i \leqslant m$. Then the first customer is served during time ξ_1 distributed in accordance with the exponential distribution with parameter $i\mu$, the next one during time $\xi_1 + \xi_2$ where ξ_2 does not depend on ξ_1 and is distributed exponentially with parameter $(i-1)\mu$, and so on; finally the $(i-j)$th customer during time $\xi_1 + \xi_2 \cdots + \xi_{i-j}$ where the parameter of ξ_{i-j} equals $(j+1)\mu$. Under the condition $v(0) = i$ the event $\{v(t) \leqslant j\}$ is equivalent to the event $\{\xi_1 + \xi_2 + \cdots + \xi_{i-j} < t\}$. Hence, for $j \leqslant i \leqslant m$

$$f_{ij}(t) = \mathsf{P}\{\xi_1 + \xi_2 + \cdots + \xi_{i-j} < t\} - \mathsf{P}\{\xi_1 + \xi_2 + \cdots + \xi_{i-j+1} < t\}. \tag{5}$$

If we set

$$\varphi_{ij}(s) = \int_0^\infty e^{-sx}\, df_{ij}(t), \quad \operatorname{Re} s \geqslant 0,$$

then it follows from formula (5) that

$$\varphi_{ij}(s) = \frac{i\mu}{s + i\mu} \cdots \frac{(j+1)\mu}{s + (j+1)\mu} - \frac{i\mu}{s + i\mu} \cdots \frac{j\mu}{s + i\mu} = \frac{\mu^{i-j} i!\, s}{j!} \prod_{r=j}^{i} \frac{1}{s + r\mu}. \tag{6}$$

The functions

$$f_{ij}(t) = C_i^j e^{-j\mu t}(1 - e^{-\mu t})^{i-j}$$

correspond to the transformations of the form (6); this result is also evident directly.

2. $i > m$. Then

$$f_{ij}(t) = \mathsf{P}\{\xi_0 + \xi_1 + \cdots + \xi_{m-j} < t\} - \mathsf{P}\{\xi_0 + \xi_1 + \cdots + \xi_{m-j+1} < t\}, \tag{7}$$

where ξ_0 is the time during which, under the condition that all the servers are busy, $i - m$ customers will be served. Analogously we have

$$\varphi_{ij}(s) = \frac{m^{i-m} m!\, \mu^{i-j} s}{j!(s + m\mu)^{i-m}} \prod_{r=j}^{m} \frac{1}{s + r\mu}. \tag{8}$$

In this problem the characteristics of the Markov chain $\{\nu_n\}$ rather than the stationary characteristics of the process $\nu(t)$ are of basic interest. Knowing the transition probabilities of this chain defined by formula (3) where $f_{ij}(t)$ are given we can solve the system of equations that determines the limit probabilities. Kendall (1960) has shown that under the condition that there exists an ergodic distribution, the formulas

$$p_k^* = \begin{cases} \mu_k, & \text{for } k \leqslant m - 2, \\ \lambda^{k-m+1} & \text{for } k \geqslant m - 1 \end{cases}$$

are valid, where μ_k are constants and λ is the unique root of the equation

$$\int_0^\infty e^{-(1-\lambda)\mu m x}\, dF(x) = \lambda$$

in the interval $0 < \lambda < 1$.

Let Q_n denote the size of the queue at time $t_n - 0$, i.e., $Q_n = \max\{0, \nu_n - m\}$ and Q denotes the random variable whose distribution coincides with the limiting distribution of Q_n $(n \to \infty)$. Then the following interesting fact is valid. If it is known that $Q > 0$, then under this condition the random variable has a geometric distribution

$$\mathsf{P}\{Q = k \mid Q > 0\} = (1 - \lambda)\lambda^{h-1} \quad (k = 1, 2, 3, \ldots). \tag{9}$$

Detailed calculations are given in the previously cited paper by Kendall as well as in a paper by Takács (1967). Here we shall present the proof of the existence of the ergodic distribution for a Markov chain $\{\nu_n\}$ under the condition

$$m\mu \int_0^\infty x\,dF(x) > 1. \tag{10}$$

First, the physical meaning of condition (10) is the same as the condition $\rho < 1$ for a single-server system with a simple incoming stream (Section 4.1). Indeed, in a unit time on the average $[\int_0^\infty x\,dF(x)]^{-1}$ customers arrive, the average duration of serving one customer equals μ^{-1} so that the inverse of the quantity on the left-hand side of inequality (10) can naturally be viewed as the load on the server.

To prove the assertion we use the ergodic theorem related to Markov chains with a finite number of states (Section 3.1). We state a frequently used particular case of this theorem.

Let $\{\nu_n\}$ *be an irreducible aperiodic Markov chain with values* 0, 1, 2, ... *Then if for some* $\varepsilon > 0$

$$\mathsf{M}\{\nu_{n+1} - \nu_n | \nu_n = i\} < -\varepsilon, \quad i > N, \tag{11}$$

$$\mathsf{M}\{\nu_{n+1} - \nu_n | \nu_n = i\} < \infty, \quad i \leqslant N, \tag{12}$$

the chain $\{\nu_n\}$ *possesses an ergodic distribution.*

In the case of the queueing system under consideration one can write

$$\nu_n - \nu_{n-1} = \beta_n,$$

where β_n is the number of customers serviced during the time interval (t_{n-1}, t_n). Let it be known that $\nu_{n-1} = i > m$ and $t_n - t_{n-1} = x$. Then for any $k \leqslant i - m$

$$\mathsf{P}\{\beta_n \geqslant k\} = e^{-m\mu x} \sum_{i=k}^{\infty} \frac{1}{i!}(m\mu x)^i$$

(as long as there is a queue the outgoing process is a Poisson process).

Hence,

$$\begin{aligned} &\mathsf{M}\{\beta_n | \nu_{n-1} = i, t_n - t_{n-1} = x\} \\ &\geqslant e^{-m\mu x} \sum_{h=1}^{i-m} \sum_{i=k}^{\infty} \frac{(m\mu x)^i}{i!} \\ &= e^{-m\mu x}\left(\sum_{k=1}^{\infty} \sum_{i=k}^{\infty} \frac{(m\mu x)^i}{i!} - \sum_{k=i-m+1}^{\infty} \sum_{i=k}^{\infty} \frac{(m\mu x)^i}{i!}\right) = m\mu x - R_{i-m+1}(x). \end{aligned}$$

Integrating over the distribution $F(x)$ we arrive at

$$\int_0^\infty m\mu x\,dF(x) = m\mu \int_0^\infty x\,dF(x) > 1;$$

$\int_0^\infty R_{i-m+1}\, dF$—being a remainder of a convergent series—can be made arbitrarily small by taking $i > N$. Whence the bound (11) follows. The bound (12) follows from the fact that $v_n - v_{n-1} \leqslant 1$ always.

4.5. $M|G|1$ System with an Unreliable and "Renewable" Server

4.5.1. *Possible Statements of the Problem*

In practical applications generalization of the queueing system to the case when the server fails and requires repair (renewal) are of great importance. Indeed it is evident a priori that there exist service systems for which the given intensity of the incoming stream, the waiting time, and queue length possess an ergodic distribution; if, however, we consider that a certain fraction of time is being used for repairs due to a systematic breakdown, then the queue indeed will grow to infinity. Therefore, it is necessary to be able to establish ergodicity conditions and to obtain various characteristics of service also for "unreliable" servers.

From a practical aspect several different mathematical models that describe the failure of a server and its renewal as well as the rules of servicing a customer who finds a server in a broken condition are of interest. First, the server can fail either only in the course of a service or during the idle time or it can fail in both of these states. Clearly, in each one of these cases a statistical law should be indicated which governs the failure of a server. In many cases the assumption that a server fails randomly, i.e., the failure during time h can occur with probability $\alpha h + o(h)$ independently of the previous history, seems to approximate the real-world situations fairly well. It may be assumed that parameter α is a random variable that is equal to α_1 if the server at a given instant is busy and equals α_0 if the server is idle.

This assumption is equivalent to the statement that the duration of reliable service of a server—whether busy or idle—is exponentially distributed.

In a more general case, the duration of reliable service is a random variable with a distribution that is not necessarily exponential. The question arises how should one reckon the duration of continuous service? In real-world systems it is sometimes of interest to consider the model according to which a server may fail only during time from t to $t + h$ with probability that depends on the duration from the beginning of the busy period still in progress at time t.

In a model considered by Mar'yanovich (1962a, b) this probability depends on another variable, namely, the time from the beginning of servicing a customer that is still in service at time t. Often the time starting from the instant of the last repair of the server is of importance.

The duration of repairs of a server after successive failures are usually assumed to be independent and identically distributed. This distribution may, however, depend on whether the server failed at the time when it was busy with a customer or whether the failure occurred during the idle time. In the majority of papers the assumption is that the repair time is exponential; for applications, it is important to relax this assumption.

Assume that a server failed during service. After the repair the customer again arrives at the server to complete the service: in the first model the service time of a customer after a repair does not depend on the time previously spent in service; in the second model this time is "stored" and the remaining service time is reduced. Intermediate situations are also possible. Of interest is the model in which a customer departs the system with some probability if during its service the server failed. This situation was studied by Mar'yanovich. We shall now consider various models previously described, starting with the simplest.

4.5.2. *Failure During Idle Time*

Assume that the server may fail only in the free state. If the busy period ends at the instant t and until $t + x$ no new customers arrive, the server may fail with probability $D_0(x) = \int_0^x d_0(t)\,dt$. In other words $D_0(x)$ is the distribution function of the service life of a "free" server. The renewal time is considered to be a random variable with distribution function $R_0(x)$ and finite mathematical expectation $\bar{\tau}_0$. We shall retain the notation of the foregoing sections for the parameter of the incoming stream, the service time, and the load on the server.

The following quantities are important service characteristics:

the distribution of waiting time at an arbitrary time, i.e., the time from instant t up to the time when the server is capable of servicing a customer arriving at t;
the probability that the server is free at the given instant;
the probability that the server is serviceable (or unserviceable) at the given instant.

All these characteristics can be studied with the aid of a suitably constructed Markov process.

First consider the following abstract servicing model. The behavior of a system is described by a regenerative random process $\gamma(t)$ with renewal times t_n, corresponding to the completion of busy periods. If $t_{n-1} + z_n$ is the time of arrival of the first customer in a given cycle, then the cycle consists of two intervals: z_n and $t_n - t_{n-1} - z_n = z_n'$. The process $\gamma(t)$ is completely defined by the random variable z_n and the random function $g(t)$, $0 < t < z_n'$, where $g(t) = \gamma(t_{n-1} + z_n + t)$ (z_n' is also a random variable).

Assume that $\mathsf{M}z_n < \infty$, $\mathsf{M}z_n' < \infty$. Now let $P(A)$ be the ergodic probability of the event $\{\gamma(t) \in A\}$. Then this probability will have the same value when $\gamma(t)$ is replaced by another process for which the distributions of $g(t)$ and z_n' are the same but that of z_n is different provided $\mathsf{M}z_n$ remains unchanged. Indeed, it follows from Smith's theorem that

$$\lim_{t\to\infty} \mathsf{P}\{\gamma(t) \in A\} = \mathsf{M}\tau(A)/\mathsf{M}\{z_n + z_n'\},$$

where $\tau(A)$ is the total time of sojourn of the function $g(t)$, $0 < t < z_n'$ and the function $g_0(t) = 0$, $0 < t < z_n$ in the set of states A. The first summand in both cases is the same, while the second equals $\mathsf{M}z_n$ or 0 depending on whether or not the set A contains the point 0.

Return now to the situation with an unreliable server. We shall interpret a failure as a customer and renewal as service time. Let $\gamma(t)$ be the virtual waiting time. Within the busy periods this process is changed in the same manner as in the case of an $M|G|1$ system, but the distribution function of servicing a customer with which the busy period starts is equal to

$$B_0(x) = \int_0^\infty e^{-\lambda y}\, dD_0(y) R_0(x) + \left(1 - \int_0^\infty e^{-\lambda y}\, dD_0(y)\right) B(x).$$

The period of free (and reliable) state of a server has the mathematical expectation

$$\lambda_0^{-1} = \int_0^\infty e^{-\lambda x} \bar{D}_0(x)\, dx.$$

In view of the preceding remark, for calculations of ergodic probabilities we shall assume that during an idle state a customer arrives in accordance with the exponential distribution with parameter λ_0. We thus have an $M|G|1$ system with the special features of arrivals and servicing of the customers stipulated previously.

Analogously to the results in Section 4.2 the following assertion is verified.

Theorem. *For $\rho = \lambda\tau < 1$, there exists a limiting distribution $F(x) = \lim_{t\to\infty} \mathsf{P}\{\gamma(t) < x\}$, which does not depend on the initial distribution and is defined by the solution of the integral equation*

$$F'(x) - \lambda \int_0^x \bar{B}(x - y)\, dF(y) + (\lambda\bar{B}(x) - \lambda_0 \bar{B}_0(x))F(+0) = 0, \quad x > 0, \quad (1)$$

where $F(+0)$ is determined from the normalizing condition $F(\infty) = 1$.

Introduce the notation

$$f(s) = \int_0^\infty e^{-sx}\, dF(x), \qquad \psi(s) = \int_0^\infty e^{-sx}\, dB(x),$$

$$\psi_0(s) = \int_0^\infty e^{-sx}\, dB_0(x).$$

Integrating both sides of (1) with the weight e^{-sx}, $0 < x < \infty$, we arrive at

$$f(s) = F(+0)\frac{s + \lambda_0(1 - \psi_0(s)) - \lambda(1 - \psi(s))}{s - \lambda(1 - \psi(s))}. \quad (2)$$

The constant $F(+0)$ is obtained by means of the limiting transition in formula (2) as $s \to 0$: on the left-hand side $f(s) \to 1$, on the right-hand side we have

$$F(+0)\frac{1 + \lambda_0\tau_0 - \lambda\tau}{1 - \lambda\tau},$$

where

$$\tau_0 = \int_0^\infty \bar{B}_0(x)\,dx.$$

Thus

$$F(+0) = \frac{1 - \rho}{1 - \rho + \lambda_0\tau_0}. \tag{3}$$

Formula (3) has the following simple interpretation. The average amount of work arriving during time dt is equal to $\lambda_0 F(+0)\tau_0\,dt + \lambda(1 - F(+0))\tau\,dt$, i.e., in an interval of unit length the server is busy on the average during the time $\lambda_0 F(+0)\tau_0 + \lambda(1 - F(+0))\tau$; on the other hand, this time is equal to $1 - F(+0)$.

Similar consideration allows us to derive the probability that a server is in working condition in a stationary state. This probability is called the *availability coefficient* (K_Γ) in reliability theory.

The times at which a server fails form a renewal stream with intensity $\lambda_0 F(+0)$. Whence

$$K_\Gamma = 1 - \lambda_0 F(+0)\bar{\tau}_0. \tag{4}$$

We suggest that the reader derive this formula from Smith's theorem for regenerative processes choosing the instants of completion of busy intervals as the renewal times.

Observe that the same mathematical equations describe a completely different physical process. In many cases when serving the first customer in the busy period an additional operation usually called "warming-up" is required. This causes $B_0(x)$ and $B(x)$ to be different functions. (In this case usually $\lambda = \lambda_0$.) Actually, this model can be adapted to the systems where both "warming-up" and failures are observed.

4.5.3. *The General Case*

It is clear that the queueing system studied in detail in the preceding subsection is very restrictive; for applications it is important to generalize it in such a manner that the possibility of failure during servicing a customer will be allowed. However, for a wide class of formulations of servicing with an unreliable server, the preceding result can be used; one needs only to interpret the service time in a different manner.

We clarify the preceding discussion by an example. Assume that in addition to the assumptions stipulated in the preceding subsection, a server can fail—in the course of servicing a customer—during a short time h with probability $\alpha_1 h + o(h)$ independently of the previous history. We assume that the renewal time is a random variable with distribution function $R_1(x)$ and finite mathematical expectation τ_1. As to failure during idle time, the assumptions are the same as in the preceding subsection. At time t_0 let the servicing of a certain customer commence. Denote by β the time from the instant t_0 up to the time when the server is ready to service the next customer. The time β may consist of the time servicing a customer who arrived at the service at time t_0 (this will occur in the case when the server did not fail during the service time); if, however, the server failed during the service n times, then β will consist of the service time and n renewal times. By the total probability formula

$$B_\beta(x) \equiv \mathsf{P}\{\beta < x\} = \sum_{n=0}^{\infty} \int_0^x R_1^{(n)}(x-y) \frac{(\alpha_1 y)^n}{n!} e^{-\alpha_1 y}\, dB(y),$$

where $R_1^{(n)}(x)$ is the distribution function of the sum of n independent random variables with the distribution function $R_1(x)$ if $n \leqslant 1$, $R_1^{(0)} \equiv 1$. Clearly, as far as waiting time of a customer is concerned, it is irrelevant what fraction of time was spent by the server on service or on repair; only the amount of time—after the beginning of a service of a customer—that is needed for a server to be able to service the next customer is of importance. Hence, if one views service time as a random variable β, then the waiting time of a customer in a steady state will be determined by formulas (2)–(3) in which $\psi(s)$ is replaced by the Laplace-Stieltjes transform $\psi_\beta(s)$ of the random variable β.

In the preceding example

$$\begin{aligned}\psi_\beta(s) &= s \int_0^\infty e^{-sx} \frac{d}{dx} \left\{ \sum_{n=0}^{\infty} \int_0^x R_1^{(n)}(x-y) \frac{(\alpha_1 y)^n}{n!} e^{-\alpha_1 y}\, dB(y) \right\} dx \\ &= \psi(s + \alpha_1 - \alpha_1 P_1(s)),\end{aligned}$$

where $P_1(s)$ is the Laplace-Stieltjes transform of the distribution of the renewal time of a server under the condition that the failure occurred during servicing a customer. A condition under which the random process $\zeta(t)$ possesses an ergodic distribution can be derived. For this purpose we compute the mathematical expectation of the random variable β evaluating the derivative of $\psi_\beta(s)$ with respect to s at $s = 0$:

$$\mathsf{M}\beta = -\psi_\beta(0) = -\psi'(0)[1 - \alpha_1 p_1'(0)] = \tau[1 + \alpha_1 \tau_1].$$

It thus follows immediately that the ergodicity condition of the process $\zeta(t)$ is given by the inequality

$$\lambda\tau[1 + \alpha_1 \tau_1] = \rho[1 + \alpha_1 \tau_1] < 1.$$

As it is to be expected for some values of parameters, the process that possesses an ergodic distribution in the case when a server is totally reliable ceases to have this property when the server is subject to a random failure.

Several formulations of this kind have been considered by Mar'yanovich, who determined, using another method, the distribution of the number of customers in the system under steady-state conditions and the probability that the server is unserviceable.

4.5.4. *The Influence of Partial Failure*

Up until now we have considered servers who may be in only two states, "serviceable" or "unserviceable." A serviceable server is capable of serving customers at any instant of time (this is reflected by the identical distributions and independence of service times for different customers). On the other hand, when the server is unserviceable, it is totally incapable of servicing and requires renewal (repair). In real-world situations, however, the server may be capable of service to some extent at different times. As an example, consider the service to be a transmission of a message by radio. With the passage of time the external noise level changes, and this results in variations in the rate of information transmission. Yet another example is a system of information transmission in which several methods of transmission, each having its own transmission rate, are available. When one piece of equipment fails, another is used.

This state of affairs substantially diversifies mathematical statements of queueing problems. We deal here only with one possible approach, which has apparently not been dealt with sufficiently in the literature as yet.

We use the interpretation of service in terms of "work": service of each customer will be understood as work performed by the server. The amounts of work corresponding to different customers will be considered independent random variables $\omega_1, \omega_2, \ldots, \omega_n, \ldots$ with a common distribution function $B(x)$. Assume, furthermore, that at almost every instant of time the server's capacity is $E(t)$, i.e., from t to $t + h$ the server can perform $E(t)h + o(h)$ units of work. If we assume that the capacity of the server is constant, $E(t) = E$, the problem is reduced to service with waiting; the distribution function of the service time will be $B(Ex)$. The preceding formulation of the problem of an unserviceable server fits well into the new framework. Indeed, when the server is unserviceable, the capacity must clearly be considered zero.

We make the following assumptions. At each instant of time the server may be in one of n states: the 1st, 2nd, ..., nth; a certain capacity E_i corresponds to the ith state. Transition from state to state occurs in a random manner; if the server is in the ith state, it may pass into the jth in the short time h with probability $q_{ij}h + o(h)$. In other words, the states of the server form a homogeneous Markov process $l(t)$.

The following characteristics of service are of interest in this formulation of the problem:

the distribution of the amount of work required starting from the instant t in order to complete servicing the customers already in the system at t;

the distribution of the waiting time of a customer arriving at time t;
the probability that at a given instant of time there are no customers in the system, and the server is in a particular state.

In studying systems with waiting, special importance is attached to the parameter

$$\rho = \lambda\tau,$$

i.e., the load on the system. In the present formulation we shall also utilize the load concept, whose value will have substantial influence on the nature of the process.

First, we note that the process $l(t)$ does not depend on the service process. Assume that all possible transitions from state i to some other state j are admissible. Then, by the ergodic theorem of Section 3.1, the process $l(t)$ possesses an ergodic distribution, defined by the probabilities

$$p_i = \lim_{t\to\infty} \mathsf{P}\{l(t) = i\}, \quad 1 \leqslant i \leqslant n,$$

which is determined by the system of equations

$$p_i \sum_{j\neq i} q_{ij} = \sum_{j\neq i} p_j q_{ji}, \quad 1 \leqslant i \leqslant n,$$

with the normalizing condition

$$\sum_{i=1}^{n} p_i = 1.$$

Now define the mean capacity of a server as:

$$\bar{E} = \sum_{i=1}^{n} E_i p_i$$

and the load on the system by the equation:

$$\rho = \frac{1}{\bar{E}} \lambda \mathsf{M}\omega_1.$$

We introduce a stochastic process $\gamma^*(t)$, defined as the amount of work required from the time t to complete servicing the customers already in the system at t (clearly if server is free at the time t, then $\gamma^*(t) = 0$). Consider now the two-dimensional stochastic process

$$\zeta^*(t) = \{l(t), \gamma^*(t)\}.$$

Clearly $\zeta^*(t)$ is a Markov process.

The distribution of the process $\zeta^*(t)$ at an arbitrary instant is given by the function

$$F_i^*(x, t) = \mathsf{P}\{l(t) = i, \gamma^*(t) \leqslant x\}.$$

Theorem. *A necessary and sufficient condition for the existence of the ergodic distribution of the process* $\zeta^*(t)$ *is*

$$\rho < 1.$$

When this condition is satisfied, the limits

$$F_i^*(x) = \lim_{t\to\infty} F_i^*(x, t)$$

satisfy the following system of integro-differential equations:

$$E_i F_i'(x) - \left(\lambda + \sum_{j\neq i} q_{ij}\right) F_i(x) + \lambda \int_0^x B(x-y)\, dF_i(y) + \sum_{j\neq i} q_{ji} F_j(x) = 0 \quad (1 \leqslant i \leqslant n). \tag{5}$$

PROOF. The derivation of (5) is analogous to that of Eq. (1) for the distribution function of the waiting time F. Only the proof that the process $\zeta^*(t)$ possesses an ergodic distribution for $\rho < 1$ is somewhat different.

Let x_n be the duration of the nth free interval, y_n be the duration of the nth busy interval. The working time of the server during time T is not less than $A(T) - E' \sum_{i=1}^{J(T)} x_i$, where $A(T) = \int_0^T E(t)\, dt$, $E' = \max E_i$, x_i is the length of the idle interval that does not intersect with $(0, T)$. On the other hand, $A(t) \leqslant \sum_{i=1}^{N(i)} \eta_i$, where $N(t)$ is the number of customers in the interval $(0, T)$ and η_i are the durations of their service. Consequently,

$$\frac{1}{T} A(T) - E' \frac{J(T)}{T} \frac{1}{J(T)} \sum_{i=1}^{J(T)} x_i \leqslant \frac{N(T)}{T} \frac{1}{N(T)} \sum_{i=1}^{N(T)} \omega_i.$$

From the ergodic theorem for semi-Markov processes, $T^{-1}A(T) \to \bar{E}$ in probability. By the property of a Poisson distribution $T^{-1}N(T) \to \lambda$ in probability. From the strong law of large numbers we have that

$$N^{-1} \sum_{i=1}^{N} \omega_i \xrightarrow[N\to\infty]{} \mathsf{M}\omega_1$$

with probability 1; hence,

$$N^{-1}(T) \sum_{i=1}^{N(T)} \omega_i \to \mathsf{M}\omega_1$$

in probability.

Finally x_i is an exponentially distributed random variable with parameter λ; $J^{-1} \sum_{i=1}^{J} x_i \to 1/\lambda$ with probability 1 and hence $J^{-1} \sum_{i=1}^{J} x_i \leqslant c_\varepsilon$ for all J simultaneously with probability at least $1 - \varepsilon$.

Thus, with probability at least $1 - \varepsilon/2$

$$\frac{J(T)}{T} \geqslant (c_\varepsilon E')^{-1}(\bar{E} - \lambda \mathsf{M}\omega_1) - \varepsilon = \delta > 0. \tag{6}$$

Furthermore, $J(T) = \sum_k J_k(T)$, where $J_k(T)$ is the number of renewal points in the interval $(0, T)$ of the renewal process of times of completion of busy periods in the state $E(t) = E_k$. When the mean length of a cycle is infinite, $T^{-1}J_k(T) \to 0$ in probability. Consequently, the average of a cycle is finite for some k or equivalently

$$\lim_{t\to\infty} F_k(t, +0) > 0. \tag{7}$$

Since $\zeta(t)$ is a regenerative process with the chosen renewal process in the form of a set of regeneration times, it follows from Smith's theorem that the process possesses a proper limit distribution. □

We now turn to the explicit solution of integro-differential equations (5). We shall utilize both analytic and probabilistic arguments.

By the method already used several times in this chapter, it is easy to establish the absolute continuity of $F_i(x)$ and thus to justify application of the Laplace transformation to (5). Denote

$$\Phi_i(s) = \int_0^\infty e^{-sx}\, dF_i(x), \quad 1 \leqslant i \leqslant n, \quad a_i = F_i(0), \quad 1 \leqslant i \leqslant n;$$

$$\psi(s) = \int_0^\infty e^{-sx}\, dB(x).$$

Applying the Laplace transform to (5) we obtain

$$\Phi_i(s)\left\{sE_i - \lambda[1-\psi(s)] - \sum_{j\neq i} q_{ij}\right\} + \sum_{j\neq i} q_{ij}\Phi_j(s) = sa_iE_i, \quad 1 \leqslant i \leqslant n. \tag{8}$$

The system of equations (8) has a unique analytic solution $\Phi_i(s)$, $1 \leqslant i \leqslant n$, for $\operatorname{Re}\{s\} > 0$, if the constants a_i appearing on the right-hand side actually correspond to their stochastic meaning, i.e., if they are equal to the probabilities $F_i(0)$. Indeed, such a solution always exists; it is the Laplace–Stieltjes transform of the distributions $F_i(x)$. We shall prove that in the neighborhood of point $s = 0$ (excluding the point $s = 0$ itself) the determinant of system (8) does not vanish; this will imply that the solution is unique in the neighborhood of zero. Next by the principle of analytic continuation the solution may then be extended uniquely over the whole right-half-plane.

Thus it is required to establish that in a neighborhood of point $s = 0$ the determinant

$$\Delta(s) = \begin{vmatrix} sE_1 - \lambda(1-\psi(s)) - \sum_{j\neq 1} q_{1j} & \cdots & q_{n1} \\ q_{12} & \cdots & q_{n2} \\ \cdots & \cdots & \cdots \\ q_{1n} & \cdots & sE_n - \lambda(1-\psi(s)) - \sum_{j\neq n} q_{nj} \end{vmatrix}$$

does not vanish. For $s = 0$, the determinant vanishes (the sum of the elements of each column is zero). Furthermore,

$$\Delta'(s) = [E_1 + \lambda\psi'(s)] \begin{vmatrix} sE_2 - \lambda[1-\psi(s)] - \sum_{j\neq 2} q_{2j} & \cdots & q_{n2} \\ \cdots & \cdots & \cdots \\ q_{2n} & \cdots & sE_n - \lambda[1-\psi(s)] - \sum_{j\neq n} q_{nj} \end{vmatrix}$$

$$+ [E_2 + \lambda\psi'(s)] \begin{vmatrix} sE_1 - \lambda[1-\psi(s)] - \sum_{j\neq 1} q_{1j} & \cdots & q_{n1} \\ q_{1n} & \cdots & sE_n - \lambda[1-\psi(s)] - \sum_{j\neq n} q_{nj} \end{vmatrix} + \cdots.$$

Substituting $s = 0$ we obtain

$$\Delta'(0) = [E_1 + \lambda\psi'(0)] \begin{vmatrix} -\sum_{j\neq 2} q_{2j} & \cdots & q_{n2} \\ \cdots & \cdots & \cdots \\ q_{2n} & \cdots & -\sum_{j\neq n} q_{nj} \end{vmatrix}$$

$$+ [E_2 + \lambda\psi'(0)] \begin{vmatrix} -\sum_{j\neq 1} q_{1j} & \cdots & q_{n1} \\ \cdots & \cdots & \cdots \\ q_{n1} & \cdots & -\sum_{j\neq n} q_{nj} \end{vmatrix} + \cdots$$

or, denoting the coefficient at $[E_i + \lambda\psi'(0)]$ in this expansion by Δ_i we have

$$\Delta'(0) = \sum_{i=1}^{n} [E_i + \lambda\psi'(0)]\Delta_i.$$

Recall that the system of equations determining p_i is given by

$$p_i \sum_{j\neq i} q_{ij} = \sum_{j\neq i} p_j q_{ji}, \quad 1 \leqslant i \leqslant n, \qquad \sum_{i=1}^{n} p_i = 1. \tag{9}$$

Since the process $l(t)$ is ergodic, this system has a unique solution. Thus, one of the first n equations of this system may be omitted so that the remaining system will possess the same unique solution. Hence,

$$p_1 : p_2 : \cdots : p_n = \Delta_1^* : (-\Delta_2^*) : \cdots : (-1)^{n+1}\Delta_n^*,$$

where Δ_i^* is the determinant of the matrix obtained by deleting the ith column and the last row in the matrix of the system of n equations under consideration. Direct comparison of Δ_i^* and Δ_i shows that either for all i between 1 to n, $\Delta_i = (-1)^n \Delta_i^*$ or, for all these i, $\Delta_i = (-1)^{n+1}\Delta_i^*$. But then $\Delta_i = \omega p_i$, where ω is a proportionality factor different from zero.

Noting that

$$\sum_{i=1}^{n} E_i p_i = \bar{E}, \qquad \sum_{i=1}^{n} p_i = 1,$$

we obtain

$$\Delta'(0) = \omega(\bar{E} + \lambda\psi'(0)) = \omega\bar{E}(1 - \rho) \neq 0.$$

Hence for sufficiently small $s \neq 0$ the determinant does not vanish, q.e.d.

For a complete solution of the problem it is required to determine the constants a_i, $1 \leqslant i \leqslant n$. To simplify the computation we shall confine ourselves to the case $n = 2$. The calculations for an arbitrary n are analogous.

For $n = 2$, the determinant $\Delta(s)$ has the form

$$\Delta(s) = \begin{vmatrix} sE_1 - \lambda(1 - \psi(s)) - q_{12} & q_{21} \\ q_{12} & sE_2 - \lambda(1 - \psi(s)) - q_{21} \end{vmatrix}.$$

Since $\Phi_1(s)$ is analytic for $\operatorname{Re}\{s\} > 0$, in the expression

$$\Phi_1(s) = \frac{\begin{vmatrix} a_1 E_1 & q_{21} \\ a_2 E_2 & sE_2 - \lambda(1 - \psi(s)) - q_{21} \end{vmatrix}}{\begin{vmatrix} sE_1 - \lambda(1 - \psi(s)) - q_{12} & q_{21} \\ q_{12} & sE_2 - \lambda(1 - \psi(s)) - q_{21} \end{vmatrix}} \tag{10}$$

if the denominator vanishes, so does the numerator. Note that

$$\Delta(0) = 0, \quad \Delta'(0) = \frac{q_{12}}{p_2}\bar{E}(1 - \rho) < 0, \quad \Delta(+\infty) = +\infty.$$

Thus there exists a positive s_0 such that $\Delta(s_0) = 0$. Hence the condition

$$\begin{vmatrix} a_1 E_1 & q_{21} \\ a_2 E_2 & s_0 E_2 + \lambda[1 - \psi(s_0)] - q_{21} \end{vmatrix} = 0,$$

or

$$a_1 E_1\{s_0 E_2 - \lambda[1 - \psi(s_0)] - q_{21}\} = a_2 E_2 q_{21}$$

is obtained.

Thus we have a relation between a_1 and a_2. A second equation can be derived from probabilistic arguments. Indeed,

$$\Phi_1(0) + \Phi_2(0) = F_1(\infty) + F_2(\infty) = 1,$$

and if we write

$$\Phi_2(s) = \frac{\begin{vmatrix} sE_1 - \lambda[1 - \psi(s)] - q_{12} & a_1 E_1 \\ q_{12} & a_2 E_2 \end{vmatrix}}{\begin{vmatrix} sE_1 - \lambda[1 - \psi(s)] - q_{12} & q_{21} \\ q_{12} & sE_2 - \lambda[1 - \psi(s)] - q_{21} \end{vmatrix}} \tag{11}$$

and find the limits of expressions (10) and (11) as $s \to 0$ we then obtain

$$\Phi_1(0) = p_1[\bar{E}(1 - \rho)]a_1 E_1 + a_2 E_2,$$

$$\Phi_2(0) = p_2[\bar{E}(1 - \rho)]a_1 E_1 + a_2 E_2,$$

whence

$$a_1 E_1 + a_2 E_2 = \bar{E}(1 - \rho).$$

The constants a_1 and a_2 are determined from these two equations as follows:

$$a_1 = \frac{q_{21}\bar{E}(1 - \rho)}{E_1\{s_0 E_2 + \lambda[1 - \psi(s_0)]\}},$$

$$a_2 = \frac{q_{12}\bar{E}(1 - \rho)}{E_2\{s_0 E_1 + \lambda[1 - \psi(s_0)]\}}.$$

Substituting these expressions into formulas (10) and (11), we obtain the final expressions for $\Phi_1(s)$ and $\Phi_2(s)$.

4.6. Mixed Service Systems

4.6.1. *Mixed System with Constant Service Rate*

The reader is already familiar with the statement of the problem according to which a customer can join the queue only if the total number of customers in the queue is less than m; any customer arriving when the queue already contains m customers does not wait for service and is lost. The following more general model is also of interest: a customer may remain in the queue or be lost with a probability that depends in an arbitrary manner on the length of the queue.

We introduce the following notation.

b_k is the probability that the customer joins the queue if k customers (including the customer being served) are in the system when he arrives;
$\nu(t)$ is the number of customers in the system at time t;
$\xi(t)$ is the time from t until the customer that is being served at time t leaves the system;

$$\zeta(t) = \{\nu(t), \xi(t)\}, \qquad \varphi_0 = \lim_{t\to\infty} \mathsf{P}\{\nu(t) = 0\},$$

$$\varphi_k(x) = \lim_{t\to\infty} \mathsf{P}\{\nu(t) = k, \xi(t) < x\}, \qquad \Phi_k(s) = \int_0^\infty e^{-sx}\, d\varphi_k(x).$$

The symbols λ, $B(x)$, and $\psi(s)$ have the same meaning as in the models considered above.

4.6.2. *Condition for Ergodicity*

Assume that the limit

$$b = \lim_{k\to\infty} b_k \tag{1}$$

exists.

Of special interest is the case

$$\lambda b\tau < 1. \tag{2}$$

However, for ergodicity of the process, a more general condition

$$\lambda\tau \varlimsup_{k\to\infty} b_k < 1 \tag{3}$$

is sufficient. This case is completely settled by the following theorem.

Theorem. *If condition (3) is satisfied, the random process $\zeta(t)$ has an ergodic distribution, which is the unique bounded absolutely continuous solution of the system of differential equations*

$$\begin{gathered}\varphi_n'(x) - \lambda b_n\varphi_n(x) + \lambda b_{n-1}\varphi_{n-1}(x) = \varphi_n'(0) - \varphi_{n+1}'(0)B(x), \quad n \geqslant 2,\\ \varphi_1'(x) - \lambda b_1\varphi_1(x) = \varphi_1'(0) - \varphi_2'(0)B(x) - \lambda b_0\varphi_0 B(x), \quad \lambda b_0\varphi_0 = \varphi_1'(0)\end{gathered} \tag{4}$$

under the additional conditions

$$\varphi_n(0) = 0, \quad n \geqslant 1, \tag{5}$$

$$\varphi_0 + \sum_{n=1}^{\infty} \varphi_n(\infty) = 1. \tag{6}$$

For the proof we first establish the existence of the ergodic distribution of the embedded Markov chain ν_n—the number of customers in the system after the service of the nth customer was terminated.

Let $\nu_{n-1} = i \geqslant 1$. Then up until the time of completion of the service that started in state i there will be at least i customers in the system and hence the probability of arrival of a customer in the interval $(t, t + dt)$ is at most $\lambda\bar{b}_i\,dt$ where $\bar{b}_i = \sup(b_1, b_{i+1}, \ldots)$. Whence

$$\mathsf{M}\{\nu_n - \nu_{n-1} | \nu_{n-1} = i\} \leqslant \lambda\bar{b}_i \int_0^\infty \bar{B}(t)\,dt - 1, \quad i \geqslant 1.$$

Under the condition (3) there exists an N such that $\lambda\tau b_i \leqslant 1 - \varepsilon$, $i \geqslant N$, and hence $\lambda\tau\bar{b}_i \leqslant 1 - \varepsilon$. Thus

$$\mathsf{M}\{\nu_n - \nu_{n-1} | \nu_{n-1} = i\} \leqslant -\varepsilon, \quad i \geqslant N.$$

At the same time we clearly have for any i

$$\mathsf{M}\{\nu_n - \nu_{n-1} | \nu_{n-1} = i\} \leqslant \lambda\tau.$$

The Markov chain $\{\nu_n\}$ is irreducible and nonperiodic since for $i \geqslant 1$

$$\mathsf{P}\{\nu_n = i - 1 | \nu_{n-1} = i\} \geqslant \int_0^\infty e^{-\lambda x}\,dB(x) > 0,$$

$$\mathsf{P}\{\nu_n = i | \nu_{n-1} = i\} \geqslant \lambda b_i \int_0^\infty x e^{-\lambda x}\,dB(x) > 0.$$

Thus all the conditions of the theorem in Section 3.1.2 are fulfilled; hence, $\{v_n\}$ is an ergodic Markov chain.

Let T_i be the time from the instant of completion of service associated with the state $v_n = i$ up to completion of the next service. We have $T_i \leqslant 1/\lambda + \tau$ uniformly in i. Thus, the average length of the cycle of the regenerative process, which describes the behavior of the system, is finite. Now the assertion of the theorem follows from Smith's theorem.

Application of the Laplace transformation to (4), taking (5) into account yields:

$$(s - \lambda b_n)\Phi_n(s) + \lambda b_{n-1}\Phi_{n-1}(s) = \varphi_n'(0) - \varphi_{n+1}'(0)\psi(s), \quad n \geqslant 2, \tag{7}$$

$$(s - \lambda b_1)\Phi_1(s) = \varphi_1'(0) - \varphi_2'(0)\psi(s) - \lambda b_0 \varphi_0 \psi(s). \tag{8}$$

Set $s = \lambda b_n$, $n = 1, 2$. Then the formulas (7) and (8) become:

$$\varphi_n'(0) - \varphi_{n+1}'(0)\psi(\lambda b_n) = \lambda b_{n-1}\Phi_{n-1}(\lambda b_n), \quad n \geqslant 2,$$

$$\varphi_1'(0) - \varphi_2'(0)\psi(\lambda b_1) = \lambda b_0 \varphi_0 \psi(\lambda b_1).$$

Thus, knowing φ_n and Φ_{n-1} we can determine $\varphi_{n+1}(0)$. Substituting the resulting expression in the right-hand side of (7), we obtain $\Phi_n(s)$ for all s, $\operatorname{Re}\{s\} > 0$. Since

$$\varphi_1'(0) = \lambda b_0 \varphi_0$$

and, consequently,

$$\Phi_1(s) = (s - \lambda b_1)^{-1}\lambda b_0 \varphi_0 \left[1 - \psi(s) - \frac{1 - \psi(\lambda b_1)}{\psi(\lambda b_1)}\psi(s) \right],$$

we can determine all $\Phi_n(s)$, $n = 2, 3, \ldots$, successively. As usual, the constant φ_0 is determined from the normalizing condition (6).

Having determined $\Phi_n(s)$, it is easy to derive the distribution of the queue. Indeed

$$\lim_{t\to\infty} \mathsf{P}\{v(t) = n\} = \int_0^\infty d\varphi_n(x) = \Phi_n(0), \quad n \geqslant 1,$$

$$\lim_{t\to\infty} \mathsf{P}\{v(t) = 0\} = \varphi_0.$$

The probability that an arbitrary customer is lost can be determined from the total probability formula, since we know that the probability of a loss under the condition that n customers are in the system is $(1 - b_n)$ and the probability of the condition is $\Phi_n(0)$. Thus, the probability of a loss of an arbitrary customer is determined by the expression

$$1 - b_0\varphi_0 - \sum_{n=1}^{\infty} b_n \Phi_n(0).$$

The waiting time characteristics of an arbitrary customer are also of interest. If the customer arrives when the number of customers in the system is n and remains in the queue, his or her waiting time until the beginning of service consists of the time $\xi(t)$ until the completion of service that is already in progress and of the remaining $n-1$ customers' service time. Since these times are independent, the Laplace-Stieltjes transform of the conditional distribution of the waiting time of a customer that arrived when there were n customers in the system and joined the queue is

$$\frac{\Phi_n(s)}{\Phi_n(0)}\psi^{n-1}(s).$$

From here the Laplace-Stieltjes transform for the unconditional distribution of the waiting time can easily be evaluated.

4.6.3. *Mixed System with Variable Service Rate*

Ivnitskiĭ has solved a number of problems pertaining to single-server systems with streams of random intensity and variable service rate. We shall present one of his formulations (1965, 1966) and present his results.

Consider the following queueing system:

At an arbitrary time t the system is characterized by the number of customers present, denoted by $\nu(t)$.

If $\nu(t) = k$, an additional customer may arrive in a short time period h with probability $\lambda_k h + o(h)$. The amount of work used for servicing the customer depends on the number of customers who are in the system immediately after its service begins. If this number equals i, then this amount of work has the distribution function $B_i(x)$.

Furthermore, if $\nu(t) = k$, $k = 1, 2, \ldots$, the service rate is α_k. For example, if at time t a customer arrives whose service involves an amount of work x, the system is free, and no other customer arrives until his or her service is completed, then this service ends at time $t + x/\alpha_1$.

Our task is to obtain the stationary distribution of the queue length

$$p_k = \lim_{t\to\infty} \mathsf{P}\{\nu(t) = k\} \quad (k = 0, 1, 2, \ldots).$$

If $\nu(t) \geqslant 1$, and hence the server is servicing a customer at the time t, we shall denote the amount of work required to complete the service after time t by $\xi(t)$. Denote further by $\zeta(t)$ the compound random process

$$\zeta(t) = \begin{cases} 0 & \text{if } \nu(t) = 0, \\ (\nu(t), \xi(t)) & \text{if } \nu(t) > 0. \end{cases}$$

This process belongs to the class of piecewise-linear processes (see Chapter 3). Denote its stationary distribution by

$$F_0 = \lim_{t\to\infty} \mathsf{P}\{\nu(t) = 0\},$$

$$F_k(x) = \lim_{t\to\infty} \mathsf{P}\{\nu(t) = k, \xi(t) < x\}, \quad k \geqslant 1.$$

In addition let

$$p_k = \lim_{t\to\infty} \mathsf{P}\{v(t) = k\}, \quad k \geqslant 0.$$

In particular, $p_0 = F_0$. (Conditions for the existence of the limiting distributions will be presented below.)

Introduce the Laplace-Stieltjes transforms

$$\psi_i(s) = \int_0^\infty e^{-sx}\, dB_i(x), \tag{9}$$

$$\varphi_i(s) = \int_0^\infty e^{-sx}\, dF_i(x). \tag{10}$$

Finally, let

$$\tau_i = \int_0^\infty x\, dB_i(x) < \infty.$$

The following theorem is valid:

Theorem. *If the process $\zeta(t)$ has a stationary distribution, then p_i and $\varphi_i(s)$ satisfy the following recurrence relations ($i \geqslant 2$):*

$$p_i = \frac{\lambda_{i-1}\left[p_{i-1} - \varphi_{i-1}\left(\dfrac{\lambda_i}{\alpha_i}\right)\right]}{\lambda_i \psi_i\left(\dfrac{\lambda_i}{\alpha_i}\right)}, \tag{11}$$

$$\varphi_i(s) = \frac{1}{\alpha_i s - \lambda_i}[\lambda_{i-1} p_{i-1} - \lambda_i p_i \psi_i(s) - \lambda_{i-1}\varphi_{i-1}(s)], \tag{12}$$

where

$$p_1 = \frac{\lambda_0 p_0\left[1 - \psi_1\left(\dfrac{\lambda_1}{\alpha_1}\right)\right]}{\lambda_1 \psi_1\left(\dfrac{\lambda_1}{\alpha_1}\right)},$$

$$\varphi_1(s) = \frac{\lambda_0 p_0\left[\psi_1\left(\dfrac{\lambda_1}{\alpha_1}\right) - \psi_1(s)\right]}{(\alpha_1 s - \lambda_1)\psi_1\left(\dfrac{\lambda_1}{\alpha_1}\right)}. \tag{13}$$

PROOF. It can be proved that the functions $F_k(x)$ satisfy the system of differential equations

$$\lambda_0 F_0 = \alpha_1 F_1'(0), \tag{14}$$

$$\alpha_1 F_1'(x) - \lambda_1 F_1(x) - \alpha_1 F_1'(0) + \alpha_2 F_2'(0) B_1(x) + \lambda_0 F_0 B_1 = 0, \tag{15}$$

$$\alpha_i F_i'(x) - \lambda_i F_i(x) - \alpha_i F_i'(0) + \alpha_{i+1} F_{i+1}'(0) B_i(x) + \lambda_{i-1} F_{i-1}(x) = 0, \quad i \geqslant 2. \tag{16}$$

Letting $x \to \infty$ in (14)–(16) we obtain

$$\lambda_0 F_0 = \alpha_1 F_1'(0),$$
$$-\lambda_1 F_1(\infty) - \alpha_1 F_1'(0) + \alpha_2 F_2'(0) + \lambda_0 F_0 = 0,$$
$$-\lambda_i F_i(\infty) - \alpha_i F_i'(0) + \alpha_{i+1} F_{i+1}'(0) + \lambda_{i-1} F_{i-1}(\infty) = 0, \quad i \geqslant 2.$$

By adding the first equation to the second and the resulting equation to the third, etc., we obtain $\alpha_i F_i'(0) = \lambda_{i-1} F_{i-1}(\infty)$. Clearly, $F_i(\infty) = p_i$. Therefore,

$$\alpha_i F_i'(0) = \lambda_{i-1} p_{i-1}.$$

After applying the Laplace transform to (16) for $i \geqslant 2$, this equation becomes:

$$(\alpha_i s - \lambda_i)\varphi_i(s) = \alpha_i F_i'(0) - \alpha_{i+1} F_{i+1}'(0)\psi_i(s)$$
$$-\lambda_{i-1}\varphi_{i-1}(s) = \lambda_{i-1} p_{i-1} - \lambda_i p_i \psi_i(s) - \lambda_{i-1}\varphi_{i-1}(s). \quad (17)$$

On the left-hand side we have $\alpha_i s - \lambda$ multiplied by a function that is obviously analytic for $\operatorname{Re}\{s\} > 0$. Consequently, for $s = \lambda_i/\alpha_i$ the left-hand side vanishes. Equating the right-hand side of the equation to zero, we obtain

$$\lambda_{i-1} p_{i-1} - \lambda_i p_i \psi_i\left(\frac{\lambda_i}{\alpha_i}\right) - \lambda_{i-1}\varphi_{i-1}\left(\frac{\lambda_i}{\alpha_i}\right) = 0,$$

or

$$p_i = \frac{\lambda_{i-1}\left[p_{i-1} - \varphi_{i-1}\left(\frac{\lambda_i}{\alpha_i}\right)\right]}{\lambda_i \psi_i\left(\frac{\lambda_i}{\alpha_i}\right)}, \quad i \geqslant 2. \quad (18)$$

Substituting this value for p_i into (17) we obtain an equation for $\varphi_i(s)$:

$$\varphi_i(s) = \frac{1}{\alpha_i s - \lambda_i}[\lambda_{i-1} p_{i-1} - \lambda_i p_i \psi_i(s) - \lambda_{i-1}\varphi_{i-1}(s)]$$
$$= \frac{\lambda_{i-1}}{\alpha_i s - \lambda_i}\left[p_{i-1}\left(1 - \frac{\psi_i(s)}{\psi_i\left(\frac{\lambda_i}{\alpha_i}\right)}\right) + \frac{\psi_i(s)}{\psi_i\left(\frac{\lambda_i}{\alpha_i}\right)}\varphi_{i-1}\left(\frac{\lambda_i}{\alpha_i}\right) - \varphi_{i-1}(s)\right]. \quad (19)$$

For $i = 1$

$$(\alpha_1 s - \lambda_1)\varphi_1(s) = \alpha_1 F_1'(0) - \alpha_2 F_2'(0)\psi_1(s) - \lambda_0 F_0 \psi_1(s)$$
$$= \lambda_0 F_0(1 - \psi_1(s)) - \lambda_1 p_1 \psi_1(s). \quad (20)$$

Whence

$$p_1 = \frac{\lambda_0 F_0\left[1 - \psi_1\left(\frac{\lambda_1}{\alpha_1}\right)\right]}{\lambda_1 \psi_1\left(\frac{\lambda_1}{\alpha_1}\right)}, \quad (21)$$

$$\varphi_1(s) = \frac{\lambda_0 \varphi_0 \left[\psi_1\left(\frac{\lambda_1}{\alpha_1}\right) - \psi_1(s) \right]}{(\alpha_1 s - \lambda_1)\psi_1\left(\frac{\lambda_1}{\alpha_1}\right)}. \tag{22}$$

The theorem is proved. □

Thus, by using the recurrence relations (18) and (19), p_i and $\varphi_i(s)$ are successively determined. For example,

$$p_2 = \frac{\lambda_1 \lambda_0 p_0}{\lambda_2 \psi_2\left(\frac{\lambda_2}{a_2}\right)\psi_1\left(\frac{\lambda_1}{a_1}\right)} \left[\frac{1 - \psi_1\left(\frac{\lambda_1}{a_1}\right)}{\lambda_1} - \frac{\psi_1\left(\frac{\lambda_1}{a_1}\right) - \psi_1\left(\frac{\lambda_2}{a_2}\right)}{\frac{a_1 \lambda_2}{a_2} - \lambda_1} \right]. \tag{23}$$

All the p_i and $\varphi_i(s)$ contain the common factor $F_0 = p_0$. This constant is determined by the normalizing condition

$$\sum_{i=0}^{\infty} p_i = 1.$$

In particular, the Pollaczek-Khinchin formula follows easily from these formulas.

Theorem. *For stationary probabilities of states the equality*

$$p_0 \lambda_0 \tau_1 = \sum_{i=1}^{\infty} p_i(\alpha_i - \lambda_i \tau_i) \tag{24}$$

is valid.

Proof. By summing Eqs. (14), (15), and (16) over i we obtain the equation

$$\sum_{i=1}^{\infty} \alpha_i F_i'(x) = \sum_{i=2}^{\infty} \alpha_i F_i'(0) \bar{B}_{i-1}(x) + \alpha_1 F_1'(0). \tag{25}$$

Since both sides of (25) contain series of nonnegative functions, we may integrate both sides termwise from 0 to ∞. Since

$$\int_0^{\infty} F_i'(x)\,dx = F_i(\infty) = p_i,$$

$$\int_0^{\infty} \bar{B}_i(x)\,dx = \tau_i,$$

the result of integration is

$$\sum_{i=1}^{\infty} \alpha_i p_i = \sum_{i=2}^{\infty} \alpha_i F_i'(0) \tau_{i-1} + \alpha_1 F_1'(0) \tau_1.$$

Since, as we have seen,

$$\alpha_i F_i'(0) = \lambda_{i-1} p_{i-1}, \quad i \geqslant 1,$$

we have

$$\sum_{i=1}^{\infty} \alpha_i p_i = \sum_{i=2}^{\infty} \lambda_i p_i \tau_i + \lambda_0 p_0 \tau_1,$$

which is clearly equivalent to the required equation (24). □

We shall now tackle the problem of the existence of the ergodic distribution for the process $\zeta(t)$. (The answer is obviously positive if there exists n such that $\lambda_i = 0$ for $i \geqslant n$ and all $\alpha_i > 0$.)

Theorem. *Let $\alpha_i > 0$ for any i. Then the ergodic distribution exists if the constant p_0 defined by* (24) *is positive and $\sum \lambda_i^{-2} = \infty$.*

Proof. If the condition of the theorem is satisfied, the sequence of functions $\{F_i(x)\}$ determined by the above recurrence relations determines a non-singular distribution function. Direct substitution into the equation of the process shows that this distribution is stationary. Since by the condition of the theorem the states communicate with each other, the distribution is ergodic.

The same result follows from the theory of regenerative random processes. Indeed, the process $\zeta(t)$ is regenerative. If $p_0 > 0$, this means that if we take $\{F_i(x)\}$ as the initial distribution, the mathematical expectation of the interval between the regeneration times will be finite. Since the stationary distribution is characterized by the fact that $p_i > 0$ as long as $\lambda_0 \lambda_1 - \lambda_{i-1} > 0$, while by the condition of the theorem a transition from states $v = i$ such that $\lambda_0 \lambda_1 \cdots \lambda_{i-1} = 0$ into a state j for which $\lambda_0 \lambda_1 \cdots \lambda_{i-1} > 0$ occurs in a finite amount of time, it follows that the mathematical expectation of the inter-regeneration interval is finite for any initial distribution. Then by Smith's theorem (taking the continuity of the distribution of the given interval into account) we verify that the ergodic distribution exists.

Denote

$$\hat{p}_i = \frac{p_i}{p_0}.$$

$\hat{p}_i$ is determined by the recurrence relations (11)–(13), substituting 1 for p_0. Thus, as it follows from the above arguments, $p_0 > 0$.

The condition $p_0 > 0$ is valid if and only if the series $\sum \hat{p}_i$ converges. □

Theorem. *The fulfillment of the following conditions*

$$\begin{aligned} &\lambda_i \leqslant c\alpha_i, \quad i \geqslant N; \\ &c\tau_i \leqslant 1 - \varepsilon, \quad i \geqslant N; \\ &\alpha_i \geqslant c_0, \quad i \geqslant N \end{aligned} \tag{26}$$

is sufficient for the existence of an ergodic distribution of the process $\zeta(t)$.

PROOF. For an embedded Markov chain of the process $\nu(t)$ the first two conditions in (26) are the same as conditions (11) and (12) in Section 4.4, which are sufficient for the existence of an ergodic distribution. From an embedded Markov chain we proceed in the usual manner to the process $\zeta(t)$; here the second and third conditions of (26) are used. □

4.6.4. *Example*

To illustrate the application of the recurrence relations (11)–(12) consider the following queueing system, which has applications in reliability theory.

Two identical devices are subject to random failures. In the case of a failure they are sent to a worker who repairs them in the order of their arrival. The duration of continuous operation of each device is independent of the state of the others and is exponentially distributed with parameter λ. η units of work are required to repair one device. Denote

$$B(x) = \mathsf{P}\{\eta < x\}, \qquad \tau = \mathsf{M}\eta, \qquad \psi(s) = \int_0^\infty e^{-sx}\,dB(x).$$

When one device is malfunctioning, the repairman repairs it at the rate α_1; when both are out of order, the repair rate is α_2. Similar situations frequently occur in practice. Denote by p_i the probability that i devices are malfunctioning (under stationary conditions). Formulas (11) and (12) then yield (setting $\lambda_0 = 2\lambda, \lambda_1 = \lambda$)

$$p_1 = \frac{\lambda_0\left[1 - \psi\left(\dfrac{\lambda_1}{\alpha_1}\right)\right]}{\lambda_1\psi\left(\dfrac{\lambda_1}{\alpha_1}\right)}p_0,$$

$$p_2 = \frac{1}{\alpha_2}\left\{\lambda_0\tau - \frac{(\alpha_1 - \lambda_1\tau)\lambda_0\left[1 - \psi\left(\dfrac{\lambda_1}{\alpha_1}\right)\right]}{\lambda_1\psi\left(\dfrac{\lambda_1}{\alpha_1}\right)}\right\}p_0,$$

$$p_0 = \left\{1 + \frac{\lambda_0\left[1 - \psi\left(\dfrac{\lambda_1}{\alpha_1}\right)\right]}{\lambda_1\psi\left(\dfrac{\lambda_1}{\alpha_1}\right)} + \frac{\lambda_0\tau}{\alpha_2} - \frac{(\alpha_1 - \lambda_1\tau)\lambda_0\left[1 - \psi\left(\dfrac{\lambda_1}{\alpha_1}\right)\right]}{\alpha_2\lambda_1\psi\left(\dfrac{\lambda_1}{\alpha_1}\right)}\right\}^{-1}$$

4.6.5. *$M|G|1|m$ System*

For the preceding example we are not going to trace the structure of the general formulas in this subsection: the most important aspect is the connection of the characteristics of an $M|G|1|m$ system with the characteristics of an $M|G|1$ system having the same λ and $B(x)$.

Let $\{\pi_k^{(m)}\}$ be the distribution of an embedded Markov chain for a limited queue and $\{\pi_k\}$ be the same for a system with an unlimited queue (provided the latter distribution exists).

We have

$$\pi_k^{(m)} = \pi_0^{(m)} f_k + \sum_{j=1}^{k+1} \pi_j^{(m)} f_{k-j+1}, \quad 0 \leqslant k \leqslant m-1. \tag{27}$$

The system (27) is of a triangular form and hence allows us to express uniquely $\pi_k^{(m)}$, $1 \leqslant k \leqslant m$, in terms of $\pi_0^{(m)}$. Since the coefficients of the equation do not depend on m, we have

$$\pi_k^{(m)} = \pi_0^{(m)} \hat{\pi}_k, \quad 0 \leqslant k \leqslant m. \tag{28}$$

As we have seen previously, π_k satisfy the same system of equations. Thus,

$$\pi_k^{(m)} = \frac{\pi_0^{(m)}}{\pi_0} \hat{\pi}_k, \quad 0 \leqslant k \leqslant m. \tag{29}$$

We have obtained an interesting result: the distribution of an embedded Markov chain for an $M|G|1|m$ system is proportional on the corresponding interval to the corresponding distribution for the system $M|G|1$. The latter, however, is determined by the generating function

$$\pi(z) = \frac{(1-\rho)(1-z)\psi(\lambda(1-z))}{\psi(\lambda(1-z)) - z} = (1-\rho)\hat{\pi}(z). \tag{30}$$

We have

$$\hat{\pi}(z) = \sum_{k=0}^{\infty} \hat{\pi}_k z^k,$$

whence

$$\hat{\pi}_k = \frac{1}{2\pi i} \oint_{|\zeta|=R} (\pi(\zeta)/\zeta^{k+1})\, d\zeta.$$

Thus

$$\sum_{k=0}^{m} \hat{\pi}_k z^k = \frac{1}{2\pi i} \oint_{|\zeta|=R} \frac{\hat{\pi}(\zeta)}{\zeta} \frac{1-(z/\zeta)^{m+1}}{1-z/\zeta}\, d\zeta, \quad |z| < R.$$

Denoting the right-hand side of this formula by $\hat{\pi}^{(m)}(z)$ we obtain

$$\sum_{k=0}^{m} \pi_k^{(m)} z^k = \hat{\pi}^{(m)}(z)/\hat{\pi}^{(m)}(1), \tag{31}$$

since obviously $\sum_{k=0}^{m} \pi_k^{(m)} = 1$.

Formula (31) is meaningful for $\rho > 1$ also, although in this case (30) is not a generating function of a probability distribution.

Stationary probabilities of states $p_k^{(m)} = \lim_{t\to\infty} \mathsf{P}\{\nu(t) = k\}$, where $\nu(t)$—the number of customers in the system at time t—is easily seen to be determined by the relations

$$p_0^{(m)} = \mu\pi_0^{(m)} \int_0^\infty e^{-\lambda t}\,dt = \mu\pi_0^{(m)}/\lambda; \tag{32}$$

$$p_k^{(m)} = \mu\pi_0^{(m)} \int_0^\infty \frac{(\lambda t)^{k-1}}{(k-1)!} e^{-\lambda t}\bar{B}(t)\,dt$$

$$+ \mu \sum_{j=1}^{k} \pi_j^{(m)} \int_0^\infty \frac{(\lambda t)^{k-j}}{(k-j)!} e^{-\lambda t}\bar{B}(t)\,dt, \quad 1 \leqslant k \leqslant m; \tag{33}$$

$$p_{m+1}^{(m)} = \mu\pi_0^{(m)} \int_0^\infty \left(1 - \sum_{i=0}^{m-1} \frac{(\lambda t)^i}{i!} e^{-\lambda t}\right) \bar{B}(t)\,dt$$

$$+ \mu \sum_{j=1}^{m} \pi_j^{(m)} \int_0^\infty \left(1 - \sum_{i=0}^{m-j} \frac{(\lambda t)^i}{i!} e^{-\lambda t}\right) \bar{B}(t)\,dt. \tag{34}$$

Here μ is the intensity of the outgoing stream, which is equal to $\lambda(1 - p_{m+1}^{(m)})$. At the same time for $\rho < 1$ $\pi_j^{(m)} = \pi_j/(\pi_0 + \cdots + \pi_m)$, $0 \leqslant j \leqslant m$. Finally,

$$\lambda\pi_0 \int_0^\infty \left(1 - \sum_{i=0}^{m-1} \frac{(\lambda t)^i}{i!} e^{-\lambda t}\right) \bar{B}(t)\,dt + \lambda \sum_{j=1}^{m} \pi_j \left(1 - \sum_{i=0}^{m-j} \frac{(\lambda t)^i}{i!} e^{-\lambda t}\right) \bar{B}(t)\,dt$$

$$= \lambda(\pi_0 + \cdots + \pi_m)\tau - (p_1 + \cdots + p_m).$$

Substituting the last relations into (34) yields

$$x = (1 - x)\left(\rho - \frac{p_1 + \cdots + p_m}{\pi_0 + \cdots + \pi_m}\right),$$

where $x = p_{m+1}^{(m)}$.

From the mathematical law for a stationary queue we have $p_j = \pi_j$. Finally we obtain

$$x = (1 - x)(\rho - 1 + p_0/(p_0 + \cdots + p_m)).$$

The root of this equation is positive since $p_0 = 1 - \rho$ and hence $\rho - 1 + p_0/(p_0 + \cdots + p_m) = (p_{m+1} + p_{m+2} + \cdots) \times (p_0 + \cdots + p_m)^{-1}$. Whence

$$p_{m+1}^{(m)} = x = \frac{p_0 - (1 - \rho)(p_0 + \cdots + p_m)}{p_0 + \rho(p_0 + \cdots + p_m)}.$$

We now obtain from (32) and (33) that

$$p_0^{(m)} = \frac{(1 - x)(1 - \rho)}{p_0 + \cdots + p_m};$$

$$p_k^{(m)} = \frac{(1 - x)p_k}{p_0 + \cdots + p_m}, \quad 1 \leqslant k \leqslant m.$$

For $\rho \geqslant 1$, formulas (32)–(34) also uniquely determine $p_k^{(m)}$, although in this case a stationary distribution $\{p_j\}$ for the system $M|G|1$ does not exist.

4.7. Systems with Restrictions

4.7.1. *Various Forms of Restrictions*

In Sections 1.8 and 1.9 we considered two models of service with restrictions, a system with limited waiting time until the beginning of servicing and a system characterized by limited holding time (waiting time plus service time). A direct generalization of each of these models is obtained if one assumes that the waiting time (or the holding time) is limited not necessarily by a constant τ but by a random variable with a distribution function $A(x)$.

We may then consider the following statement of the problem. Assume that the server possesses a certain effective range and can serve the customers only when they are within this range. Customers move through this range at a constant, say unit, speed. When service begins, the speed of the customer becomes α. It is easy to see that $\alpha = 0$ corresponds to service with limited waiting time (the customer "pauses" until the service is completed) and $\alpha = 1$ corresponds to limited holding time.

4.7.2. *Formulation of Restrictions*

When customers are serviced in the order of their arrival, it is possible to describe in a compact manner numerous possible forms of restrictions introducing the process $\gamma(t)$, which denotes the time from the instant t up to the time when the system is disengaged from all the customers arriving up to time t (in the case when the customers have departed from the system before time t we set $\gamma(t) = 0$). Let the customers form a simplest stream with parameter λ. Denote by η_y the value of the jump at the time of arrival of a customer which found the process in state y; denote $B_y(x) = \mathsf{P}\{\eta_y < x\}$ assuming that $B_y(x)$ is a distribution function for any y and measurable in y for any x. Thus, we assume that the virtual waiting time is described by a homogeneous Markov process $\gamma(t)$ whose transitions during time dt are described by the stochastic differential equation

$$d\gamma(t) = -\operatorname{sign}\gamma(t)\,dt + \eta_{\gamma(t),t}\,dN(t),$$

where $\eta_{y,t}$ are independent for different (y, t) random variables with the distribution $B_y(x)$, $N(x)$ is a Poisson process with parameter λ.

We shall interpret the function $\bar{B}_y(x) = 1 - B_y(x)$ for different restriction models. In all cases it is assumed that the time necessary for servicing a customer is a random variables with the distribution function $B(x)$.

1. The waiting time for service is bounded by a random variable ω with the distribution function $H(x)$. We have

$$\bar{G}_y(x) = \mathsf{P}\{\eta \geqslant x, \omega \geqslant y\} = \bar{B}(x)\bar{H}(y).$$

2. The sojourn time of a customer in the system is bounded by a random variable ω with the distribution function $H(x)$. In this case

$$\bar{G}_y(x) = \mathsf{P}\{\eta \geqslant x, \omega \geqslant x + y\} = \bar{B}(x)\bar{H}(x + y).$$

3. The variable $\omega + \alpha\omega'$, where ω is the waiting time of a customer, ω' is the service time bounded by a random variable ω with the distribution function $H(x)$. We have

$$\bar{G}_y(x) = \mathsf{P}\{\eta \geqslant x, \omega \geqslant y + \alpha x\} = \bar{B}(x)\bar{H}(y + \alpha x).$$

4.7.3. *Existence of the Ergodic Distribution*

Denote

$$F(t, x) = \mathsf{P}\{\xi(t) < x\}, \qquad F(x) = \lim_{t \to \infty} F(t, x).$$

Theorem. *If the total time of sojourn of a customer in a system is bounded by a constant T, then the process $\xi(t)$ has an ergodic distribution.*

Proof. The proof is almost trivial. Indeed the condition implies

$$\mathsf{P}\{\xi(t) \leqslant T\} = 1.$$

Hence, if no customer arrives between t and $t + T$, we must have $\gamma(t + T) = 0$. Since the probability that no new customers arrive in the interval $(t, t + T)$ is $e^{-\lambda T}$, we have the bound

$$\mathsf{P}\{\gamma(t + T) = 0 | \gamma(t) = x\} \geqslant e^{-\lambda T}, \quad 0 \leqslant x \leqslant T.$$

This means that the mathematical expectation of the number of renewals of the regenerative process $\gamma(t)$ in unit time is at least $\lambda e^{-\lambda T} > 0$. This is possible only if the mathematical expectation of the interrenewal interval of the process is finite. It is also clear that this interval possesses a density. Hence, by Smith's theorem, the ergodic distribution exists. □

We shall now present, in a modified form, a more refined criterion for the existence of the ergodic distribution, due to Afanas' eva (1965).

Theorem. *If for $y \geqslant y_0$*

$$B_y(x) \geqslant M(x),$$

while for $y < y_0$ and some $c > 0$

$$B_y(x) \geqslant M(cx),$$

where $M(x)$ is a distribution function satisfying

$$\lambda \int_0^\infty [1 - M(x)]\, dx < 1,$$

then the process $\gamma(t)$ has an ergodic distribution.

Proof. Denote by $\tau_y(y \geqslant y_0)$ the mathematical expectation of the time elapsing from the instant at which $\gamma(t) = y$ until the instant when the process assumes the value y_0 for the first time. It follows from the inequality

$$F_y(x) \geqslant M(x)$$

that

$$\tau_y \leqslant T_y,$$

where T_y is a variable defined analogously to τ_y but for a single-server system *with waiting*, with a simple stream with parameter λ and service time distribution $M(x)$.

We shall derive an equation for T_y. If $\gamma(t) = y$, then with probability $(1 - \lambda h) + o(h)$ we have $\gamma(t + h) = y - h$ and $\gamma(t + h) = y + \eta + o(h)$ is valid with probability $\lambda h + o(h)$. Here η is a random variable with distribution function $M(x)$. This argument leads to the relation

$$T_y = h + (1 - \lambda h) T_{y-h} + \lambda h \int_0^\infty T_{y+x}\, dM(x) + o(h),$$

or, equivalently, to the equation

$$\frac{1}{h}[T_y - T_{y-h}] + \lambda T_{y-h} = \lambda \int_0^\infty T_{y+x}\, dM(x) + o(h).$$

Approaching h to zero, we obtain

$$\frac{dT_h}{dy} + \lambda T_y = \lambda \int_0^\infty T_{y+x}\, dM(x),$$

which is an integro-differential equation of the convolution type on a half-line (cf. Krein (1961)). This equation means that either $T_y = \infty$ for every y, or it has a unique continuous solution such that $T_0 = 0$. The first case is impossible, since $\int_0^\infty T_x\, dM(x)$ is the mathematical expectation of the busy period of the server, equal to

$$\int_0^\infty \overline{M}(x)\, dx \bigg/ \left(1 - \lambda \int_0^\infty \overline{M}(x)\, dx\right) < \infty.$$

By direct substitution one verifies that the solution of the integro-differential equation is the function

$$T_y = \frac{y}{1 - \rho}, \quad y \geqslant 0,$$

where

$$\rho = \lambda \int_0^\infty \overline{M}(x)\, dx.$$

We shall now investigate the mathematical expectation of the length of the interval for which $\xi(t) > y_0$. We have

$$\tau = \int_{y_0}^\infty \tau_x\, dR(x),$$

where $R(y)$ is the distribution function of the random variable $\gamma(t + 0)$ under the conditions that $\gamma(t - 0) \leqslant y_0$ and $\gamma(t + 0) > y_0$.

From the condition

$$B_y(x) \geqslant M(cx)$$

it follows that

$$1 - R(x) \leqslant \sup_{y \leqslant y_0} [1 - B_y(x - y_0)] \leqslant 1 - M(c(x - y_0)),$$

so that

$$\int_{y_0}^{\infty} x \, dR(x) < \infty.$$

Now the bound obtained implies that

$$\tau = \int_{y_0}^{\infty} \tau_x \, dR(x) \leqslant \int_{y_0}^{\infty} T_x \, dR(x) = \frac{1}{1-\rho} \int_{y_0}^{\infty} x \, dR(x).$$

Thus, the mathematical expectation of the holding time of the process $\gamma(t)$ above the level y_0 is finite.

Let τ_0 be the mathematical expectation of the length of the interval for which $\gamma(t) \leqslant y_0$. If during time t after the beginning of this interval no customers arrive, this interval is not completed. Hence,

$$\tau_0 \geqslant \int_0^{\infty} e^{-\lambda t} \, dt = 1/\lambda.$$

We have

$$\lim_{t \to \infty} \mathsf{P}\{\gamma(t) \leqslant y_0\} = \frac{\tau_0}{\tau + \tau_0} = \beta > 0.$$

At the same time the probability of beginning of the busy interval during the time from t to $t + dt$ is not less than the probability of the event $\{\gamma(t - y_0) \leqslant y_0$; during the time interval from $t - y_0$ to t no customers arrived; in the interval $(t, t + dt)$ one customer arrived$\}$. Thus, the intensity of the stream of busy intervals is

$$\mu \geqslant \beta e^{-\lambda y_0} \lambda > 0,$$

and hence the average value of the busy interval is finite. Since the process $\gamma(t)$ is regenerative (regeneration times are the instants of beginning of busy intervals) by the Smith theorem, the process $\gamma(t)$ possesses an ergodic distribution. □

4.7.4. *Equation for the Stationary Distribution*

We shall assume that the process $\gamma(t)$ has a stationary distribution (this is, of course, possible even if the condition of the theorem in the preceding subsection is not satisfied). Assume that at some instant t the distribution of $\gamma(t)$ coincides with this stationary distribution. Expressing the probability of the event $\{\gamma(t+h) < x\}$ in terms of the distribution of $\gamma(t)$ and the probabilities of the events that may occur in the interval $(t, t+h)$, we obtain

$$\bar{F}(x) = (1 - \lambda h)\bar{F}(x+h) + \lambda h \int_0^{x+h} \bar{B}_y(x - y + \theta h)\, dF(y) + o(h), \tag{1}$$

where $\theta = \theta(y, x)$, $0 \leqslant \theta \leqslant 1$.

The first term on the right-hand side corresponds to no customer arriving between t and $t + h$. The integral corresponds to the case when exactly one customer arrives in this time interval. The last term is $o(h)$ due to the orderliness of a simple stream.

Theorem. *If the stationary distribution exists, it is determined by a constant A and a function $p(x)$:*

$$F(x) = A + \int_0^x p(t)\, dt, \quad x > 0,$$

where A and $p(x)$ satisfy the integral equation

$$p(x) - \lambda \int_0^x \bar{B}_y(x - y) p(y)\, dy = \lambda A \bar{B}_0(x) \tag{2}$$

valid for almost all $x > 0$.

PROOF. The integrand in (1) does not exceed 1, therefore

$$|\bar{F}(x) - (1 - \lambda h)\bar{F}(x+h)| \leqslant \lambda h + o(h),$$

or

$$\frac{F(x+h) - F(x)}{h} \leqslant 2\lambda + o(h),$$

implying the absolute continuity of the function $F(x)$ for $x > 0$. Thus, there exist an A and $p(x) \leqslant 2\lambda$ such that for $x > 0$

$$F(x) = A + \int_0^x p(t)\, dt.$$

Equation (2) then follows in the limit from (1) by approaching $h \to 0$. □

We note that for fixed A Eq. (2) is a Volterra equation with a bounded kernel; it is well known that such an equation can have only one integrable solution in any interval $(0, T)$. Since the probabilistic solution required for our problem is obviously integrable, this solution is determined up to a constant factor A. The latter is determined by the normalizing condition

$$A + \int_0^\infty p(t)\, dt = 1.$$

4.7.5. *Embedded Markov Chain*

Let t_n be the time of arrival of the nth customer, $\gamma_n = \gamma(t_n - 0)$. Then $\{\gamma_n\}$ is a homogeneous Markov chain.

We shall now prove an interesting property.

Theorem. *Let $F(x)$ be a distribution function, $F(0) = 0$, $F(+0) = A$, $F(x) = A + \int_0^x p(t)\,dt$, $x > 0$, where A and $p(x)$ satisfy the integral equation (2) for $x > 0$. Then $F(x)$ is a stationary distribution of a Markov chain $\{\gamma_n\}$.*

Proof. It can be assumed without loss of generality that $\lambda = 1$. Let γ_1 have the distribution $F(x)$. It is sufficient to prove that γ_2 has the same distribution. Denote $dG(x) = \mathsf{P}\{x < \gamma_2 < x + dx\}$. The following stochastic relation is valid:

$$\gamma_2 = (\gamma_1 + \eta_{\gamma_1} - \xi)^+.$$

Here η_{γ_1} is a random variable with the distribution function $B_y(x)$ for $y = \gamma_1$, ξ is an independent from $(\gamma_1, \eta_{\gamma_1})$ random variable with the density e^{-t}, $t > 0$. For $x > 0$

$$\begin{aligned} dG(x) &= \int_0^\infty e^{-t}\,dt\,d\mathsf{P}\{\gamma_1 + \eta_{\gamma_1} < x + t\} \\ &= \int_0^\infty e^{-t}\,dt\left\{A\,dB_0(x + t) + \int_0^{x+t} p(y)\,dy\,dB_y(x + t - y)\right\}. \end{aligned}$$

Whereas it follows from Eq. (2) that

$$\int_0^z p(y)\,dy\,dB_y(z - t) = p(z)\,dz - dp(z) - A\,dB_0(z).$$

Substituting this identity into the preceding relation we obtain

$$dG(x) = \int_0^\infty \{p(x + t)\,dx - dp(x + t)\}e^{-t}\,dt = p(x)\,dx$$

(here integration by parts was used). Since obviously $G(x)$ is a distribution function, we have $G(+0) = 1 - \int_0^\infty p(x)\,dx$. Thus, $G(x) = F(x)$. □

In terms of the distribution $F(x)$ one can express various service characteristics: distribution function of the waiting time of a customer in the system before the service begins; distribution function of the holding time of a customer in the system; distribution function of the "degree of service," i.e., the ratio of the actual service time to the time required for complete service; the probability that a customer is completely served if the time required for complete service is x; the probability of complete loss (a customer leaves the system before the service begins); the probability of partial loss (a customer is being served but departs before the service is completed). All these characteristics are obtained in the same manner. Let $\alpha(y)$ be the "individual" characteristic of a customer under the condition that at the instant of his or her arrival $\gamma(t) = y$. Then the average characteristic is

$$\bar{\alpha} = \int \alpha(y)\, dF(y).$$

4.8. Priority Service

4.8.1. *Assumptions and Notation*

In Section 1.7 we explained the practical importance of queueing systems characterized by priority service to one type of customers before others. We have also indicated the main statements of the problems of priority service. In this section we shall deal with more general formulations, along the lines of this chapter whose purpose is to extend the analytic results of Chapter 1 to cases in which the service time has an arbitrary distribution.

We shall consider three different statements of priority service problems.

1. When a customer of the first type arrives, service of a customer of the second type is interrupted. When all the available customers of the first type have been served, the server resumes the interrupted service, and the remaining service time of the customer of the second type is decreased by the amount of time spent on his or her service before the arrival of the customer of the first type.
2. As above, except that upon resumption of service to the customer of the second type the amount of time previously spent on his or her service is not taken into account; the service starts "from scratch."
3. When a customer of the first type arrives, service to a customer of the second type is completely discontinued and the customer is lost.

We shall also assume that in all three models of priority service the arriving customers of both types constitute independent simple streams with parameters λ_1 and λ_2, respectively. The service time for a customer of the ith type $(i = 1, 2)$ is a random variable with distribution function $B_i(x)$ and Laplace–Stieltjes transform

$$\psi_i(s) = \int_0^\infty e^{-sx}\, dB_i(x).$$

Denote by τ_i the mathematical expectation of the service time of a customer of the ith type; we shall assume that τ_1 and τ_2 are finite.

Let $\gamma_i(t)$ be the waiting time for a customer of the ith type who arrives at time t, and $\gamma_i^*(t)$ be the same customer's holding time in the system, i.e., the time from t until the customer leaves the system.

We introduce the following functions:

$$F_i(x) = \lim_{t\to\infty} \mathsf{P}\{\gamma_i(t) < x\}, \quad i = 1, 2;$$

$$F_i^*(x) = \lim_{t\to\infty} \mathsf{P}\{\gamma_i^*(t) < x\}, \quad i = 1, 2.$$

These functions will be determined in terms of their Laplace–Stieltjes transforms $\varphi_i(s)$, $\varphi_i^*(s)$. In the third model another quantity is also of interest, namely the probability that an arbitrary customer of the second type is lost.

We shall not indicate to which of the models—1, 2, or 3—our notation refers; the following presentation will preclude any possibility of confusion.

4.8.2. *Service of Customers of the First Type*

Customers of the first type are served completely independently of the customers of the second type; thus, the random process $\gamma_i(t)$ is of the same nature as the process $\gamma(t)$ studied extensively in the beginning of this chapter. By Khinchin's formula (23) (Section 4.2.6.)

$$\varphi_1(s) = (1 - \lambda_1\tau_1)\bigg/\left(1 - \lambda_1\frac{1 - \psi_1(s)}{s}\right),$$

provided only

$$\lambda_1\tau_1 < 1.$$

The holding time (waiting time until the completion of service) in the system is the sum of the waiting and the service times; since these two variables are independent, the Laplace-Stieltjes transform of the sum is the product of transforms of the summands, i.e.,

$$\varphi_1^*(s) = \frac{(1 - \lambda_1\tau_1)\psi_1(s)}{1 - \dfrac{\lambda_1}{s}(1 - \psi_1(s))}.$$

For customers of the second type, the corresponding characteristics are more complicated; we shall handle these problems in the following subsections.

4.8.3. *The Method of Investigation*

Essentially we already have the mathematical tools required for investigating the characteristics of service to customers of the second type, namely, the theory of servicing systems with unreliable servers developed in detail in Section 4.5. Indeed, servicing customers of the first type is equivalent to a failure of the server. Thus, instead of considering service to customers of two types we can consider service to customers of the second type only, interpreting service to customers of the first type as failure of a server.

Let β be the time from the beginning of servicing a customer of the second type until the time when the server is available for servicing the next customer of the second type,

$$\psi_\beta(s) = \int_0^\infty e^{-sx}\, d\mathsf{P}\{\beta < x\}.$$

Under the condition

$$\lambda_2 \mathsf{M}\beta < 1$$

we have

$$\varphi_2(s) = \frac{(1 - \lambda_1\tau_1)(1 - \lambda_2\mathsf{M}\beta)\{s + \lambda_1[1 - P_0(s)]\}}{s - \lambda_2[1 - \psi_\beta(s)]}. \tag{1}$$

Clearly the mathematical expectation of the period in which the server remains "unserviceable" provided that the failure occurs when no customers (of the second type) are in the system is

$$\tau_3 = \frac{\tau_1}{1 - \lambda_1\tau_1}.$$

Since this random period coincides with the period in which the server is busy with customers of the first type, it follows from the same considerations that $P_0(s)$ is equal to the Laplace-Stieltjes transform of the period during which the server is occupied with customers of the first type. $P_0(s)$ is defined as the unique analytic solution, real for positive s, of the functional equation

$$P_0(s) = \psi_1[s + \lambda_1 - \lambda_1 P_0(s)]. \tag{2}$$

4.8.4. *Determination of the Function* $\psi_\beta(s)$

As a result of the calculations presented previously the function $\psi_\beta(s)$ is now the only unknown function in the formula (1) for $\varphi_2(s)$. We shall deal separately with each one of the above servicing models.

Model 1. If we interpret the beginning of the period in which the server is busy with customers of the first type as a failure of the server and the end of this period as a renewal, then this model corresponds exactly to the model with failures discussed in Section 4.5.3. Using the formula derived therein we obtain

$$\psi_\beta(s) = \psi_2(s + \lambda_1 - \lambda_1 P_0(s)),$$

where $P_0(s)$ as above is determined by formula (2). In particular, if $\psi_1(s) = \psi_2(s)$, we obtain from the same formula

$$\psi_\beta(s) = P_0(s).$$

Model 2. Here the period β whose distribution we are seeking may be represented as follows. Let $\{\gamma_n\}$, $\{\zeta_n\}$, $\{\xi_n\}$ be three independent sequences of independent random variables where the Laplace-Stieltjes transforms of the distributions of all the variables γ_i and ζ_i are $\psi_2(s)$ and $P_0(s)$, respectively, and all the ξ_i are exponentially distributed with parameter λ_1.

If $\gamma_1 \leqslant \xi_1$, then $\beta = \gamma_1$ (while a customer of the second type is being served, no customers of the first type arrive; in this case the customer of the second type leaves the system after a period exactly equal to the duration of service).

If $\xi_1 < \gamma_1, \gamma_2 \leqslant \xi_2$, then $\beta = \zeta_1 + \xi_1 + \gamma_2$ (while a customer of the second type is being served, a customer of the first type arrives during a period ξ_1 from the beginning of the service; a period ζ_1 is then devoted to servicing the customers of the first type and servicing of customers of the second type is then resumed and is completed before the arrival of any customer of the first type). Analogously if $\xi_1 < \gamma_1, \xi_2 < \gamma_2, \ldots, \xi_n < \gamma_n, \gamma_{n+1} \leqslant \xi_{n+1}$, then $\beta = \xi_1 + \zeta_1 + \xi_2 + \zeta_2 + \cdots + \xi_n + \zeta_n + \gamma_{n+1}$. Thus, by the total probability formula we have

$$\begin{aligned} \mathsf{P}\{\beta < x\} = \sum_{n=0}^{\infty} \mathsf{P}\{&\xi_i < \gamma_i, 1 \leqslant i \leqslant n; \gamma_{n+1} \leqslant \xi_{n+1}; \\ &\xi_1 + \zeta_1 + \xi_2 + \zeta_2 + \cdots + \xi_n + \zeta_n + \gamma_{n+1} < x\}. \end{aligned} \tag{3}$$

Noting that

$$\begin{aligned} \int_0^\infty e^{-sx}\, d_x \mathsf{P}\{\xi_i < x, \xi_i < \gamma_i\} &= \lambda_1 \int_0^\infty e^{-sx}(1 - B_2(x)) e^{-\lambda_1 x}\, dx \\ &= \frac{\lambda_1}{s + \lambda_1}[1 - \psi_2(s + \lambda)], \end{aligned}$$

$$\int_0^\infty e^{-sx}\, d_x \mathsf{P}\{\gamma_i < x, \gamma_i \leqslant \xi_i\} = \int_0^\infty e^{-(s+\lambda_1)x}\, dB_2(x) = \psi_2(s + \lambda_1),$$

formula (3) after an application of the Laplace-Stieltjes transform becomes:

$$\begin{aligned} \psi_\beta(s) &= \sum_{n=0}^{\infty} \left\{\frac{\lambda_1}{s + \lambda_1}[1 - \psi_2(s + \lambda_1)] P_0(s)\right\}^n \psi_2(s + \lambda_1) \\ &= \frac{(s + \lambda_1)\psi_2(s + \lambda_1)}{s + \lambda_1 - \lambda_1[1 - \psi_2(s + \lambda_1)] P_0(s)}. \end{aligned}$$

Model 3. Utilizing the random variables γ_i, ζ_i and ξ_i introduced previously, we have

$$\beta = \begin{cases} \gamma_1, & \text{if } \gamma_1 \leqslant \xi_1, \\ \xi_1 + \zeta_1, & \text{if } \gamma_1 > \xi_1. \end{cases}$$

It is then easy to see that

$$\psi_\beta(s) = \psi_2(s + \lambda_1) + \frac{\lambda_1}{s + \lambda_1}[1 - \psi_2(s + \lambda_1)] P_0(s).$$

4.8.5. *Determination of the Function $\Phi_2^*(s)$*

For models 1 and 2 the time from t until the completion of service to a customer of the second type is equal to $\gamma_2(t) + \beta$. Since these random variables are independent, we have in both cases the relation

$$\Phi_2^*(s) = \Phi_2(s)\psi_\beta(s).$$

In model 3 customers of the second type may be lost. Clearly, a customer will be lost if and only if at least one customer of the first type arrives during his or her service; whence the probability of loss is

$$\mu = \mathsf{P}\{\xi_1 < \gamma_1\} = 1 - \psi_2(\lambda_2).$$

It is also clear that for this model

$$\gamma_2^*(t) = \gamma_2(t) + \min\{\xi_1, \gamma_1\}.$$

Since

$$\int_0^\infty e^{-sx}\, d_x \mathsf{P}\{\min\{\xi_1,\gamma_1\} = x\} = \psi_2(s+\lambda_1) + \frac{\lambda_1}{s+\lambda_1}[1 - \psi_2(s+\lambda_1)],$$

it follows that

$$\Phi_2^*(s) = \Phi_2(s)\left\{\psi_2(s+\lambda_1) + \frac{\lambda_1}{s+\lambda_1}[1 - \psi_2(s+\lambda_1)]\right\}.$$

4.8.6. *The Ergodicity Condition*

All the formulas derived in the preceding three subsections are valid only under the assumption that the process has an ergodic distribution. In view of (1) the existence of such a distribution is equivalent to the validity of the inequalities

$$\lambda_1\tau_1 < 1, \qquad \lambda_2 M\beta < 1.$$

Using the expressions obtained for the Laplace-Stieltjes transform of the distribution of the random variable β, we obtain for each of the three models:

for model 1

$$\mathsf{M}\beta = \tau_2/(1 - \lambda_1\tau_1),$$

the ergodicity condition is $\lambda_2\tau_2 < 1 - \lambda_1\tau_1$;

for model 2

$$\mathsf{M}\beta = [1 - \psi_2(\lambda_1)]/\lambda_1(1 - \lambda_1\tau_1)\psi_2(\lambda_1),$$

the ergodicity condition is $\lambda_2[1 - \psi_2(\lambda_1)] < \lambda_1(1 - \lambda_1\tau_1)\psi_2(\lambda_1)$;

for model 3

$$\mathsf{M}\beta = [1 - \psi_2(\lambda_1)]/\lambda_1(1 - \lambda_1\tau_1),$$

the ergodicity condition is $\lambda_2[1 - \psi_2(\lambda_2)] < \lambda_1(1 - \lambda_1\tau_1)$.

4.9. The Generalized Scheme of Priority Service with a Limited Queue

4.9.1. *Statement of the Problem*

Consider a queueing system that at any time t is characterized by a state $v(t)$ from a finite set X where the element 0 is singled out. If $v(t) = 0$, then there is no customer in the system in this state; for $v(t) \neq 0$ there are customers and one of them is serviced. A server is available that services a customer with the rate α_v in the state $v(t) = v \neq 0$: if the remaining amount of work necessary for servicing this customer equals $\xi(t)$, then $\xi'(t) = \alpha_{v(t)}$. At the beginning of the service at time t if $v(t + 0) = v$ the amount of work for servicing this customer is defined as an independent—from the previous history of the process—random variable with distribution function $B_v(x)$ and Laplace–Stieltjes transform $\psi_v(s)$.

If $v(t) = v$, then during time dt with probability $\lambda_{v\mu}^{(0)}\,dt$, independent of the past, a transition of $v(t)$ into the state μ may occur while the process $\xi(t)$ does not undergo a jump; only the rate of service will be changed. With probability $\lambda_{v\mu}^{(1)}\,dt$ a jump occurs after which $v(t + dt) = \mu$ and servicing of a new (priority) customer commences.

The customer whose service has been interrupted remains in the system; the work required for completion of his or her service is "stored" and when the service resumes $\xi(t)$ retains the same value as at the time when the service was interrupted.

If at a given time t the service of a customer in the state v is completed, i.e., $v(t - 0) = v$, $\xi(t - 0) = 0$, then with probability $p_{v\mu}$ the process $v(t)$ proceeds into state μ. If, however, $\mu \neq 0$, then either a service of a new customer begins or the previously interrupted service resumes. We assume that the customer whose servicing commences by the system in the new state is uniquely determined by the values of v and μ. If at time t the transition of $v(t)$ from $v \neq 0$ into the state 0 occurred, then the idle interval (of the server) commences.

We shall assume the following:

1. $\alpha_v > 0$, $v \neq 0$.
2. The average amount of work for servicing any customer is finite.
3. For any states v and μ there exists a chain of states $v = v_0, v_i, \ldots, v_n = \mu$ such that if $\alpha = v_i$, $\beta = v_{i+1}$, then either $\lambda_{\alpha\beta}^{(0)} > 0$ or $p_{\alpha\beta} > 0$ or $\lambda_{\alpha\beta}^{(1)} > 0$.
4. Let A and B be any operations of a server. Then if at the time of the beginning of operation B, there is already a customer for operation A, operation A may not commence until B is completed. This assumption will be called an *ordering (regulating) condition.*

We also denote $\lambda_v = \sum_\mu (\lambda_{v\mu}^{(0)} + \lambda_{v\mu}^{(1)})$.

In the assumptions stated above from general ergodic considerations the existence of an ergodic distribution

$$p_v = \lim_{t \to \infty} \mathsf{P}\{v(t) = v\}$$

follows independently of the initial state of the system. Algorithms for determining this distribution are the most interesting analytic part of the problem, and we shall dwell on this problem.

4.9.2. *The Structure of the Process*

Denote by $|v|$ the number of interrupted services at a given instant provided $v(t) = v$ $(v \neq 0)$. Clearly the possible values of $|v|$ are $0, 1, 2, \ldots, l$. If $v = 0$, i.e., there are no customers in the system, we set $|v| = -1$.

The structure of the process is as follows (Fig. 6): busy intervals ($|v(t)| \geqslant 0$) are replaced by idle intervals ($|v(t)| = -1$). Each busy interval consists of intervals (a, b) (in the Figure these are separated by vertical lines) where a and b are the times of the beginning and completion of servicing of the same customer. Intervals of the type (a, b) adjoining each other form a cycle of order 0. Thus, if (a_0, b_0) is a cycle of order 0, then

$$|v(a_0 - 0)| = |v(b_0 + 0)| = -1,$$

$$|v(a_0 + 0)| = |v(b_0 - 0) = 0.$$

Inside the cycle of order 0 cycles of order 1 are located, for example, (a_1, b_1) and (a_1', b_1') in Fig. 6. Each one of these intervals starts with a transition of $|v(t)|$ from the state 0 into state 1 and is completed by the reverse transition. Analogously, we define cycles of order k, where $1 < k \leqslant l$, as the intervals of sojourn of $|v(t)|$ in the states $\geqslant k$. Each cycle of order k consists of one or several intervals that start and end with servicing a customer.

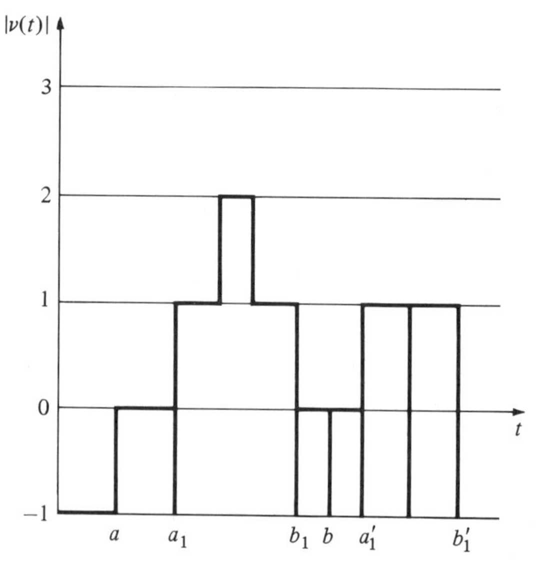

FIGURE 6

4.9.3. *Basic Equations*

We introduce the following notation:

$L_{\nu\mu}\,dt$ is the probability of occurrence during time dt in the state ν of a cycle of order $|\nu| + 1$, after completion of which the process proceeds into the state μ ($|\mu| = |\nu|$); $T_{\nu\mu}^{\nu'}$ is the mathematical expectation of the duration of the process $\nu(t)$ in the state ν' ($|\nu'| > |\nu|$) during the cycle of order $|\nu| + 1$ that starts with the departure of the process from the state ν and is completed with the transition into the state μ ($|\mu| = |\nu|$);

$l_{\nu\mu}$ is the probability that if at the beginning of the service the customer was in the state $\nu(t + 0) = \nu$, then immediately after completion of the service of this customer the process proceeds into the state μ ($|\mu| = |\nu|$ or $|\nu| = -1$);

$t_{\nu\mu}^{\nu'}$ is the conditional mathematical expectation of the sojourn of the process in the state ν' ($|\nu'| > |\nu|$) during the interval (a, b), where $\nu(a + 0) = \nu$, $\nu(b + 0) = \mu$; a and b are the times of beginning and completion of the service of the same customer where $|\mu| = |\nu|$ or $|\nu| - 1$.

The equation

$$p_\nu = \lambda_0 p_0 T_{00}^{\nu}, \quad \nu \neq 0 \tag{1}$$

follows from general ergodic considerations (or from Smith's theorem for regenerative processes). Indeed in the interval between the beginnings of two consecutive cycles of order 0 the process $\nu(t)$ is in the state ν on the average T_{00}^{ν} units of time and in the state 0 on the average λ_0^{-1} units of time.

We observe that constants $T_{00}^{\nu'}$ completely determine the stationary distribution provided we adjoin to (1) the normalizing condition

$$\sum_{\nu} p_\nu = 1,$$

whence

$$p_0^{-1} = 1 + \lambda_0 \sum_{\nu' \neq 0} T_{00}^{\nu'}. \tag{2}$$

A method for a recurrent definition of constants $T_{\nu\mu}^{\nu'}$ will be given below. First,

$$l_{\nu\mu} = \sum_{|\mu'| = |\nu|} p_{\mu'\mu} \int_0^\infty l_{\nu\mu'}(x)\,dB_\nu(x), \tag{3}$$

where $l_{\nu\mu}(x)$ is the conditional probability that the service that commenced in state ν is completed in state μ under the condition that the amount of work required for completion of a given service constitutes x units.

We shall construct a system of differential equations satisfied by functions $l_{\nu\mu}(x)$.

The variable x plays the role of "time"; it shows the amount of work that was already completed in servicing a given customer. Cycles of higher orders occur instantaneously in the given units of time; during this time the variable x ceases to increase. The occurrence of a cycle of order $|\nu| + 1$ in pseudotime

x is equivalent to a spontaneous change of the parameter of the process $v(x)$. Thus, the relation

$$l_{v\mu}(x+h) = l_{v\mu}(x)\left(1 - \frac{\lambda_\mu - L_{\mu\mu}}{\alpha_\mu} h\right) + h \sum_{|\mu'|=|v|} \frac{\lambda_{\mu'\mu}^{(0)} + L_{\mu'\mu}}{\alpha_{\mu'}} l_{v\mu'}(x) + o(h)$$

is easily deduced; whence in the usual manner we obtain

$$l'_{v\mu}(x) + \frac{\lambda_\mu - L_{\mu\mu}}{\alpha_\mu} l_{v\mu}(x) = \sum_{\substack{|\mu'|=|v| \\ \mu' \neq \mu}} \frac{\lambda_{\mu'\mu}^{(0)} + L_{\mu'\mu}}{\alpha_{\mu'}} l_{v\mu'}(x). \tag{4}$$

Functions $l_{v\mu}(x)$ satisfy the initial condition $l_{v\mu}(0) = \delta_{v\mu}$, where $\delta_{v\mu}$ is the Kronecker symbol. In Laplace transforms Eq. (4) is of the form

$$\left(s + \frac{\lambda_\mu - L_{\mu\mu}}{\alpha_\mu}\right)\tilde{l}_{v\mu}(s) = \sum_{\substack{|\mu'|=|v| \\ \mu' \neq \mu}} \frac{\lambda_{\mu'\mu}^{(0)} + L_{\mu'\mu}}{\alpha_{\mu'}} \tilde{l}_{v\mu'}(s) + \delta_{v\mu}, \quad |v| - 1 \leqslant |\mu| \leqslant |v|, \tag{5}$$

where

$$\tilde{l}_{v\mu}(s) = \int_0^\infty e^{-sx} l_{v\mu}(x)\, dx. \tag{6}$$

For a fixed v and for $|\mu| = |v|$ or $|v| - 1$ the system of equations (5) has the determinant

$$\Delta_{|v|}(s) = s^{q_{|v|}} + o(s^{q_{|v|}}), \quad s \to \infty, \quad \operatorname{Re} s \geqslant \varepsilon > 0,$$

where $q_{|v|}$ is a positive number. Therefore, the given system possesses a unique solution that is a rational function of the variable s and the constants $\lambda_{\mu'\mu}/\alpha_{\mu'}$, $L_{\mu'\mu}/\alpha_{\mu'}$.

Consequently, if the functions $l_{v\mu}(x)$ are represented in the form

$$l_{v\mu}(x) = \sum_r \alpha_{v\mu r} x^{b_{v\mu r}} e^{-\rho_{v\mu r} x}, \tag{7}$$

where $b_{v\mu r}$ are nonnegative integers. Boundedness of $l_{v\mu}(x)$ for $x > 0$, which is evident from probabilistic considerations, implies the inequality $\operatorname{Re} \rho_{v\mu r} \geqslant 0$. Thus the formula (3) yields the equality

$$l_{v\mu} = \sum_{|\mu'|=|v|} p_{\mu'\mu} \sum_r a_{v\mu r} (-1)^{b_{v\mu r}} \psi_v^{(b_{v\mu r})}(\rho_{v\mu r}). \tag{8}$$

In place of representation (7) one can also use Parseval's equality: if

$$\tilde{l}(s) = \int_0^\infty e^{-st} l(t)\, dt, \quad \psi(s) = \int_0^\infty e^{-st}\, dH(t),$$

then

$$\int_0^\infty l(t)\,dH(t) = \frac{1}{2\pi}\int_{-\infty}^{\infty} \tilde{l}(ix)\psi(-ix)\,dx. \tag{9}$$

Then instead of (8) we obtain an integral of the characteristic function with rational weights. Both the residual method and the quadrature methods are used for calculation of these integrals.

The constants $t_{\nu\mu}^{\nu'}$ can be obtained as follows. Let $\nu(x)$ be the value of the process $\nu(t)$ at the instant when x units of work for servicing a customer that started at the state ν is completed. If $\nu(x) = \nu'$, then from x to $x + h$, where h is a small quantity, $p/\alpha_{\nu'}$ units of time of the sojourn of the process in state ν' will expire (with probability $1 + o(1)$).

Next, between x and $x + h$ a cycle of order $|\nu(x)| + 1$ may take place; if $\nu(x) = \nu_1$, then the probability of occurrence of the cycle after which there will be $\nu(x + h) = \nu_2$ is equal to $L_{\nu_1\nu_2}h/\alpha_{\nu_1} + o(h)$; during such a cycle the process will be in the state ν' on the average $T_{\nu_1\nu_2}^{\nu'}$ units of time. Note furthermore that $t_{\nu\mu}^{\nu'} l_{\nu\mu}$ is the mathematical expectation of the sojourn time in state ν', multiplied by the indicator of the event consisting of the transition after completion of the service in the state μ. Consequently,

$$t_{\nu\mu}^{\nu'} l_{\nu\mu} = \sum_{|\mu'|=|\nu|} p_{\mu'\mu}\int_0^\infty dB_\nu(x)\int_0^x \left\{\frac{1}{\alpha_{\nu'}} l_{\nu\nu'}(y) l_{\nu'\mu'}(x-y) + \sum_{|\nu_1|=|\nu_2|=|\nu|} \frac{L_{\nu_1\nu_2}T_{\nu_1\nu_2}^{\nu'}}{\alpha_{\nu_1}} l_{\nu\nu_1}(y) l_{\nu_2\mu'}(x-y)\right\} dy. \tag{10}$$

We now express $L_{\nu\mu}$ and $T_{\nu\mu}^{\nu'}$ in terms of $l_{\nu'\mu'}$ and $t_{\nu'\mu'}^{\nu''}$. This is simpler than the derivation of formula (10):

$$L_{\nu\mu} = \sum_{|\mu'|=|\nu|+1} \lambda_{\nu\mu'} K_{\mu'\mu}, \tag{11}$$

where $K_{\nu\mu}$ is the probability that the cycle of order ν which started in state ν will be completed with the transition to the state μ, $|\mu| = |\nu| - 1$.

To determine $K_{\nu\mu}$ we have the system of equations

$$K_{\nu\mu} = l_{\nu\mu} + \sum_{|\mu'|=|\nu|} l_{\nu\mu'} K_{\mu'\mu}, \quad |\mu| = |\nu| - 1 \tag{12}$$

which in accordance with the theory of Markov chains has a unique solution.

In the same manner

$$L_{\nu\mu} T_{\nu\mu}^{\nu'} = \sum_{|\mu'|=|\nu|+1} \lambda_{\nu\mu'} S_{\mu'\mu}^{\nu'}, \tag{13}$$

where the constants $S_{\nu\mu}^{\nu'}$ are determined by the system of equations

$$K_{\nu\mu} S_{\nu\mu}^{\nu'} = l_{\nu\mu} t_{\nu\mu}^{\nu'} + \sum_{|\mu'|=|\nu|} l_{\nu\mu'} K_{\mu'\mu}(t_{\nu\mu'}^{\nu'} + S_{\mu'\mu}^{\nu'}), \quad |\mu| = |\nu| - 1. \tag{14}$$

We have thus obtained a recurrent process: $l_{\nu\mu}$ and $t_{\nu\mu}^{\nu'}$ are determined by means of the formulas (8), (10) in terms of $L_{\nu_1\mu_1}$ and $T_{\nu_1\mu_1}^{\nu'}$ where $|\nu_1| = |\nu|$; formulas (11)–(14) allow us to determine $L_{\nu\mu}$ and $T_{\nu\mu}^{\nu'}$ in terms of $l_{\nu_1\mu_1}$ and $T_{\nu_1\mu_2}^{\nu'}$ where $|\nu_1| = |\nu| + 1$.

4.9.4. *Remarks*

Among the books on the priority system, we mention Jayswal (1968) and Gnedenko et al. (1973). Some important results contaned in the last book are due to Klimov, Danielyan, and others. Priority system with bounded queue were studied intensively by G.P. Basharin and his students: Basharin (1967a, 1967b), Bocharev (1968), and Bocharev and Lysenko (1969). Multilinear priority systems were studied by Williams (1967). Schrage (1968) and Schrage and Miller (1966) in their important papers proved the optimality of priority with respect to the shortest remaining service time. Special types of priority service associated with methods of operation of computers are discussed in Artamonov and Brekhov (1979) and Kleinrock (1975). The current state of the problem of deriving formulas for characteristics of one-server priority queueing systems is depicted in the works of Franken et al. (1984) and Gnedenko and König (1984).

Multiphase service systems in which the customers pass sequentially through two or several servers is a difficult problem analytically. We note the works of Neuts (1968), Klimov (1970), Rosental' and Tolmachev (1975), and Taylor (1972) in this connection.

Numerous papers have been devoted to systems with a nonordinary ("group") incoming stream. Among these we mention the papers by Shakhbazov and Samandarov (1964) and Mills (1980).

Substantial literature devoted to outgoing streams in queueing systems and loss streams is available. We shall note the papers by Brémaud (1978), Yarovitskii (1961), Simonova (1969), Rudlovchak (1967), Viskov and Ismailov (1970), and Bokuchava et al. (1969). In a number of investigations the problem of feasibility of reconstructing the characteristics of a system from the observations of the outgoing stream is investigated. Thus in the paper by Kovalenko (1965d) it was shown that in a $M|G|1$ system for $\rho < 1$ one can always reconstruct $B(x)$ from the observations of the outgoing process except in the case when $B(x) = 1 - e^{-\mu x}$, $x \geqslant 0$: in this case, as it follows from Berk's theorem, the outgoing stream is simple with parameter λ for any $\mu \geqslant \lambda$ and hence the value of μ cannot be reconstructed. Ivnitskiĭ solved many of these problems; we mention especially the results presented in (1969) and (1977).

Methods described in this chapter allow us to study systems with an incoming stream and sequences of duration of service in the form of a semi-Markov process. We shall mention Çinlar's paper (1967) in this connection.

Relations between the distribution of the queue length and the waiting time were studied by Haji and Newell (1971), Stidham (1974), and Franken et al. (1984).

5
Application of More General Methods

In this chapter we shall discuss solutions of a number of problems in queueing theory that require more advanced methods of investigation than in the preceding chapter.

5.1. The $GI|G|1$ System

5.1.1. *Basic Recurrence Relations*

Consider a single-server queueing system with waiting. Assume that the incoming stream has limited aftereffects. At the initial time the system is empty. The arrival times are $t_1 \leqslant t_2 \cdots \leqslant t_n \leqslant \cdots$, the variables $z_n = t_n - t_{n-1}$ are jointly independent and have the same distribution

$$A(x) = \mathsf{P}\{z_n < x\}, \quad n \geqslant 2. \tag{1}$$

The service times are jointly independent random variables η_n with distribution

$$B(x) = \mathsf{P}\{\eta_n < x\}, \quad n \geqslant 1. \tag{2}$$

Also let $\zeta_n = \eta_{n-1} - z_n$,

$$K(x) = \mathsf{P}\{\zeta_n < x\} = \int_0^\infty \bar{A}(y - x)\, dB(y). \tag{3}$$

It is assumed that the sequences $\{z_n\}$ and $\{\eta_n\}$ are mutually independent.

Let w_n denote the nth customer's waiting time. Clearly, $w_1 = 0$.

We shall consider, following Lindley (1952) how the successive values of w_n are formed.

If the nth customer arrives immediately after the $(n-1)$th, he or she will have to wait $w_{n-1} + \eta_{n-1}$ units of time; in time z_n this quantity is decreased by z_n, i.e., the nth customer's waiting time will be $w_{n-1} + \eta_{n-1} - z_n = w_{n-1} + \zeta_n$ units of time. However, the following should be noted. If z_n is sufficiently large, $w_{n-1} + \zeta_n$ may be negative. It is easy to see that the actual waiting time of the nth customer is then zero.

We thus obtain the recurrence relation

$$w_n = \max\{w_{n-1} + \zeta_n, 0\}, \tag{4}$$

or in symbolic notation

$$w_n = (w_{n-1} + \zeta_n)^+. \tag{5}$$

Denote

$$F_n(x) = \mathsf{P}\{w_n < x\}. \tag{6}$$

Then relation (4) or (5) may be expressed in terms of the distribution functions as follows:

$$F_{n+1}(x) = \begin{cases} \int_{-\infty}^{x} F_n(x-y)\,dK(y), & \text{if } x > 0,\ n \geqslant 2, \\ 0, & \text{if } x \leqslant 0,\ n \geqslant 1. \end{cases} \tag{7}$$

Formulas (7), together with the obvious relation

$$F_1(x) = \begin{cases} 0, & \text{if } x \leqslant 0, \\ 1, & \text{if } x > 0, \end{cases} \tag{8}$$

enable us in principle to obtain recursively distribution of the waiting time for any customer.

Lindley made an important observation that was due to an appropriate choice of a random sequence.

Formula (7) also retains its meaning when $\{\eta_n\}$ and $\{z_n\}$ are dependent sequences; the essential point is that the random variables $\{\zeta_n\}$ are jointly independent.

Following Lindley we shall give two interpretations of the preceding recurrence formulas.

The first interpretation is a random walk with an impenetrable barrier at the origin.

Imagine a moving particle whose coordinate is x at time n. At the initial time $n = 1$, $x_1 = 0$. At each subsequent instant n the particle is displaced by ζ_n, where ζ_n is independent of $\zeta_1, \zeta_2, \ldots, \zeta_{n-1}$ and has distribution $K(x)$. If at any step the particle moves into the negative coordinate domain, it returns immediately to the origin. Then the distribution of the coordinate of the particle at the instant n coincides with that of the nth customer's waiting time. This follows directly from (5), which is a formal representattion of a random walk with an impenetrable barrier at the origin.

The second interpretation that leads to the distribution of the waiting time is that of a random walk with an absorbing barrier. Again we consider a moving particle with coordinate x_n at the nth step; $x_1 = 0$. Let $x_n = x_{n-1} + \zeta_n$, $n \geqslant 2$. An absorbing barrier is located at level x. If at any instant $x_n \geqslant x$, the particle is absorbed.

We assert that the probability that the particle is absorbed during n steps coincides with $F_n(x) = \mathsf{P}\{w_n < x\}$. In other words, for an unrestricted random walk

$$x_n = x_{n-1} + \zeta_n, \quad n \geqslant 2, x_1 = 0,$$
$$\mathsf{P}\{w_n < x\} = \mathsf{P}\left\{\max_{1 \leqslant i \leqslant n} x_i < x\right\},$$

i.e., w_n may be interpreted as $\max_{1 \leqslant i \leqslant n} x_i$. This assertion is proved by inducation. For $n = 1$ it is trivially valid. For $n > 1$ we have

$$\max_{1 \leqslant i \leqslant n} x_i = \begin{cases} \max(0, x_2) & \text{if } \max_{3 \leqslant i \leqslant n} (x_i - x_2) \leqslant 0, \\ \max(0, x_1) + \max_{3 \leqslant i \leqslant n} [0, (x_i - x_1)] & \text{if } \max_{3 \leqslant i \leqslant n} (x_i - x_1) > 0. \end{cases} \tag{9}$$

However, the distribution of $\max_{3 \leqslant i \leqslant n}(x_i - x_2)$ coincides with that of the random variable $\max_{1 \leqslant i \leqslant n-1} x_i$ (due to the homogeneity of the walk). Denote $L_n(x) = \mathsf{P}\{\max_{1 \leqslant i \leqslant n} x_i < x\}$. Then (9) yields the formulas:

$$L_n(x) = \begin{cases} \displaystyle\int_{-\infty}^{x} L_{n-1}(x - y)\, dK(y) & x > 0, \\ 0 & x \leqslant 0. \end{cases}$$

These formulas are equivalent to (7) and since the latter uniquely determine $F_n(x)$, we have $L_n(x) = F_n(x)$ for all $n \geqslant 1$.

5.1.2. *The Integral Equation; the Existence of the Limiting Distribution*

We have just proved the result that is equivalent to

$$F_n(x) = \mathsf{P}\left\{\max_{1 \leqslant i \leqslant n} x_i < x\right\}; \tag{10}$$

whence it follows that for every $n \geqslant 2$

$$F_n(x) \leqslant F_{n-1}(x);$$

thus, the limit

$$F(x) = \lim_{n \to \infty} F_n(x) \tag{11}$$

exists in the sense of weak convergence. The definition of $F(x)$ can be completed by means of continuity from the left. By Lebesgue's convergence theorem we can let n approach infinity in (7), leading to the integral equation

$$F(x) = \begin{cases} \displaystyle\int_{-\infty}^{x} F(x - y)\, dK(y), & x > 0, \\ 0, & x \leqslant 0. \end{cases} \tag{12}$$

Equation (12) is an integral equation on a half-line with a kernel depending on the difference of the arguments. A general theory of such equations has been developed by Krein (1961). The solution is found by factorization of the Fourier transform of the kernel of this integral equation.

We shall investigate whether $F(x)$ is a distribution function, or, equivalently, we shall check the value of $F(\infty) = \lim_{x\to\infty} F(x)$.

Theorem. *Let the mathematical expectation*

$$\alpha = \mathsf{M}\zeta_n$$

be finite or infinite, but of definite sign. Then

1. *if* $\alpha < 0$ $F(\infty) = 1$,
2. *if* $\alpha > 0$ $F(\infty) = 0$,
3. *if* $\alpha = 0$ $F(\infty) = 0$, *unless when* $\zeta_n = 0$ *with probability* 1.

Proof. One can write

$$F(x) = \mathsf{P}\left\{\max_{1\leqslant i\leqslant\infty} x_i < x\right\}, \tag{13}$$

which also inclues the case

$$\max_{1\leqslant i\leqslant\infty} x_i = \infty.$$

Consider the first case: $\alpha < 0$. In view of the strong law of large numbers only finitely many x can be positive. Thus, the maximum is also finite with probability 1.

In the second case, we again appeal to the strong law of large numbers: $x_n \xrightarrow[n\to\infty]{} \infty$ with probability 1; this implies that $\max_{1<i\leqslant\infty} x_i = \infty$ and $F(x) = 0$ for every x.

The third case is the most difficult one. There are two possibilities, $F(+0) = 0$ or $F(+0) > 0$. If $F(+0) = 0$, we obtain from Eq. (12)

$$\int_{-\infty}^{0} F(-y)\,dK(y) = -\int_{0}^{\infty} F(y)\,dK(-y) = 0. \tag{14}$$

Again there are two possibilities, $\zeta_n = 0$ with probability 1 or $\mathsf{P}\{\zeta_n \neq 0\} > 0$. In the first case all the w_n coincide.

In the second case, since $\mathsf{M}\zeta_n = 0$ there exists an $\varepsilon > 0$ such that $\int_{-\varepsilon}^{0} dK(y) > 0$. However, then equality (14) implies that $F(y) = 0$ for $y < \varepsilon$. Substituting $x = \varepsilon' < \varepsilon$ into (12) we obtain $F(y) = 0$ for $y < \varepsilon + \varepsilon'$; repeating this procedure we arrive at $F(y) = 0$. Now let $F(0) = \delta > 0$ and $w_{i_1} = w_{i_2} = \cdots = w_{i_n} = \cdots = 0$ where the intermediate w_k are positive. Then the differences $i_2 - i_1$, $i_3 - i_2$, ... are independent identically distributed random variables. It follows from the theory of Markov chains that $F(0)$ can be valid only if $\mathsf{M}\{i_n - i_{n-1}\} < \infty$. The strong law of large numbers implies that, if m_n is the number of w_k equal to zero ($1 \leqslant k \leqslant n$), then $m_n/n \to \delta$. Let ξ_n denote the time between $t_{i_{n-1}}$ and t_{i_n} during which the server is free. All the ξ_n are identically distributed and (for $\zeta \not\equiv 0$) are positive with positive probability. Hence,

$$\sum \xi_i > cn, \quad c > 0, \quad n \geqslant n_0. \tag{15}$$

However, $w_n = w_1 + \sum_{i=2}^n \zeta_i + \sum \xi_i$. By the strong law of large numbers

$$\sum_{i=2}^{n} \zeta_i = o(n), \quad n \to \infty;$$

in conjuction with the bound (15), this implies that

$$w_n > \frac{c}{2} n, \quad n \geqslant n_1,$$

hence $F(x) = 0$ for every x. The theorem is proved.

We shall now establish the ergodicity of the limiting distribution: for each y

$$F_n(x|y) = \mathsf{P}\{w_n < x | w_1 = y\} \xrightarrow[n\to\infty]{} F(x).$$

Denote by $w_n(y)$ the solution of the recurrence system of equations $w_n(y) = (w_{n-1}(y) + \zeta_n)^+$ under the initial condition $w_1(y) = y$. Then since the function x^+ is monotonically nondecreasing in x we have

$$w_n(y) \leqslant w_n(z), \quad y \leqslant z. \tag{16}$$

Suppose that $\alpha < 0$. Let γ_n be the idle time of a server in the interval $(0, t_n)$. If at the initial time an additional work that equals γ_n was assigned to the server, it would have completed it together with the other work, i.e.,

$$w_n(\gamma_n) = w_n, \tag{17}$$

Hence, in view of (16)

$$\mathsf{P}\{w_n < x\} - \mathsf{P}\{\gamma_n < y\} \leqslant \mathsf{P}\{w_n(y) < x\} \leqslant \mathsf{P}\{w_n < x\}. \tag{18}$$

Furthermore,

$$\gamma_n \geqslant t_n - (\eta_1 + \cdots + \eta_{n-1}) \geqslant -(\zeta_2 + \cdots + \zeta_n)$$

and in view of the law of large numbers $\gamma_n \to \infty$ in probability. Thus, $P\{\gamma_n < y\} \xrightarrow[n\to\infty]{} 0$, and it follows from (18) that $F_n(x|y) - F_n(x) \xrightarrow[n\to\infty]{} 0$, $F_n(x|y) \xrightarrow[n\to\infty]{} F(x)$. In the case when $\alpha > 0$ and in the case when $\alpha = 0$, $\mathsf{P}\{\zeta_n \neq 0\} = 0$ we obtain from (16) that

$$\lim_{n\to\infty} F_n(x|y) \leqslant \lim_{n\to\infty} F_n(x) = 0.$$

Only in the case when $\alpha = 0$, $\mathsf{P}\{\zeta_n = 0\} = 1$ does ergodicity fail to hold: $w_n(y) = y$, $n \geqslant 1$.

Denote

$$\rho = \mathsf{M}\eta_{n-1}/\mathsf{M}z_n. \tag{19}$$

Then

$$\rho - 1 = \frac{\alpha}{Mz_n},$$

hence $\rho < 1$, $\rho = 1$, or $\rho > 1$ according to whether $\alpha < 0$, $\alpha = 0$, or $\alpha > 0$. Thus, the ergodic distribution of the random sequence $\{w_n\}$ exists for $\rho < 1$; it evidently does not exist for $\rho > 1$, and exists only in the degenerate case $\mathsf{P}\{\zeta_n = 0\} = 1$ for $\rho = 1$. The constant ρ defined by (19) has the same probabilistic meaning as $\lambda\tau$ in a system with a simple incoming stream: it is the load on the system. □

5.1.3. *Analytic Methods*

Let ξ_n, $n \geqslant 1$, be independent random variables with the distribution function $F(x)$ and characteristic function $\psi(t)$, $s_n = \xi_1 + \cdots + \xi_n$, $s_0 = 0$; $F_n(x) = \mathsf{P}\{s_n < x\}$. For $|z| < 1$ and real t

$$1 - z\psi(t) = \exp\{\ln(1 - z\psi(t))\},$$

$$\ln(1 - z\psi(t)) = -\sum_{k=1}^{\infty} \frac{1}{k} z^k \psi^k(t) = -\sum_{k=1}^{\infty} \frac{1}{k} z^k \int e^{itx}\, dF_k(x)$$

$$= u(z, t) + v(z, t) + w(z),$$

where

$$u(z, t) = -\sum_{k=1}^{\infty} \frac{1}{k} \int_{+0}^{\infty} e^{itx}\, dF_k(x),$$

$$v(z, t) = -\sum_{k=1}^{\infty} \frac{1}{k} z^k \int_{-\infty}^{0} e^{itx}\, dF_k(x),$$

$$w(z) = -\sum_{k=1}^{\infty} \frac{1}{k} z^k (F_k(+0) - F_k(0)).$$

The function $u(z, t)$ is analytic in the upper half-plane in the complex variable t ($\operatorname{Im} t > 0$) and is continuous in the region $\{\operatorname{Im} t \geqslant 0\}$. Analogously, $v(z, t)$ is analytic in the lower half-plane and continuous for $\operatorname{Im} t \leqslant 0$. These properties are preserved also for the functions $w_{z+}(t) = \exp\{-u(z, t)\}$, $w_{-z}(t) = \exp\{v(z, t)\}$. Moreover, $|w_{z\pm}(t)| \geqslant \varepsilon > 0$ in the corresponding half-planes and $w_{z+}(i\infty) = 1$. Denoting $\varphi(z) = \exp\{w(z)\}$, we obtain the representation

$$1 - z\psi(t) = \varphi(z) w_{z+}(t)/w_{z-}(t),$$

which is called *factorization*. We now disregard that the functions $\varphi(z)$, $w_{z\pm}(t)$ possess a definite probabilistic meaning and assume that only the factorization equation and the previously stated analytic properties of φ, $w_{z\pm}$ are known. These properties uniquely determine the factorization components φ, w_{z+}, and w_{z-} up to a constant factor.

It turns out that various characteristics of random walks are expressed in terms of the factorization components of the corresponding function; in particular, the stationary distribution of the queue length, waiting time, and virtual waiting in the system $GI|G|1$. An important example is Spitzer's identity: if $y_n = \max\{0, s_1, \ldots, s_n\}$, $\varphi_n(t) = Me^{ity_n}$, then

$$\sum_{n=0}^{\infty} z^n \varphi_n(t) = w_{z+}(t)/((1-z)w_{1+}(t)), \quad |z| < 1.$$

If $M|\xi_1| < \infty$, $M\xi_1 < 0$, $y_\infty = \sup_{n \geqslant 0} s_n$, $\varphi(t)$ is a characteristic function of the random variable y_∞, then

$$\varphi(t) = w_{1+}(0)/w_{1+}(t).$$

Choosing for $F(x)$ the distribution function $K(x)$ of the random variables ζ_n, we obtain an expression for the characteristic function of a stationary distribution of the waiting time in the system $GI|G|1$.

Since it is difficult to solve analytically the equations that determine the characteristics of the systems $GI|G|1$, $GI|G|m$, and similar systems, various inequalities between characteristics of these systems and corresponding characteristics of simpler systems have acquired substantial importance. These are based on stochastic ordering. We shall present an example of such a result. (Stoyan (1977)).

Let $F_1(t)$, $F_2(t)$ be distribution functions. We define two relations of stochastic ordering:

$$F_1 \overset{(1)}{\leqslant} F_2 \Leftrightarrow F_1(t) \geqslant F_2(t), \qquad -\infty < t < \infty;$$

$$F_1 \overset{(2)}{\leqslant} F_2 \Leftrightarrow \int_{-\infty}^{t} \bar{F}_1(x)\,dx \leqslant \int_{-\infty}^{t} \bar{F}_2(x)\,dx, \quad -\infty < t < \infty.$$

If $F(t)$ possesses a finite moment m_F and $F(0) = 0$, we define

$$F_R(t) = \frac{1}{m_F} \int_0^t \bar{F}(x)\,dx, \quad t \geqslant 0.$$

We say that F is of $NBUE$ ($NWUE$) type if $F_R \overset{(1)}{\leqslant} F$ (respectively, $F \overset{(1)}{\leqslant} F_R$). If in a $GI|G|1$ system the distribution $A(x)$ of the interarrival time between the customers is of the type $NBUE$ ($NWUE$), then

$$F \overset{(1)}{\leqslant} F_M \quad (F_M \overset{(1)}{\leqslant} F),$$

where F is the stationary distribution of the waiting time in the given system while F_M is the corresponding distribution of the system $M|G|1$ with the same load and the service time distribution.

5.2. $GI|G|m$ Systems

5.2.1. *Multidimensional Random Walk*

In this section we shall deal with a queueing system with waiting and with m servers. The incoming stream has limited aftereffects; $B(x)$ is the distribution function of η_n, where η_n is the nth customer's service time; $A(x) = \mathsf{P}\{z_n < x\}$, where z_n is the length of the interval between the arrival of the $(n-1)$th and nth customers. We shall assume that the random variables $\{\eta_n\}$ and $\{z_n\}$ are jointly independent.

Denote by w_n the nth customer's waiting time until the service begins ($n \geqslant 1$). When $m > 1$, there is no direct connections between w_n and w_{n-1}. We shall therefore consider a multidimensional random walk.

Consider the operation of the ith server separately. Assume that the incoming stream for this server is the same as for the whole system and the service time of the nth server is equal to η_{ni}, where $\eta_{ni} = \eta_n$ in the case when the given customer actually arrived at this server and $\eta_{ni} = 0$ otherwise. We introduce the random variable w_{ni}—the time from the instant t_n up until the time of the ith server's disengagement from the customers that appeared before time t_n, or 0 if all of them were served before time t_n. Then for $\{w_{ni}\}$ the recurrent relation analogous to (5) presented in Section 5.1, namely,

$$w_{ni} = (w_{n-1,i} + \eta_{n-1,i} - z_n)^+ \tag{1}$$

is fulfilled. At the same time

$$w_n = \min_{1 \leqslant i \leqslant m} w_{ni}, \tag{2}$$

since any customer arrive at a server with minimal waiting time. The random variables η_{ni} are determined by the formula

$$\eta_{ni} = \begin{cases} \eta_n, & w_n = w_{ni}, \\ 0, & w_n < w_{ni}, \end{cases}$$

in the case that among the numbers $w_{ni}, \ldots, w_{nm}$ only one is minimal. However, if, for example, $w_n = w_{ni_1} = \cdots = w_{ni_k}$ and all the other $w_{ni} > w_n$, then a rule must be determined to choose from the indices $i_1, \ldots, i_k$. Since this rule does not affect the distribution of $\{w_n\}$, we shall assume that from the set $\{i_1, \ldots, i_k\}$ an i is selected—in accordance with an equiprobable distribution—such that $\eta_{ni} = \eta_n$; for all other j it is assumed that $\eta_{nj} = 0$.

The random variables $\bar{w}_n = (w_{n1}, \ldots, w_{nm})$ determine an m-dimensional random walk.

5.2.2. *Kiefer and Wolfowitz's Ergodic Theorem* (1955)

Theorem. *Let*

$$\tau = \mathsf{M}\eta_n < \infty, \quad \lambda^{-1} = \mathsf{M}z_n, \quad \rho = \lambda\tau.$$

Then,

1. *if $\rho < m$, the random sequence $\{w_n\}$ has an ergodic distribution;*
2. *if $\rho > m$, then for any $x, \bar{y} = (y_1, y_2, \ldots, y_m)$*

$$\mathsf{P}\{w_n < x \mid \bar{w}_1 = \bar{y}\} \xrightarrow[n\to\infty]{} 0; \tag{3}$$

3. *if* $\rho = m$ *and* $\eta_n \neq mz_n$ *with positive probability, condition* (3) *holds;*
4. *if* $\rho = m$ *and* $\eta_n = mz_n$ *with probability* 1, *the variables* w_n *are deterministic and bounded;* w_n *is periodically dependent (with period* m*) on* n.

We shall prove only that part of the theorem that deals with the case $\rho < m$. First we shall prove several lemmas. The inequality $\bar{x} < \bar{y}$ for vectors $\bar{x}$ and $\bar{y}$ will be interpreted as a system of inequalities $x_i < y_i$ for the corresponding components of this vector.

Let $\bar{g}_n = (g_{n1}, \ldots, g_{nm})$ be a vector whose components are ordered in a nondecreasing order of components of the vector $\bar{w}_n$:

$$F_n(\bar{x}|\bar{y}) = \mathsf{P}\{\bar{g}_n < \bar{x}|\bar{g}_1 = \bar{y}\}, \quad F_n(\bar{x}) = F_n(\bar{x}|0).$$

Also denote $\bar{f}(z, \eta|\bar{y})$ as the value of the vector $\bar{g}_n$ the fixed values $\bar{g}_{n-1} = \bar{y}$, $z_n = z$, $\eta_{n-1} = \eta$. (It is easy to see that the form of function f does not depend on n.)

Lemma 1. *If* $\bar{y} \leqslant \bar{y}'$, $\eta \leqslant \eta'$, $z \geqslant z'$, *then,*

$$\bar{f}(z, \eta|\bar{y}) \leqslant \bar{f}(z', \eta'|\bar{y}'). \tag{4}$$

Proof. Denote by $n_x(\bar{x})$ the number of components of vector $\bar{x}$ smaller than x; $E(x) = 0$ for $x \leqslant 0$, $E(x) = 1$ for $x > 0$. Then

$$n_x(\bar{f}(z, \eta|\bar{y})) = E(x - y_1 - \eta + z) + \sum_{i=2}^{m} E(x - y_i + z). \tag{5}$$

Since the function $E(x)$ is nondecreasing, it follows from (5) that $n_x(\bar{f}(z, \eta|\bar{y}) \geqslant n_x(\bar{f}(z', \eta'|\bar{y}'))$. This means that each component of $\bar{f}(z, \eta|\bar{y})$ does not exceed the corresponding component of $\bar{f}(z', \eta'|\bar{y}')$. □

Corollary. *Let* $\{\bar{g}_n\}$ *be constructed based on the data* η_i, z_i, $\bar{y} = \bar{g}_1$, $\{\bar{g}_n'\}$ *be an analogous sequence constructed based on* η_i', z_i', *and* $\bar{y}'$, where $z_i \geqslant z_i'$, $\eta_i \leqslant \eta_i'$, *and* $\bar{y} \leqslant \bar{y}'$. *Then for any* $n \geqslant 1$

$$\bar{g}_n \leqslant \bar{g}_n'.$$

To prove the corollary it is sufficient to iterate the inequality (4). Whence

$$F_n(\bar{x}) \geqslant F_n(\bar{x}|\bar{y}). \tag{6}$$

Lemma 2. *For any* $n \geqslant 2$

$$F_n(\bar{x}) \leqslant F_{n-1}(\bar{x}).$$

Proof. One can write

$$\bar{g}_3 = \bar{f}(z_3, \eta_2|\bar{g}_2), \ldots, \bar{g}_n = \bar{f}(z_n, \eta_{n-1}|\bar{g}_{n-1}).$$

In the same manner, we shall write the formulas for $\bar{g}_2'$, ..., $\bar{g}_{n-1}'$ where, however, in place of z_i and η_i we shall write z_{i+1}, η_{i+1}; this naturally does not alter the probability distribution:

$$\bar{g}'_2 = \bar{f}(z_3, \eta_2 | 0), \ldots, \bar{g}'_{n-1} = \bar{f}(z_n, \eta_{n-1} | \bar{g}'_{n-2}).$$

Comparing the last two formulas and applying Lemma 1 we obtain

$$\begin{aligned} \bar{g}_3 &= \bar{f}(z_3, \eta_2 | \bar{g}_2) \geqslant \bar{f}(z_3, \eta_2 | 0) = \bar{g}'_2, \\ \bar{g}_4 &= \bar{f}(z_4, \eta_3 | \bar{g}_3) \geqslant \bar{f}(z_4, \eta_3 | \bar{g}'_2) = \bar{g}'_3, \\ &\vdots \\ \bar{g}_n &= \bar{f}(z_n, \eta_{n-1} | \bar{g}_{n-1}) \geqslant f(z_n, \eta_{n-1} | \bar{g}'_{n-2}) = \bar{g}'_{n-1}. \end{aligned}$$

Thus

$$\bar{g}_n = \bar{g}_n(z_2, \ldots, z_n; \eta_1, \ldots, \eta_{n-1}) \geqslant \bar{g}_{n-1}(z_3, \ldots, z_n; \eta_2, \ldots, \eta_{n-1}). \tag{7}$$

Let $I_{\bar{x}}(\bar{g}_n)$ be a random variable which takes on value 1 for $\bar{g}_n < \bar{x}$ and 0 otherwise. Then we have from (7) that $I_{\bar{x}}(\bar{g}_n) \leqslant I_{\bar{x}}(\bar{g}'_{n-1})$. Applying mathematical expectations we arrive at

$$F_n(\bar{x}) \leqslant F_{n-1}(\bar{x}). \qquad \square$$

Corollary. *There exists* $F(\bar{x}) = \lim_{n\to\infty} F_n(\bar{x})$.

We shall consider an auxiliary "cyclic" service system in which the allocation of customers to servers is carried out in a cyclic order: if $n \equiv i \pmod m$, then nth customer is assigned to the ith server. The random variables related to the cyclic system will be denoted in the same manner as those related to the system with the common queue using, however, capital letters. We shall assume that in both systems the sequences $\{z_n\}$ and $\{\eta_n\}$ are the same.

Let a server serve a customer in the interval $(t, t + \eta_k)$ in a cyclic system and start servicing the next customer at time $t + \eta_k + \varepsilon_k$. In this case set $\eta'_k = \eta_k + \varepsilon$. The transition from η_k to η'_k can evidently only increase the values of G_{ni}. For the new values of service times there will be no interruptions in the service and the system will become a system with a common queue, i.e., $\bar{G}'_n = \bar{g}'_n$, where $\bar{G}'_n$ and $\bar{g}'_n$ are the variables obtained from $\bar{G}_n$ and $\bar{g}_n$ when η_k becomes η'_k. By the corollary to Lemma 1 under the inverse transition the components of $\bar{g}_n$ can only be decreased. Thus,

$$\bar{g}_n \leqslant \bar{G}'_n. \tag{8}$$

Let $G_{nm+i,i} = G_n^{(i)}$. The sequence $\{G_n^{(i)}\}$ has the same meaning as $\{w_n\}$ for the system $GI|G|1$ with the distribution $B(x)$ for the service time and distribution $A^{*(n)}(x)$ for the interarrival times of customers. Thus, its load is less than 1, whence

$$\mathsf{P}\{G_n^{(i)} < x\} \xrightarrow[n\to\infty]{} \Phi(x), \tag{9}$$

where $\Phi(x)$ is a distribution function.

Let $n = km + i_0$. Then

$$G'_{ni_0} = \max\{G'_{ni_0}, z_{n+1} + \cdots + z_{n+m}\},$$

$$G'_{n,i_0+1} = \max\{(G_{n-m+1,i_0+1} - (z_{n-m+2} + \cdots + z_n))^+, z_{n+1}\}$$

and so on. Hence in view of (9) $\sup_n \mathsf{P}\{G'_{ni} \geqslant x\} \xrightarrow[x\to\infty]{} 0$ and then in view of (8) this property is also fulfilled for $\{g_{ni}\}$. It thus follows that $F(\bar{x}) = \lim_{n\to\infty} F_n(\bar{x})$ is a distribution function. Since $\{G_n^{(i)}\}$ are ergodic Markov chains, it follows, in view of the relations between $\bar{g}_n$ and $\{G_n^{(i)}\}$ previously derived, that there exists x such that the number of returns of $\bar{g}_n$ into the set of states $\{g_{nm} < x\}$ is infinite with probability 1.

If $\rho < m$, this means that $\eta < z_1 + \cdots + z_m$ with positive probability. There exist $\varepsilon > 0$ and $\delta > 0$ such that

$$\mathsf{P}\{\eta_n \leqslant ma - \varepsilon\} \geqslant \delta, \qquad \mathsf{P}\{z_n \geqslant a\} \geqslant \delta. \tag{10}$$

Now assume that $g_{ni} < x$, $1 \leqslant i \leqslant m$ and that η_s and z_{s+1} satisfy (10) for $s \geqslant n$. Then after $[mx/\varepsilon] + 1$ steps the queue disappears; moreover, the distribution of the random vector g_N, $N = n + [mx/\varepsilon] + 1$ ceases to depend on g_n.

Furthermore, it is obvious that

$$F_n(x_1,\ldots,x_m|y_1,\ldots,y_n) \geqslant F_n\left(x_1 - \sum_{i=1}^m y_i,\ldots,x_m - \sum_{i=1}^m y_i\right);$$

hence we have

$$F_n(x_1,\ldots,x_m|y_1,\ldots,y_m) \geqslant c > 0 \tag{11}$$

for fixed y_i uniformly in n for sufficiently large x. Denote by A_m the event

$$\left\{g_{ni} \leqslant x, 1 \leqslant i \leqslant m; \eta_s \leqslant ma - \varepsilon, z_{s+1} \geqslant a, n \leqslant s \leqslant \frac{mx}{\varepsilon} + 1\right\}.$$

The events $\{A_m\}$ are recurrent (see Feller (1950)) so one can write

$$\begin{aligned} &F_n(x_1,\ldots,x_m|y_1,\ldots,y_m) \\ &\quad = \sum_{k=1}^n \mathsf{P}\{\bar{A}_1\bar{A}_2\ldots\bar{A}_{k-1}A_k|y_1,\ldots,y_m\} \\ &\quad\quad \times \int F_{m-l}(x_1,\ldots,x_m|y'_1,\ldots,y'_m)\,d\Phi(y'_1,\ldots,y'_m) + \theta\mathsf{P}\{\bar{A}_1\ldots\bar{A}_n\}, \end{aligned} \tag{12}$$

where

$$l = [mx/\varepsilon] + 1, \quad 0 \leqslant \theta \leqslant 1,$$

$\Phi(x_1,\ldots,x_m)$ is the conditional distribution of the random vector $w^* = (w_i^*,\ldots,w_m^*)$, where w_i^* are the random variables

$$w_i = \eta_i - \sum_{k=1}^i z_{k+1}, \quad 1 \leqslant i \leqslant m, \tag{13}$$

arranged in ascending order, under the condition that for all η_i and z_{k+1} in (14), relation (10) holds.

For some a and sufficiently small ε in view of

$$F_n(x_1,\ldots,x_m) \geqslant F_{n+1}(x_i,\ldots,x_m)$$

the sequence

$$\{F_{n-l}(\ldots)d\Phi(\ldots)\}$$

on the right-hand side of (12) is nonincreasing. In view of (10) and (11)

$$\mathsf{P}\{\bar{A}_1\cdots\bar{A}_n\}\xrightarrow[n\to\infty]{}0$$

and thus

$$\lim_{n\to\infty} F_n(x_1,\ldots,x_m|y_1,\ldots,y_m)$$
$$=\lim_{n\to\infty}\int F_n(x_1,\ldots,x_m|y_1',\ldots,y_m')d\Phi(y_1',\ldots,y_m').$$

In particular, putting $y_1 = y_2 = \cdots = y_m = 0$ we have

$$\lim_{n\to\infty} F_n(x_1,\ldots,x_m|y_1,\ldots,y_m) = F(x_1,\ldots,x_m).$$

This shows that the limiting distribution is ergodic.

5.3. The $M|G|m|0$ System

5.3.1. *The Ergodic Theorem*

Consider the system $M|G|m$ with the distribution function $B(x)$ of service time and parameter λ of the incoming stream.

The result presented subsequently states that Erlang's formula (Section 1.4) is valid for arbitrary $B(x)$ where $\rho = \lambda\tau$, $\tau = \int_0^\infty \bar{B}(t)\,dt$.

Let $\nu(t)$ denote the number of servers that are in service at time t. Assume that at a certain instant of time $\nu(t)$ takes on the value k $(1 \leqslant k \leqslant m)$ [hence $\nu(t-0) \neq k$]. We randomly assign indices from 1 to k to those servers who are busy at time t. This can be envisioned as follows: k balls are placed in an urn with the numbers of the busy servers indicated on them. Then one ball is selected randomly: if its number is j_1, the index 1 is assigned to the j_1th server and the ball is not returned to the urn. The index 2 will correspond to the second ball selected and so on until all the balls are selected. The assigned indices are thus valid as long as $\nu(t)$ does not attain a new value. Now denote by $\xi_j(t)$ the time from instant t up to the instant when the server with index j completes the service that was in progress at time t. The random process

$$\zeta(t) = \{\nu(t); \xi_1(t),\ldots,\xi_{\nu(t)}(t)\}$$

is the basis for our discussion below.

Introduce the notation

$$F_h(t; x_1, x_2, \ldots, x_k) = \mathsf{P}\{v(t) = k; \xi_1(t) < x_1, \xi_2(t) < x_2, \ldots, \xi_k(t) < x_k\},$$
$$0 \leqslant k \leqslant n. \tag{1}$$

Clearly

$$p_k(t) = F_k(t; \infty, \infty, \ldots, \infty),$$

and hence

$$p_k = \lim_{t \to \infty} F_k(t; \infty, \infty, \ldots, \infty), \tag{2}$$

so that the determination of function $F_k(\ldots)$ will solve the problem at hand.

Theorem. *If $\tau < \infty$, then the random process $\zeta(t)$ possesses an ergodic stationary distribution.*

PROOF. The previously introduced random process is a regenerative one. First we shall prove that regeneration times that are the times when $v(t-0) = 0$, $v(t+0) = 1$ form an infinite sequence or equivalently each busy period of at least one server is finite with probability 1.

Let t_0 be regeneration time, ω be the duration of the busy period that started at this instant. Then for $a > 0$, $n \geqslant 1$

$$\mathsf{P}\{\omega > na\} \leqslant \mathsf{P}\{v(t_0 + a) > 0, \ldots, v(t_0 + na) > 0\}.$$

Denote by $v_0(t)$ the number of servers that are busy at time t whose total service time is not less than a units of time; $v_1(t) = v(t) - v_0(t)$. Then

$$\mathsf{P}\{v(t_0 + a) > 0, \ldots, v(t_0 + na) > 0\}$$
$$\leqslant \mathsf{P}\{v_0(t_0 + a) > 0, \ldots, v_0(t_0 + na) > 0\} + \sum_{k=1}^{n} \mathsf{P}\{v_1(t_0 + ka) > 0\}. \tag{3}$$

We take $k = 1$, $t_0 = 0$. If at no interval $((i-1)/N, i/N)$, $\Lambda \leqslant i \leqslant [Na] + 1$, a customer with a service time longer than $a - (i/N)$ arrived we have $v_0(a) = 0$. Whence

$$\mathsf{P}\{v_0(a) = 0\} \geqslant \prod_{i=1}^{[Na]+1} \left(1 - \frac{\lambda}{N} \bar{B}\left(a - \frac{i}{N}\right) + o\left(\frac{1}{N}\right)\right). \tag{4}$$

The same inequality is also fulfilled for $\mathsf{P}\{v_0(ka) = 0 | v_0(a| > 0, \ldots, v_0((k-1)a) > 0\}$. Utilizing the formula $1 - z = \exp\{-z + o(z)\}$ for each term of the product on the right-hand side of (4) and N approaching infinity we obtain

$$\mathsf{P}\{v_0(a) = 0\} \geqslant \exp\left\{-\lambda \int_0^a \bar{B}(t)\, dt\right\} \geqslant e^{-\rho}. \tag{5}$$

In order that the inequality $v_1(t) > 0$ be valid, it is necessary that in some interval $(x, x + dx)$ a customer arrives whose service time is longer than

$\max\{a, t - x\}$. Whence

$$\mathsf{P}\{v_1(t) > 0\} \leqslant \lambda \int_0^t \bar{B}(\max\{a, x - t\})\, dx \leqslant \lambda \left(a\bar{B}(a) + \int_a^\infty \bar{B}(x)\, dx \right) = \beta_a. \tag{6}$$

With the aid of (3) we obtain from the bounds (5) and (6)

$$\mathsf{P}\{\omega > na\} \leqslant (1 - e^{-\rho})^n + n\beta_a. \tag{7}$$

We have

$$a\bar{B}(a) = a \int_a^\infty dB(x) \leqslant \int_a^\infty x\, dB(x) \xrightarrow[a \to \infty]{} 0; \int_a^\infty \bar{B}(x)\, dx \xrightarrow[a \to \infty]{} 0$$

being a remainder of a convergent integral. Therefore, $\beta_a \to 0$, $a \to \infty$. Now the bounds (7) easily implies that $\mathsf{P}\{\omega > z\} \to 0$, $z \to \infty$, i.e., ω is a proper random variable. □

The average duration of the interval in which $v(t) = 0$ is equal to $1/\lambda$; at the same time for $p_0(t) = \mathsf{P}\{v(t) = 0\}$ we have analogously to (5) the bound

$$p_0(t) \geqslant e^{-\rho}. \tag{8}$$

We thus obtain that the mathematical expectation of a busy interval is finite. Ergodicity of the process $\zeta(t)$ now follows from Smith's theorem on regenerative processes.

5.3.2. *Proof of Sevast'yanov's Formula*

We shall present here a proof of Sevast'yanov's formula based on an embedded Markov chain.

Denote by t_n the instant of arrival into the system of the nth customer ($n \geqslant 1$). Let $F^n_{k;i_1,\ldots,i_k}(x_1, \ldots, x_k)$ denote the probability of the following event. At time $t_n - 0$ k servers are busy, namely, the servers with the numbers $i_1, \ldots, i_k$; at least $x_1, \ldots, x_k$ time units are still required until the completion of the current services.

We shall prove that $F^n_{k;i_1,\ldots,i_k}(x_1, \ldots, x_k)$ possess limits that are independent of the initial state.

It follows from the mathematical law of a stationary queue that the stationary distributions of the Markov chain $\zeta_n = \{v(t_n - 0); \xi_1(t_n - 0), \xi_2(t_n - 0), \ldots\}$ and of the random process $\zeta(t)$ coincide.

Let χ be the number of customers serviced during the busy interval ω. Then $\chi + 1$ is the return time of ζ_n into the state 0. We have

$$\mathsf{P}\{\chi > n\} = \mathsf{P}\{z_1 + \cdots + z_n \leqslant \omega\}, \tag{9}$$

where z_k are the interarrival intervals. Next

$$\mathsf{P}\{z_1 + \cdots + z_n \leqslant \omega\} \leqslant \mathsf{P}\{z_1 + \cdots + z_n \leqslant n/(2\lambda)\} + \mathsf{P}\{\omega > n/(2\lambda)\}. \tag{9}$$

If we prove that the right-hand side of (10) is a general term of a convergent series, then in view of (9) we shall arrive at

$$\mathsf{M}\chi = \sum_{n=0}^{\infty} \mathsf{P}\{\chi > n\} < \infty. \tag{11}$$

First, $\sum_n \mathsf{P}\{\omega > n/(2\lambda)\} < \infty$ since $\mathsf{M}\omega$ is finite. Furthermore, for any $s > 0$

$$\begin{aligned}\mathsf{P}\{z_1 + \cdots + z_n \leqslant n/(2\lambda)\} &\leqslant e^{sn/(2\lambda)}(\mathsf{M}e^{-sz_1})^n \\ &= (e^{s/(2\lambda)}/(1 + s/\lambda))^n = \exp\left\{-n\left(\frac{s}{2\lambda} + O(s^2)\right)\right\}\end{aligned}$$

(as $s \to 0$).

Thus $\mathsf{P}\{z_1 + \cdots + z_n \leqslant n/(2\lambda)\} \leqslant \theta^n$, $0 < \theta < 1$. Hence, the average return time of ζ_n into the state 0 is finite. Also a transition from 0 into 0 in one step [with probability $\int_0^\infty e^{-\lambda x}\,dB(x)$] is possible. Hence, $\{\zeta_n\}$ possesses an ergodic distribution.

We introduce the notation $P_n(A) = \mathsf{P}\{\zeta_n \in A\}$. Since the distribution is ergodic to prove Sevast'yanov's formula, it is sufficient to show that, if $\{P_{n-1}(A)\}$ is given by this formula, then $\{P_n(A)\}$ is also of the same form.

Denote

$$p^n_{k;i_1,\ldots,i_k}(x_1,\ldots,x_k) = \frac{\partial^k}{\partial x_1 \cdots \partial x_k} F^n_{k;i_1,\ldots,i_k}(x_1,\ldots,x_k)$$

and set

$$p^{n-1}_{k;i_1,\ldots,i_k}(x_1,\ldots,x_k) = \frac{1}{m!}(m-k)!\,\lambda^k \prod_{i=1}^{k} \bar{B}(x_i).$$

Since the interarrival time is exponentially distributed with parameter λ,

$$p^n_{k;i_1,\ldots,i_k}(x_1,\ldots,x_k) = \lambda \int_0^\infty e^{-\lambda t}\varphi_k(t;x_1,\ldots,x_k)\,dt,$$

where $\varphi_k(t;x_1,\ldots,x_k)$ is the conditional probability of the event under consideration given that $t_n - t_{n-1} = t$.

The formula of the total probability yields the equality

$$\begin{aligned}&m!(\lambda p^n_0)^{-1} p^n_{k;i_1,\ldots,i_k}(x_1,\ldots,x_k) \\ &\quad = (m-k)! \int_0^\infty e^{-\lambda t} L^{m-k}\Pi\,dt + \sum_{r=k}^{m-1} \frac{\lambda^r}{(r-k)!} \int_0^\infty e^{-\lambda t} B(t) L^{r-k}\Pi\,dt,\end{aligned}$$

where for abbreviation we set

$$L = \int_0^t \bar{B}(x)\,dx, \qquad \Pi = \prod_{i=1}^{k} \bar{B}(t + x_i).$$

Integration by parts results in

$$\int_0^\infty e^{-\lambda t} B(t) L^k \Pi \, dt$$

$$= \int_0^\infty e^{-\lambda t} L^k \Pi \, dt - \frac{1}{k+1} \int_0^\infty e^{-\lambda t} L^{k+1} \Pi \left[\lambda \, dt + \sum_{j=1}^k dB(t + x_j)/\bar{B}(t + x_j) \right].$$

Substituting the last equality into the initial formula we obtain

$$m!(\lambda p_0^n)^{-1} p_{k;i_1,\ldots,i_k}^n(x_1, \ldots, x_k)$$

$$= (m-k)! \, \lambda^{k-1} \int_0^\infty e^{-\lambda t} \Pi \left[\lambda \, dt + \sum_{j=1}^k dB(t + x_j)/\bar{B}(t + x_j) \right]$$

$$= -(m-k)! \, \lambda^{k-1} \int_0^\infty d(e^{-\lambda t} \Pi)$$

$$= -(m-k)! \, \lambda^{k-1} e^{-\lambda t} \prod_{i=1}^k \bar{B}(t + x_i)|_{t=0}^\infty = (m-k)! \, \lambda^{k-1} \prod_{i=1}^k \bar{B}(x_i).$$

Sevast'yanov's formula is now obtained by summing over $i_1, \ldots, i_k$ and integrating with respect to $x_1, \ldots, x_k$.

5.4. More Complex Systems with Losses

We shall consider several more complex systems with losses all having the following common features:

1. The probability that a customer arrives in the interval $(t, t + dt)$ depends on the "qualitative" state of the process at the given time and does not depend on prehistory.
2. Servicing of a customer is characterized by a distribution function of the amount of work and the speed of executing this work in a particular state.
3. Probabilities of "qualitative" states (values of a discrete component of a Markov process) in a stationary mode do not depend on the form of the distribution of the amount of work (the service time) provided the mathematical expectation is fixed.
4. If it is known that at a given instant (in the stationary mode) the system is in a particular qualitative state and specific operations are performed where $\xi_1, \ldots, \xi_k$ are the remaining amounts of work (service times), then $\xi_1, \ldots, \xi_k$ are independent and possess the density

   $$p_i(x) = \bar{B}_i(x)/\tau_i, \quad 1 \leqslant i \leqslant k,$$

   where $B_i(x)$ are the distribution functions of amounts of work (service times) realized at the times of arrival of customers of a particular type.

Proofs of formulas will not be presented. Basically they fit into the framework presented in Section 5.3 in the course of the proof of Sevast'yanov's theorem although in many cases ingenuity is required to arrive at the final formulas. Original proofs obtained by various authors are based as a rule on differential equations for multidimensional Markov processes with a qualitative component and with additional components that vary linearly.

5.4.1. *Reliability of a Renewable System*

Consider the following model due to Mar'yanovich (1961b).

Let a system be given with elements (units) subject to random failures. A malfunctioning unit is immediately sent for repair. The renewal time is assumed to be a random variable with the distribution function $B(x)$ and a finite mathematical expectation τ. Given that k units failed the probability that during time dt yet another unit will fail equals $\lambda_k\,dt$. We shall assume that the sequence $\{\lambda_k\}$ is bounded.

Stationary probabilities of the states of the system are of the form

$$p_k = \frac{1}{k!}\lambda_0 \ldots \lambda_{k-1}\tau^k \Big/ \left(\sum_{i=0}^{\infty} \frac{1}{i!}\lambda_0 \ldots \lambda_{i-1}\tau^i\right), \quad k \geqslant 0.$$

Here p_k is the probability that at a given instant k units are being repaired.

Based on Mar'yanovich (1961b) we shall consider the following example. A system consists of $n + m + r$ units of which n are "main" units, m constitute active and r passive redundant elements. Failed main units are replaced by the active redundant units and these in turn by passive ones. The "main" or active redundant units fail with intensity λ; Units that are in a passive redundant state do not fail. In our example

$$\lambda_k = \begin{cases} \lambda(n+m), & k \leqslant r, \\ \lambda(n+m+r-k), & k > r\ ; \end{cases}$$

$$p_k = \frac{1}{k!}(\lambda(n+m)\tau)^k p_0, \quad k \leqslant r+1;$$

$$p_k = \frac{1}{k!}(\lambda\tau)^k (n+m)^{r+1}(n+m-1)\cdots(n+m+r-k+1)p_0,$$

$$2 \leqslant k - r \leqslant n + m.$$

The availability coefficient is computed by the formula

$$K_\Gamma = \sum_{k=0}^{m+r} p_k.$$

The well-known Erlang formula (1.4.15) for the distribution of the number of serviceable devices, which was derived under the assumption of exponential distribution of the service time, follows from Mar'yanovich's formula as a particular case; one should only replace the renewal intensity by a quantity that is the reciprocal of the mean renewal time.

Let T_0 and T_1 be the mean durations of the system in serviceable and unserviceable states, respectively. If the system is in a stable state, then the times of transition from serviceable to nonserviceable states form a stationary stream of homogeneous events with intensity

$$\mu = (T_0 + T_1)^{-1} = \lambda_{m+r} p_{m+r}.$$

The second equation is obtained also from ergodic considerations:

$$T_0/(T_0 + T_1) = K_\Gamma.$$

Whence

$$T_0 = K_\Gamma/(\lambda_{m+r} p_{m+r}),$$

$$T_1 = (1 - K_\Gamma)/(\lambda_{m+r} p_{m+r}).$$

5.4.2. *A Renewable System with a Variable Renewal Rate*

Unlike in the Mar'yanovich model, let $B(x)$ be interpreted as the amount of work required for the renewal (repair) of a unit, and the rate of renewal, i.e., the rate of decrease in the remaining amount of work is equal to $\alpha_k > 0$ given that k units have failed.

Let $\nu(t)$ be the number of failed units at time t. Performing a random substitution of time

$$s = \int_0^t \alpha_{\nu(u)}\, du$$

we shall denote $\nu^*(s) = \nu(t)$. In the new time the renewal will occur with a unit rate and λ_k should be replaced by $\lambda_k^* = \lambda_k/\alpha_k$.

Thus

$$\lim_{s\to\infty} \mathsf{P}\{\nu^*(s) = k\} = \frac{1}{k!}\lambda_0^* \cdots \lambda_{k-1}^* \tau^k \Bigg/ \left(\sum_{i=0}^{\infty} \frac{1}{i!}\lambda_0^* \cdots \lambda_{i-1}^* \tau^i\right).$$

For the initial process

$$p_k = \lim_{t\to\infty} \mathsf{P}\{\nu(t) = k\} = \frac{p_k^*}{\alpha_k} \Bigg/ \sum_{i=0}^{\infty} \frac{p_i^*}{\alpha_i}.$$

This formula may be verified directly from the system of equations for stationary probabilities. Its ergodic interpretation is of interest: if the fractions of the s-time that the process spends in the states k are equal to p_k^*, then in t-time they are proportional to p_k^*/α_k.

5.4.3. *Incompletely Accessible Service System*

In telephone communication the following model is used. The subscribers are subdivided into two groups and the connecting lines into three sets L_1, L_2, L_{12} so that only subscribers of the first group can use the lines belonging to

L_1, only subscribers of the second group have access to L_2, while the lines in the set L_{12} are accessible to the subscribers of both groups provided the lines of "their own" subset are occupied.

An analogous problem arises in reliability theory. We present Yaroshenko's (1962) result.

A system consists of two subsystems; let the parameters of the stream of failures be λ_1 and λ_2 and the distribution functions of the renewal times be $H_1(x)$ and $H_2(x)$; let $\tau_l = \int_0^\infty x\,dH_l(x)$, $l = 1, 2$. n_1 workers are assigned to the first subsystem and n_2 to the second; n_{12} workers can service both subsystems. The workers of the third group are assigned to the lth subsystem only when all of the corresponding n_l workers are busy ($l = 1, 2$). Denote by $F_{ij}(x_1, \ldots, x_i; y_1, \ldots, y_j)$ the probability of the following compound event; i workers are servicing the first subsystem, j workers the second, and the repairs in progress at the given instant will be completed after at most $x_1, \ldots, y_j$ time units, respectively. If this event occurs at time t, then between the times t and $t + h$ service in the first subsystem can commence only if

$$i < n_1 \quad \text{or} \quad i \geqslant n_1 \quad \text{and} \quad n_1 + n_{12} - i - \max[0, j - n_2] > 0. \tag{1}$$

In the same manner the condition to resume service in the second subsystem will be

$$j < n_2 \quad \text{or} \quad j \geqslant n_2 \quad \text{and} \quad n_2 + n_{12} - j - \max[0, i - n_1] > 0. \tag{2}$$

Denote

$$\Delta_1(i, j) = \begin{cases} 1, & \text{if condition (1) is satisfied} \\ 0, & \text{otherwise;} \end{cases}$$

$$\Delta_2(i, j) = \begin{cases} 1, & \text{if condition (2) is satisfied} \\ 0, & \text{otherwise.} \end{cases}$$

For the negative value of at least one of the arguments we define $\Delta_l(i, j) \equiv 0$.

The stationary condition is of the form:

$$F_{ij}(x_1, \ldots, x_i; y_1, \ldots, y_j)$$
$$= \frac{\lambda_1^i \lambda_2^j}{i!\,j!} \prod_{i'=1}^{i} \int_0^{x_{i'}} \bar{B}_1(u)\,du \prod_{j'=1}^{j} \int_0^{y_{j'}} \bar{B}_2(u)\,du \prod_{i'=0}^{i-1} \Delta_1(i', 0) \prod_{j'=0}^{j-1} \Delta_2(i, j') F_{00},$$
$$0 \leqslant i \leqslant n_1 + n_{12}, \quad 0 \leqslant j \leqslant n_2 + n_{12}.$$

It follows from conditions (1) and (2) defining the functions $\Delta_l(i, j)$ that

$$\prod_{i'=0}^{i-1} \Delta_1(i', 0) = \Delta_1(i - 1, 0); \qquad \prod_{j'=0}^{j-1} \Delta_2(i, j') = \Delta_2(i, j - 1);$$
$$\Delta_1(i - 1, 0)\Delta_2(i, j - 1) = \Delta_2(i, j - 1).$$

Substituting these values in the preceding formula and approaching all $x_{i'}$ and $y_{j'}$ to infinity we obtain an expression for the probability that i workers are servicing the first subsystem and j workers the second:

$$\frac{(\lambda_1\tau_1)^i(\lambda_2\tau_2)^j\Delta_2(i,j-1)}{i!j!}F_{00}.$$

The constant F_{00} is determined from the normalizing condition.

5.4.4. *A Necessary and Sufficient Condition for Solvability of the State Equations of a System in Constants*

In all the examples presented previously we had the following situation: the stationary probabilities of the states of the system were determined by the intensity of the stream and the mean service time. An interesting question is, for which systems are the stationary probabilities invariant with respect to the form of the service time distribution, if the mean waiting time is fixed?

Kovalenko (1962) dealt with this question for a general scheme. His analytical result may be formulated in the terminology of reliability theory.

Consider a complex device consisting of s groups of units that may fail. If a unit fails, its repair begins immediately. The distribution function $B_i(x)$ of the renewal time may depend on the serial number i of the group to which the serviced unit belongs.

The failure law is as follows. Let at some instant t k_1 units in the first group, k_2 in the second, ..., and k_s in the sth malfunction. Then the probability that between t and $t+h$ another unit from the jth group fails is of the form

$$\lambda_j(k_1,k_2,\ldots,k_s)h+o(h);$$

the probability that two or more units fail during this period is $o(h)$.

Thus, in general, the reliability of the units of any group may be influenced by the number of malfunctioning units in both the given and the other groups. If the total number of units is finite, it is easy to prove the existence and ergodicity of the stationary distribution of the process

$$v(t)=\{v_1(t),v_2(t),\ldots,v_s(t)\},$$

where $v_j(t)$ is the number of units in the jth group malfunctioning at the time t.

Denote

$$p(k_1,k_2,\ldots,k_s)=\lim_{t\to\infty}\mathsf{P}\{v(t)=\{k_1,k_2,\ldots,k_s\}\}.$$

Theorem. *The probabilities $p(k_1,k_2,\ldots,k_s)$ are independent of the functional form of $B_1(x)$, $B_2(x)$, ..., $B_s(x)$, where*

$$\tau_i=\int_0^\infty x\,dB_i(x),\quad 1\leqslant i\leqslant s,$$

is fixed, if and only if for any N, for any ordered N-tuple of natural numbers

$$(m_1, m_2, \ldots, m_N),$$

where all $m_i \leqslant s$, *and for any permutation of this N-tuple*

$$(m_{l_1}, m_{l_2}, \ldots, m_{l_N})$$

the equality

$$\prod_{\mu=1}^{N} \lambda_{m_\mu}\left(\sum_{\nu=1}^{\mu-1} \delta_{1,m_\nu}, \sum_{\nu=1}^{\mu-1} \delta_{2,m_\nu}, \ldots, \sum_{\nu=1}^{\mu-1} \delta_{s,m_\nu}\right)$$
$$= \prod_{\mu=1}^{N} \lambda_{m_{l_\mu}}\left(\sum_{\nu=1}^{\mu-1} \delta_{1,m_{l_\nu}}, \sum_{\nu=1}^{\mu-1} \delta_{2,m_{l_\nu}}, \ldots, \sum_{\nu=1}^{\mu-1} \delta_{s,m_{l_\nu}}\right) \tag{3}$$

is valid. *If this condition is satisfied,* $\mathsf{P}(k_1, k_2, \ldots, k_s)$ *is determined by the formula*

$$\begin{aligned} p(k_1, \ldots, k_s) = {} & \frac{1}{k_1! k_2! \ldots k_s!} \lambda_1(0,0,\ldots,0)\lambda_1(1,0,\ldots,0)\cdots \\ & \times \lambda_1(k_1 - 1, 0, \ldots, 0)\lambda_2(k_1, 0, \ldots, 0)\lambda_2(k_1, 1, 0, \ldots, 0)\cdots \\ & \times \lambda_2(k_1, k_2 - 1, 0, \ldots, 0)\lambda_3(k_1, k_2, 0, \ldots, 0)\cdots \\ & \times \lambda_s(k_1, k_2, \ldots, k_{s-1}, k_s - 1)p(0,0,\ldots 0). \end{aligned} \tag{4}$$

The cumbersome condition of this theorem has a very simple interpretation. Its meaning is that in any situation the probability that a unit of the ith group fails in the interval $(t - h, t)$ and a unit of the jth group fails in $(t, t + h)$ coincides up to terms of order $o(h^2)$ with the probability that a unit of the jth group fails in $(t + h, t)$ and a unit of the ith group fails in $(t, t + h)$. The reader is invited to verify that formulations in Sections 5.4.1–5.4.3 are particular cases of this scheme.

We shall prove the above theorem, which is due to Kovalenko (1962), (1963) and independently to König.

Necessity. Assume that for some fixed $\{\lambda_j(k_1, k_2, \ldots, k_s)\}$ and $\{\tau_j\}$, $p(k_1, k_2, \ldots, k_s)$ does not depend on the form of the distribution $\{B_j(x)\}$ as long as the mathematical expectations

$$\int_0^\infty \bar{B}_j(x)\,dx = \tau_j, \quad 1 \leqslant j \leqslant s$$

are fixed. We express $\{B_j(x)\}$ in the following form:

$$\begin{aligned} & B_j(x) = 1 - \exp\{-x/\tau_j\}, \quad j \neq i; \\ & B_i(x) = 1 - \mu \exp\{-\mu x/\tau_i\}, \quad 0 < \mu < 1, x \geqslant 0. \end{aligned} \tag{5}$$

In other words, the renewal times η_j for the units of all groups except the ith are exponentially distributed, while η_i is a mixture of an exponential and a degenerate random variable: the renewal time vanishes with a positive probability, while when it is positive its conditional distribution is exponential.

For brevity, we shall introduce the vector notation. Denote the k-tuple $\{k_1, \ldots, k_s\}$ by k, writing $\lambda_i(k) = \lambda_i(k_1, k_2, \ldots)$, $p(k) = p(k_1, \ldots)$. We denote by the symbol e_j a vector of the form

$$\{\overbrace{0, 0, \ldots, 0, 1}^{j}, 0, \ldots, 0\}.$$

Vector addition will have the usual meaning—namely, addition of the corresponding components.

In view of (5) the process $v(t)$ is Markovain, and the usual methods yield equations that relate the probabilities of its different states. One should remember that in this case two transitions of the process from one state to another may occur at the same instant: a unit fails, but its renewal time may be zero. However, when this happens the process returns to its original state and, hence, such transitions may be ignored. The steady-state equations are of the form

$$\left[\frac{\mu k_i}{\tau_i} + \sum_{j \neq i} \frac{k_j}{\tau_j} + \lambda_i(k)\mu + \sum_{j \neq i} \lambda_j(k)\right] p(k)$$
$$= \frac{\mu(k_i + 1)}{\tau_i} p(k + e_i) + \sum_{j \neq i} \frac{k_j + 1}{\tau_j} p(k + e_j)$$
$$+ \lambda_i(k - e_i)\mu p(k - e_i) + \sum_{j \neq i} \lambda_j(k - e_j) p(k - e_j). \tag{6}$$

Here it should be assumed that

$$\lambda_j(k)p(k) = 0 \quad \text{if } k = \{k_1, k_2, \ldots, k_s\} \text{ and } \min_{1 \leqslant j \leqslant s} k_j < 0.$$

This corresponds to the probabilistic meaning of the constants $p(k)$.

Comparing the coefficients of μ at both sides of (6), we obtain

$$\left[\frac{k_i}{\tau_i} + \lambda_i(k)\right] p(k) = \frac{k_i + 1}{\tau_i} p(k + e_i) + \lambda_i(k - e_i) p(k - e_i). \tag{7}$$

Fixing k_j for $j \neq i$, the solution of the system is

$$p(k) = \frac{\tau_i^{k_i}}{k_i!} \prod_{r=1}^{k_i} \lambda_i(k - re_i) p(k - k_i e_i). \tag{8}$$

Now fix in (8) all k_j except for k_i and $k_{i'}$; we may then express $p(k)$ in terms of $p(k - k_i e_i - k_{i'} e_{i'})$, etc. Finally, we obtain

$$p(k) = \prod_{i=1}^{s} \frac{\tau_i^{k_i}}{k_i!} \prod_{l=1}^{k_1 + \cdots + k_s} \lambda_{i_l}\left(\sum_{j=1}^{s} e_j \sum_{m=1}^{l-1} \delta_{j, i_m}\right) p(0), \tag{9}$$

where $\{i_1, i_2, \ldots, i_{k_1 + \ldots + k_s}\}$ is a sequence of integers such that exactly k_j of them equals j, $1 \leqslant j \leqslant s$. Equal terms of this sequence are located one next to the other. Since i, i', etc., may be chosen in an arbitrary order, Eq. (9) [provided $p(0)$ is not zero which is obviously the case] is consistent if and only if the system of equations (3) is satisfied.

Sufficiency. Consider another model in which it is assumed that all the elements of the system are of different types, i.e., the components of the state vector $v(t) = (v_1(t), \dots, v_s(t))$ have only zero or one components; otherwise all the assumptions stipulated at the beginning of this subsection remain unchanged. It turns out that the new model is as general as the original one. Indeed since the total number of elements is finite, one may correspond to the state $(k_1, \dots, k_s)$ of the initial vector an arbitrary vector of the form

$$(x_1, \dots, x_{N_1}; x_{N_1+1}, \dots, x_{N_1+N_2}; \dots; x_{N_1+\cdots+N_{s-1}+1}, \dots, x_{N_1+\cdots+N_s})$$

whose components 0 and 1 are such that among the first N_1 components there are exactly k_1 ones, among the next N_2 components there are k_2 ones, and so on. We assume (and this is an essential assumption!) that, when an element of a certain type fails, each one of the zero components of the corresponding array becomes one with equal probability. It is easy to verify that under this assumption condition (3) is fulfilled for the process in the new space of states. It is therefore sufficient to prove the sufficiency of the condition (3) for the new model; we shall then obtain the same assertion by appropriately summing up the probabilities of "microstates."

Denote by $F_k(x_{i_1}, \dots, x_{i_k})$ the stationary probability that at a given time the elements whose numbers correspond to the unit components of vector k are malfunctioning (and only these elements) and until the completion of the renewal of each one of them less than $x_{i_1}, \dots, x_{i_k}$ time units remain, respectively. Equating the distributions related to the times t and $t + dt$ results in a system of differential equations

$$\sum_{j=1}^{s} \frac{\partial F_k}{\partial x_j} - \lambda(k) F_k + \sum_{j=1}^{s} F_{k-e_j} \lambda_j(k - e_j) B_j(x_j)$$
$$= \sum_{j=1}^{s} \frac{\partial F_k}{\partial x_j}\bigg|_{x_j=0} - \sum_{j=1}^{s} \frac{\partial F_{k+e_j}}{\partial x_j}\bigg|_{x_j=0}, \tag{10}$$

where F_{k-e_j} depends on the same variables x_i as F_k except for x_j; $F_{k \pm e_j} = 0$ if the vector $k \pm e_j$ contains components that are different from 0 and 1; $\lambda(k) = \sum_{j=1}^{s} \lambda_j(k)$. Direct substitution of (9) into (10) verifies that the distribution is stationary. One can show that the process possesses an ergodic distribution. An alternative proof is to use the same arguments as in the proof of Sevast'yanov's theorem. Observe that if s_k is the nth (in ascending order) time of renewal of an element, then $\zeta(s_n - 0)$ forms an embedded Markov chain whose probability distribution is determined by the functions $(\partial F_k/\partial x_j)|_{x_j=0}$. Each equation of the system can be integrated, which will result in an integral equation in the functions $(\partial F_k/\partial x_j)|_{x_j=0}$, i.e., a system of equations for the ergodic distribution of an embedded Markov chain is obtained.

5.4.5. *Further Generalizations*

König and Matthes discovered a very interesting fact: Sevast'yanov's formula remains valid if we relax the assumption that the service times of different customers are mutually independent. These times may be arbitrarily dependent random variables possessing an ergodic distribution.

Next König and Matthes introduced the notion of "marked (graded) service": a service for each customer consists of several stages each having a definite characteristic. Thus, the formula for the state probabilities—taking the characteristics of serviced customers into account—depends only on the mathematical expectation of the duration of separate stages of the individual service. Formula (4) [under condition (3)] may also be generalized to the case when the durations of service are arbitrarily dependent (with ergodicity conditions) and the service is subdivided into a number of stages.

The proof of these assertions is quite cumbersome; it involves operations with invariant measures in the space of infinite sequences.

We shall consider the following particular case, which has numerous interpretations.

In the first of the previously stated models we shall omit the condition of independence of durations of service. Let there be given s independent Markov chains on the phase space $(X, \mathfrak{G})$ with the ergodic distribution $\pi(dz)$ and the transition function $g(dz|z_0)$. Transitions of the ith Markov chain are associated with the times of the beginning of renewal of the ith element. We define the random processes $z_1(t), \ldots, z_s(t)$, where $z_i(t)$ is the state of the ith Markov chain after the transition associated with the last (before time t) failure of the ith element. If at time t the failure of the ith element occurred and the process $z_i(t)$ proceeds into the state z, then the renewal time η_i of this element possesses the distribution function $B_i(x|z)$ with continuous in x and π-measurable with respect to z density $b_i(x|z)$, where

$$\tau_i = \int_X \left\{\int_0^\infty x b_i(x|z)\, dz\right\} \pi(dz) < \infty.$$

It is assumed that in this situation η_i does not depend on the whole prehistory of the process; thus, the Markov chains, in a certain sense, absorb into themselves the possible dependence among the service times of an element.

Denote by $F_k(dz_1, \ldots, dz_s; x_{i_1}, \ldots, x_{i_l})$ the stationary probability of the event that $\nu(t) = k = e_{i_1} + \cdots + e_{i_l}$ with $z_i(t) \in dz_i$, $1 \leqslant i \leqslant s$; the remaining renewal durations are less than $x_{i_1}, \ldots, x_{i_l}$;

$$f_k(\cdots) = \frac{\partial^{(l)} F_k(\cdots)}{\partial x_{i_1} \cdots \partial x_{i_l}}.$$

From (10) we obtain

$$\sum_{j=1}^{s} \frac{\partial F_k}{\partial x_j} - \lambda(k) f_k + \sum_{j=1}^{s} \lambda_j(k - e_j) \int_X f_{k-e_j}(\cdots du \cdots) g(dz_j|u)\, du b_j(x_j|z_j)$$

$$+ \sum_{j=1}^{s} f_{k+e_j}|_{x_j=0} = 0. \tag{11}$$

We verify that the following distribution

$$f_k(dz_1,\ldots,dz_s;x_{i_1},\ldots,x_{i_l})$$
$$= p_0\Lambda_k\pi(dz_1)\cdots\pi(dz_s)\bar{B}_{i_1}(x_{i_1}|z_{i_1})\cdots\bar{B}_{i_l}(x_{i_l}|z_{i_l}), \tag{12}$$

where

$$\Lambda_k = \lambda_{i_1}(0)\lambda_{i_2}(e_{i_1})\cdots\lambda_{i_l}(e_{i_1}+\cdots+e_{i_{l-1}})$$

satisfies the system of equations (11) under condition (3).

The key step in the proof is that, when (12) is substituted into (11), the variables on the right-hand side are separated and the integral

$$\int g(dz_j|u)\pi(du) = \pi(dz_j)$$

is singled out, which allows us to cancel $\pi(dz_1)\cdots\pi(dz_s)$ on both sides of the equation.

We thus obtain a system of equations, which in addition to the variables contains only $\bar{B}_i(x)$ and $b_i(x)$. It is easy to verify that the system obtained is an identity.

Consider a simple case when under the conditions of the theorem of the preceding subsection the service time η_j of a customer of the jth type is comprised of several stages.

Let $(T, T+\eta_j)$ be the interval of servicing a customer of the jth type. Assume that for all t in this interval a process $\alpha(t)$ with values $\alpha_1, \alpha_2, \ldots, \alpha_m, \ldots$ is defined where τ_{jm} ($\sum_{m=1}^{\infty}\tau_{jm} = \tau_j$) is the mathematical expectation of the sojourn time in the state α_m.

Formula (4) gives the total probability of servicing customers of different types. Let it be known that a customer of the jth type is serviced. Then in accordance with (10) the conditional distribution of the remainder of the service time will have the density

$$\bar{B}_j(x)/\tau_j, \quad x > 0. \tag{13}$$

We shall calculate the conditional probability (under the stated condition) that $\alpha(t) = m$ ($m = 1, 2, \ldots$). They simplest way to do this is to use renewal theory.

It is known that for a renewal process with $F(x) = B(x)$ the probability density of the time until the next renewal is of the form (13). Hence, the probability of falling into a certain section η_j is the same as if the renewal process

$$t_1 = \eta_j^1, \quad t_2 = \eta_j^1 + \eta_j^2, \ldots, \quad t_n = \eta_j^1 + \eta_j^2 + \cdots + \eta_j^n, \ldots$$

were valid where η_j^k are independent realizations of the random variable η_j. In that case, however, the probability of falling into a point for which $\alpha(t) = m$ is equal to the ratio of the mean duration in this state to the total time:

$$\mathsf{P}\{\alpha(t) = m\} = \frac{\tau_{jm}}{\tau_j}.$$

Thus one can obtain the overall service probabilities as well as the conditional probabilities of sojourn in various states $\alpha_1, \alpha_2, \ldots$. The problem is thus completely solved.

EXAMPLE. Let there be a queueing system with losses, m servers, and a simple incoming stream with parameter λ. The service time η_0 of a customer is a random variable.

If at time t servicing of a customer commences, then under the assumption that $\eta_0 > x$ the server fails before the time $t + x$ with a positive probability. (We shall denote the failure-free service time by η_1.) In this case the customer is lost and the server is being repaired. The duration of the repair is a random variable η_2. How should one calculate the probabilities of various states?

Set

$$\eta = \begin{cases} \eta_0 & \text{for } \eta_0 < \eta_1, \\ \eta_1 + \eta_2 & \text{for } \eta_0 \geqslant \eta_1. \end{cases}$$

Then η is the busy time of the server. Denote

$$\tau = \mathsf{M}\eta.$$

Under this condition the probability that k servers are busy is obtained from Sevast'yanov's formulas

$$p_k = \frac{(\lambda\tau)^k}{k!} \Bigg/ \sum_{i=0}^{m} \frac{(\lambda\tau)^i}{i!}, \quad 0 \leqslant k \leqslant m.$$

Now the conditional probability that out of k busy servers i are in service and $k - i$ are in repair is equal to

$$C_k^i a^i (1 - a)^{k-i},$$

where

$$a = \frac{\mathsf{M}\{\min(\eta_0, \eta_1)\}}{M\eta},$$

$\min(\eta_0, \eta_1)$ is the time during which the server is servicing a customer. If η_0, η_1, and η_2 are given by their distributions, it is easy to obtain integral expression for a and τ thus completely solving the problem.

This example is due to Mar'yanovich (1960).

5.4.6. *The Problem of Redundancy with a Redistributed Load*

We shall now illustrate the possibility of deriving explicit formulas in the case when the distribution associated with the arrival of customers is of a general form.

Let two elements with the reliability distribution function $A(x)$ be given. If at a given instant both elements are functioning, the reliability decreases with a unit rate and will have the interpretation of the remaining time of faultless performance; if one element is in the state of failure, the reliability of the other decreases with the rate $a > 0$. The renewal of the elements is carried out independently; $B(x)$ is the distribution function of the renewal time.

Consider the random process $\zeta(t) = (v_1(t), v_2(t); \xi_1(t), \xi_2(t))$, where $v_i(t)$ is the indicator function of the failed state of the ith element, $\xi_i(t)$ $(i = 1, 2)$ are its reliability for $v_i(t) = 0$ and the duration until its renewal for $v_i(t) = 1$, respectively. The process is a piecewise-linear Markov process. Its additional components being positive decrease with the unit rate in any case except for the case when $v_i(t) = 0$ and $v_{3-i}(t) = 1$. In the last case $\xi_i(t)$ decreases with the rate a, $\xi_{3-i}(t)$ with the rate 1. As the component $\xi_i(t)$ reaches the zero value, the state $v_i(t)$ changes and the new value of $\xi_i(t)$ is a random variable with the distribution $A(x)$ or $B(x)$ depending on the new value of $v_i(t)$.

Let $F_{ij}(x, y) = \mathsf{P}\{v_1 = i, v_2 = j, \xi_1 < x, \xi_2 < y\}$ where (v_1, v_2, ξ_1, ξ_2) is a random vector corresponding to the steady state of the process $\zeta(t)$. The system of equation

$$\frac{\partial F_{00}}{\partial x} + \frac{\partial F_{00}}{\partial y} - \left.\frac{\partial F_{00}}{\partial x}\right|_{x=0} - \left.\frac{\partial F_{00}}{\partial y}\right|_{y=0} + \left.\frac{\partial F_{10}}{\partial x}\right|_{x=0} A(x) + \left.\frac{\partial F_{01}}{\partial y}\right|_{y=0} A(y) = 0,$$

$$\frac{\partial F_{10}}{\partial x} + a\frac{\partial F_{10}}{\partial y} - \left.\frac{\partial F_{10}}{\partial x}\right|_{x=0} - a\left.\frac{\partial F_{10}}{\partial y}\right|_{y=0} + \left.\frac{\partial F_{00}}{\partial x}\right|_{x=0} B(x) + \left.\frac{\partial F_{11}}{\partial y}\right|_{y=0} A(y) = 0,$$

$$a\frac{\partial F_{01}}{\partial x} + \frac{\partial F_{01}}{\partial y} - a\left.\frac{\partial F_{01}}{\partial x}\right|_{x=0} - \left.\frac{\partial F_{01}}{\partial y}\right|_{y=0} + \left.\frac{\partial F_{11}}{\partial x}\right|_{x=0} A(x) + \left.\frac{\partial F_{00}}{\partial y}\right|_{y=0} B(y) = 0,$$

$$\frac{\partial F_{11}}{\partial x} + \frac{\partial F_{11}}{\partial y} - \left.\frac{\partial F_{11}}{\partial x}\right|_{x=0} - \left.\frac{\partial F_{11}}{\partial y}\right|_{y=0} + a\left.\frac{\partial F_{01}}{\partial x}\right|_{x=0} B(x) + a\left.\frac{\partial F_{10}}{\partial y}\right|_{y=0} B(y) = 0$$

is established in the usual manner.

We introduce the notation

$$A^*(x) = \int_0^x \bar{A}(t)\,dt, \qquad B^*(x) = \int_0^x \bar{B}(t)\,dt$$

and attempt to satisfy this system of equations by setting

$$F_{00}(x, y) = \alpha A^*(x) A^*(y),$$

$$F_{10}(x, y) = F_{01}(y, x) = \beta B^*(x) A^*(y),$$

$$F_{11}(x, y) = \gamma B^*(x) B^*(y),$$

where α, β, and γ are constants. In other words we are assuming that for fixed

$v_1(t)$, $v_2(t)$ the "remaining" variables are independent and are distributed as the value of a jump (a transition) in a renewal process model.

Substitution into the first equation of the system yields the equality

$$(\beta - \alpha)(A(x)A^*(y) + A^*(x)A(y)) = 0.$$

Set $\alpha = \beta$. Then the second equation of the system (as also the third equation symmetric to it) results in the equality

$$\gamma - a\alpha = 0.$$

Setting $\gamma = a\alpha$ we obtain that the fourth equation yields an identity. Utilizing the normalization condition we obtain

$$1/\alpha = \tau_0^2 + 2\tau_0\tau_1 + a\tau_1^2,$$

where

$$\tau_0 = \int_0^\infty \bar{A}(x)\,dx, \qquad \tau_1 = \int_0^\infty \bar{B}(x)\,dx.$$

Stationary probabilities of the consolidated states are

$$p_{00} = \alpha\tau_0^2, \quad p_{10} = p_{01} = \alpha\tau_0\tau_1, \quad p_{11} = \alpha a\tau_1^2.$$

The second method of derivation is based on ergodic considerations. Let $\sigma(t) = a$ if $v_1(t) + v_2(t) \geqslant 1$ and $\sigma(t) = 1$ otherwise. We carry out the time substitution $s = \int_0^t \sigma(u)\,du$ and set $\zeta^*(s) = \zeta(t)$. In the s-time the random processes $(v_1^*(s), \xi_2^*(s))$ and $(v_2^*(s), \xi_2^*(s))$ are independent. Denote $F_j^*(x) = \mathsf{P}\{v_i^*(s) = j, \xi_i^*(s) < x\}$ assuming the stationary distribution. Then

$$F_0^{*\prime}(x) - F_0^{*\prime}(0) + \frac{1}{a}F_1^{*\prime}(0)A(x) = 0,$$

$$\frac{1}{a}F_1^{*\prime}(x) - \frac{1}{a}F_1^{*\prime}(0) + F_0^{*\prime}(0)B(x) = 0,$$

since in s-time the rate of renewals equals $1/a$. The solution of the last system of equations is of the form

$$F_0^*(x) = A^*(x)/(\tau_0 + a\tau_1), \qquad F_1^*(x) = aB^*(x)/(\tau_0 + a\tau_1).$$

Whence

$$p_0^* = F_0^*(\infty) = \tau_0/(\tau_0 + a\tau_1), \qquad p_1^* = F_1^*(\infty) = a\tau_1/(\tau_0 + a\tau_1).$$

The stationary distribution $(v_1^*(s), v_2^*(s))$ is of the form

$$p_{00}^* = \tau_0^2/(\tau_0 + a\tau_1)^2, \qquad p_{10}^* = p_{01}^* = a\tau_0\tau_1/(\tau_0 + a\tau_1)^2,$$
$$p_{11}^* = a^2\tau_1^2/(\tau_0 + a\tau_1)^2.$$

Transition to the t-time is carried out according to the formulas

$$p_{00} = ap_{00}^*/(ap_{00}^* + p_{10}^* + p_{01}^* + p_{11}^*) = \tau_0^2/(\tau_0^2 + 2\tau_0\tau_1 + a\tau_1^2),$$
$$p_{10} = p_{01} = \tau_0\tau_1/(\tau_0^2 + 2\tau_0\tau_1 + a\tau_1^2),$$
$$p_{11} = a\tau_1^2/(\tau_0^2 + 2\tau_0\tau_1 + a\tau_1^2).$$

Now it is sufficient to observe that

$$F_{ij}(x, y) = p_{ij}[F_i^*(x)/F_i^*(\infty)][F_j^*(y)/F_j^*(\infty)].$$

For example,

$$F_{00}(x, y) = A^*(x)A^*(y)/(\tau_0^2 + 2\tau_0\tau_1 + a\tau_1^2).$$

The intensity μ of the stream of failures of the system interpreted as falling into the state (1, 1) is determined by the formula

$$\mu = (F_{11}'(x, \infty) + F_{11}'(\infty, x))|_{x=0} = 2a\tau_1/(\tau_0^2 + 2\tau_0\tau_1 + a\tau_1^2).$$

The average time T_0 of a "serviceable" and T_1 of an "unserviceable" state of the system is determined by the formulas $T_0 = p_{00}/\mu$, $T_1 = (1 - p_{00})/\mu$.

5.5. Ergodic Theorems

5.5.1. *Sevast'yanov's Theorem*

Among the ergodic theorems used in queueing theory the following theorem due to Sevast'yanov is one of the most fruitful.

Theorem. *Let $\xi(t)$ be a homogeneous Markov process in the measurable space $(X, \mathfrak{G})$, where $\mathfrak{G}$ is a class of Borel processes with measurable with respect to x transition function $P_x(t, A)$, x is the initial state, $t \geqslant 0$, $A \in \mathfrak{G}$; X_n, $n \geqslant 1$, are selected compact sets in $\mathfrak{G}$, φ is a probability measure on $(X, \mathfrak{G})$ for which the following properties are satisfied:*

1.
$$\liminf_{n\to\infty} \lim_{x} \lim_{t\to\infty} P_x(t, X_n) = 1. \tag{1}$$

2.
$$P_x(T_n, A) \geqslant \rho_n \varphi(A), \quad x \in X_n, \quad A \subset X_n, \tag{2}$$

for some T_n and $\rho_n > 0$.

3. *For any initial distribution P_0 there exist $t_0 = t_0(n)$ and $K = K_n$ such that*

$$\overline{\lim_{n\to\infty}} \sup_{A\subset X_n} [P\{\xi(t) \in A\} - K\varphi(A)] \leqslant 0. \tag{3}$$

Then there exists an ergodic distribution $\pi = \{\pi(A), A \in \mathfrak{G}\}$ of the process $\xi(t)$:

$$\sup_{A\in\mathfrak{G}} |P_x(t, A) - \pi(A)| \xrightarrow[t\to\infty]{} 0, \quad x \in X. \tag{4}$$

We shall focus on the basic steps of the proof. We shall prove in full the important fact that as $t \to \infty$ the distributions of the process corresponding to different initial distributions converge. This part of the proof is based on the method of innovation moments developed by Borovkov (1980).

Let ψ be an arbitrary probability measure on $(X, \mathfrak{G})$. Denote $P_\psi(t, A) = \int P_x(t, A)\psi(dx)$; let T_ψ be a number such that for all $t \geqslant T_\psi$, $P_\psi(t, X_0) \geqslant \lambda$, $0 < \lambda < 1$, X_0 is an arbitrary X_n such that

$$\inf_x \lim_{t\to\infty} P_x(t, X_n) > \lambda.$$

[The existence of T_ψ follows from (1) and the measurability of the transition function.]

Assume that $\mathsf{P}\{\xi(0) \in A\} = \psi_0(A)$, $A \in \mathfrak{G}$, and set $t_1 = t_1(\psi_0) = T_{\psi_0} + T$, $T = T_n$. Then

$$P_{\psi_0}(t_1, A) \geqslant P_{\psi_0}(T_{\psi_0}, X_0) \inf_{x \in X_0} P_x(T, A) \geqslant \theta\varphi(A), \quad A \in \mathfrak{G}, \tag{5}$$

where $\theta = \lambda\rho$, $\rho = \rho_n$, or, equivalently,

$$P_{\psi_0}(t_1, A) = \theta\varphi(A) + (1 - \theta)\psi_1(A), \tag{6}$$

where ψ_1 is a probability measure.

Equation (6) can be interpreted as follows. A random trial is performed with the probability of success θ. If the trial is successful, $\xi(t_1)$ has the distribution φ, otherwise the distribution is ψ_1. If the trial is successful, we say that t_1 is an innovation time. Otherwise the same construction can be applied to ψ_1 in place of ψ_0 with a time shift in the amount t_1. We then obtain that $t_2 = t_1 + T_{\psi_1} + T$; this time may be an innovation time with probability θ. This process can be continued indefinitely.

As a result we obtain

$$\begin{aligned} P_{\psi_0}(t, A) &= \theta P_\varphi(t - t_1, A) + (1 - \theta)\theta P_\varphi(t - t_2, A) + \cdots \\ &\quad + (1 - \theta)^{k-1}\theta P_\varphi(t - t_k, A) + (1 - \theta)^k Q_{\psi_0}(t, A), \quad t > t_k, \end{aligned} \tag{7}$$

where $Q(t, A)$ is a probability distribution.

Now let ψ be an arbitrary probability measure on $(X, \mathfrak{G})$.

Clearly the sequence $\{t_k\}$ may be selected to be the same for the initial distributions ψ_0 and ψ. Then Eq. (7) will be valid for $P_\psi(t, A)$ also with ψ_0 replaced by ψ, whence

$$|P_\psi(t, A) - P_{\psi_0}(t, A)| \leqslant (1 - \theta)^k, \quad t > t_k,$$

and hence

$$\sup_A |P_\psi(t, A) - P_{\psi_0}(t, A)| \xrightarrow[t\to\infty]{} 0. \tag{8}$$

We introduce a distance on the space $\mathfrak{M}$ of probability measures on $(X, \mathfrak{G})$ to be the *distance in variation*:

$$d(\varphi, \psi) = \int_X |\varphi(dx) - \psi(dx)| = 2 \sup_{A \in \mathfrak{G}} |\varphi(A) - \psi(A)|. \tag{9}$$

Then $\mathfrak{M}$ becomes a complete metric space. Conditions (1)–(3) imply that for a given x_0 there exists a sequence $\{T_k\}$ such that

$$d(P_x(T_k, \cdot), \pi(\cdot)) \xrightarrow[k\to\infty]{} 0, \tag{10}$$

where $\{\pi(A)\}$ is a probability measure on $(X, \mathfrak{G})$. We shall denote by ψ_0 the measure concentrated at x_0, $\psi(A) = P_{x_0}(t, A)$. Then

$$P_{\psi_0}(t + T_k, A) = \int P_x(t, A) P_{\psi_0}(T_k, dx)$$

$$= \int P_x(t, A)\pi(dx) + \int P_x(t, A)[P_{\psi_0}(T_k, dx) - \pi(dx)]. \tag{11}$$

In view of (10) the last summand is infinitesimal. On the other hand, in view of (8)

$$P_{\psi_0}(t + T_k, A) = P_{\psi}(T_k, A) \xrightarrow[k\to\infty]{} \pi(A).$$

Now we obtain from (11)

$$\pi(A) = \int P_x(t, A)\pi(dx), \tag{12}$$

i.e., π is the stationary distribituon of the process $\xi(t)$. Relation (4) follows from the fact that

$$P_x(t, A) - \pi(A) = P_x(t, A) - P_\pi(t, A),$$

and it remains to apply (8) once again.

5.5.2. *Construction of Innovation Times*

In queueing systems in which periodical clearing of the system occurs, the "clearing off" times are characterized by the fact that for the subsequent course of the process the whole prehistory is irrelevant—one should only know the times of arrival of customers in the future and durations of their service. Thus, if these times are independent of the prehistory, then the arrival time that follows the "clearing off" time will be an innovation time. Hence, all the systems with a simple incoming stream possess innovation times. The system $GI|G|m$ for $m \geqslant 2$ also possesses "innovation" times for ρ sufficiently small but in a deterministic stream with an interarrival intervals Δ and service time uniformly distributed in the interval $((m-1)\Delta, m\Delta)$ there will always be customers in the system. Nevertheless, for $\rho < m$ it is possible to construct innovation times: if interarrival intervals are larger and the service times are smaller than certain fixed values, then in a finite number of steps the queue will disappear and then all the subsequent course of the process is independent of the past. We have utilized this device in the proof of the Kiefer-Wolfowitz theorem for a $GI|G|m$ system (See section 5.2).

Now let a service system with several types of customers and service operations be given. The operation of the system is determined by the sequence $\{z_k^{(r)}, \eta_k^{(r)}\}$ where $z_k^{(r)}$ are the interarrival intervals, $\eta_k^{(r)}$ are the service times, k is the (serial) number, and r is the type of customer (service). Assume that there exists a "nominal" mode of operation of the system characterized by the "tolerance" $z_k^{(r)} \leqslant a^{(r)}$, $\eta_k^{(r)} \geqslant b^{(r)}$, violation of these inequalities is viewed as a disturbance. Any reasonably constructed system is capable of "working out" disturbances: if starting with some instant of time the elements of the "controlling" sequences $\{z_k^{(r)}, \eta_k^{(r)}\}$ are within the tolerance limits, then in a finite number of steps the process will settle into a "nominal" mode. If the latter is characterized by the stationarity of the distribution of the basic Markov process, then the time of arrival at this mode will be an innovation time. It is only important to verify that the probability that the "tolerance" inequalities are satisfied is positive.

As an example consider a multiphase system of service with blocking

$$GI|G|m_1|r_1 \to |G|m_2|r_2 \to |\cdots \to |G|m_s|r_s,$$

consisting of many-server subsystems with a limited number of waiting locations. Customers that form a renewal process form a queue at the first subsystem provided there are already $m_1 + r_1$ customers in it. If in the jth subsystem ($2 \leqslant j \leqslant s$) there are $m_j + r_j$ customers, then after completion of the service in the ($j - 1$)th subsystem this customer continues to occupy the server ("blocking" it) as long as the approach to the jth subsystem is blocked.

Denote by z_k the interarrival intervals, $\eta_k^{(i)}$ durations of servicing customers in the ith subsystem. Also let $\gamma(t)$ be total work to be done after time t for completion of servicing customers present in the system at time t. As long as there is only at least one customer in the system, $\gamma'(t) \leqslant -1$; hence, if sufficiently often we have consecutively $\eta_k^{(1)} + \cdots + \eta_k^{(s)} \leqslant z_{k+1} - \varepsilon$, $\varepsilon > 0$, then a "clean-up" time will occur [for any given current value of $\gamma(t)$].

Consider the method due to Kovalenko and Kuznetsov (1980), which utilizes the existence of an absolutely continuous component of distribution of the random variables that determine the service process. Our exposition will be sufficiently general to clarify the basic idea.

Let the behavior of the system be described by a homogeneous Markov process $\zeta(t) = (\zeta_1(t), \ldots, \zeta_m(t))$ where $\zeta_i(t)$ are numerical variables and for some point $x^0 = (x_1^0, \ldots, x_m^0)$, the process with probability 1 returns infinitely often to a neighborhood of this point $U_\varepsilon(x^0) = \{(x_1, \ldots, x_m)\colon |x_i - x_i^0| < \varepsilon, 1 \leqslant i \leqslant m\}$.

It is assumed that up to the instant of a jump, the process follows a determined trajectory and the distance between the possible trajectories is not increased: if $\zeta(t)$ and $\zeta^0(t)$ are trajectories of the process corresponding to the initial states x and x^0 at time t_0, then it follows from $x \in U_\varepsilon(x^0)$ that

$$\zeta(t) \in U_\varepsilon(\zeta^0(t)), \quad t > t_0.$$

The time of a jump for the first trajectory is τ, for the second τ^0 where $|\tau - \tau^0| \leqslant c_0 \varepsilon$. [If in the intervals between the jumps the process follows the differential equation $\zeta'(t) + \alpha(\zeta(t)) = 0$, $\alpha(x) = (\alpha_1(x), \ldots, \alpha_m(x))$, $\alpha_i(x) \geqslant \alpha' > 0$, and the jumps occur at the instants at which some component of the process reaches a fixed level, then it is easy to verify that $c \leqslant 1/\alpha'$.] Denote by $H(\bar{x}|\bar{y})$ the distribution of $\zeta(t+0)$ where t is the instant of a jump given that $\zeta(t-0) = \bar{y}$. For some m-dimensional set $V(\bar{y}^0)$ of a positive measure $|V|$, let $dH(\bar{x}|\bar{y}) \geqslant \beta\, dx_1 \cdots dx_m$, $\beta > 0$, for any $\bar{y} \in U_\varepsilon(\bar{y}^0)$ and $\bar{x} \in V$. Then after each hit of $\zeta(t)$ into $U_\varepsilon(x^0)$ an innovation time occurs at which the value of $\zeta(t)$ is uniformly distributed on the set $V(\bar{y}^0)$. (We note that the innovation times corresponding to different trajectories are shifted in time one relative to the other.)

Often a situation arises when only one component is "innovated" at a jump receiving a nonzero increment. In such a case one must anticipate a time when all the components are "innovated".

EXAMPLE. Let there be m independent renewal processes with the distribution $B(x)$ of interrenewal times and $\zeta_i(t)$ be the time until the next renewal of the ith one of these processes. Assume that $dB(t) \geqslant \beta\, dt$ in the interval $(a, a + \Delta)$, where $a > 0$, $\Delta > 0$. We fix an ε, $0 < \varepsilon < \Delta$, $\varepsilon < a$, and let t be the time when some $\zeta_i(t-0) = 0$ and all the others $\zeta_j(t-0) < \varepsilon$. Then during the time ε a renewal of all these processes will occur. If $\zeta(t-0) = \bar{y}$,

$$dF_\Delta(\bar{x}|\bar{y}) = d\mathsf{P}\{\zeta(t+\Delta) < \bar{x} | \zeta(t-0) = \bar{y}\},$$

then

$$dF_\Delta(\bar{x}|\bar{y}) \geqslant \beta\, dx_1 \cdots dx_m,\ a < x_i < a + \Delta - \varepsilon,\ 1 \leqslant i \leqslant m.$$

Consequently the renewal will occur at time $t + \Delta$ with probability $\beta(\Delta - \varepsilon)^m$; at that instant the distribution of the state of the process is uniform in the hypercube

$$a < x_i < a + \Delta - \varepsilon, \quad 1 \leqslant i \leqslant m.$$

5.5.3. *Stability of Queueing Systems*

Let the operation of the system be determined by the recurrence relation

$$z(n+1) = f(z(n), \xi(n)), n \geqslant 0, \tag{13}$$

where $\alpha = (z(0), \xi(0), \xi(1), \ldots)$ is the so-called controlling sequence, $z(n)$ is the state of the system at the n-step. Along with (13) consider the sequence $(z^*(n))$ determined by the same relation but with the controlling sequence

$$\alpha^* = (z^*(0), \xi^*(0), \xi^*(1), \dots).$$

The stability (continuity) theorems establish the proximity between the sequences $(z(n))$ and $(z^*(n))$ for sufficiently small deviations of α^* from α. The proximity is determined in terms of a metric, i.e., the distance

$$\rho(\alpha, \alpha^*) = \max\{\|z(0) - z^*(0)\|, \sup_{n \geqslant 0} \|\xi(n) - \xi^*(n)\|\}, \tag{14}$$

where $\|x\|$ is the norm of the element x [the length of the vector if $z(n)$ and $\xi(n)$ are the vectors in the Euclidean space]. It is possible to consider the analog of (14) when $\|\cdots\|$ is replaced by $\mathsf{M}\|\cdots\|^r$, $r \geqslant 1$. The theory of stability of the systems described by eq. (13) for i.i.d. $\xi(n)$ was devised by Kalashnikov (1978), who developed the method of sample functions that generalizes the direct method of Lyapunov.

A Markov chain $[z(n), n \geqslant 0]$ with values in a metric space Z is called *stable on the average in time* if for any $\varepsilon > 0$ a constant $\delta > 0$ is found such that for

$$\mathsf{M}\rho_z(z(0), z^*(0)) < \delta, \qquad \mathsf{M}\rho_\xi(\xi(n), \xi^*(n)) < \delta, \quad n \geqslant 0,$$

where ρ_z and ρ_ξ are distances in the corresponding spaces the inequality

$$\overline{\lim_{n\to\infty}} \frac{1}{n} \sum_{k=0}^{n-1} \mathsf{P}\{\rho_z(z(n), z^*(n)) \geqslant \varepsilon\} < \varepsilon \tag{15}$$

is fulfilled. Let μ_1 be the distance between the distributions $z(0)$, $z^*(0)$, and μ_2 be the distance between the distributions $\xi(1)$ and $\xi^*(1)$. Let $\pi(Q, Q^*)$ be the Levy-Prokhorov distance between the distributions in Z, i.e., the minimal value of $\varepsilon > 0$ for which for any measurable $B \subset Z$

$$Q(B) \leqslant Q^*(B_\varepsilon) + \varepsilon, \qquad Q^*(B) \leqslant Q(B) + \varepsilon,$$

where B_ε is an ε-neighborhood of B. It is assumed that for any admissible distributions of $z(n)$ and $\xi(0)$ there exists the limiting distribution $Q(B)$ of the random variable $z(n)$ $(n \to \infty)$.

A Markov chain $\{z(n)\}$ is weakly stable in the limit if $\pi(Q, Q^*) < \varepsilon$ for $\mu_1 \leqslant \delta(\varepsilon)$, $\mu_2 \leqslant \delta(\varepsilon)$. It was proved by Kalashnikov that (15) implies the weak stability of $\{\xi(n)\}$ in time.

To establish the weak stability Kalashnikov considers a sequence $x(n) = (z(n), z^*(n))$ and its generating operator

$$AW(x) = \mathsf{M}\{W(x(1)) | x(0) = x\}, \tag{16}$$

where $W(x)$ is a nonnegative function.

Under certain conditions on $AW(x)$ for the corresponding function $W(x)$, the Markov chain $\{z(n)\}$ is stable on the average in time. The most important condition here is the inequality

$$AW(x) \leqslant -\varepsilon < 0$$

for $\rho(z, z^*) \geqslant \delta$.

5.6. Heavily Loaded Queueing Systems

5.6.1. *Limit Theorem for Distribution of Waiting Time in a GI|G|1 System*

A general limit theorem of probability theory will be required in what follows; the theorem was proved by Khinchin (1932) in connection with the second diffusion problem.

Let a sequence of random variables $s_0, s_1, s_2, \ldots, s_n, \ldots$ forming a Markov chain with the distribution of one-step transition probabilities

$$F(x, y) = \mathsf{P}\{s_n < y | s_{n-1} = x\}$$

and a certain initial state s_0, $a \leqslant s_0 \leqslant b$ be given. Furthermore, let two barriers at levels a and b be available.

The question is: What is the probability $F(x)$ that given $s_0 = x_0$ the upper barrier ($x = b$) will be the first one attained?

Theorem. *Let $F(x, y)$ depend on a parameter ε where for $\varepsilon \to 0$ the limiting relations (Lindeberg conditions)*

$$\begin{gathered} \mathsf{M}\{s_n - s_{n-1} | s_{n-1} = x\} = \varepsilon\alpha(x) + o(\varepsilon), \\ \mathsf{M}\{(s_n - s_{n-1})^2 | s_{n-1} = x\} = \varepsilon\beta(x) + o(\varepsilon), \\ \mathsf{M}\{(s_n - s_{n-1})^2[E(s_n - s_{n-1} - \tau) + E(\tau - s_n + s_{n-1})] | s_n = x\} = o(\varepsilon) \end{gathered} \tag{1}$$

for any $\tau > 0$ are valid where

$$E(x) = \begin{cases} 1 & x > 0 \\ 0 & x < 0, \end{cases}$$

$\alpha(x)$ and $\beta(x)$ are continuous functions and $o(\varepsilon)$ is uniformly in x. Then $F(x)$ as $\varepsilon \to 0$ converges to the solution of the differential equation

$$\alpha(x)\frac{dv}{dx} + \frac{1}{2}\beta(x)\frac{d^2v}{dx^2} = 0 \tag{2}$$

with the following boundary conditions:

$$v(a) = 0, \qquad v(b) = 1. \tag{3}$$

We note that a similar result is valid also in the case when the lower barrier goes to $-\infty$ provided only $\alpha(x)$ and $\beta(x)$ satisfy some additional analytic conditions [restriction of the growth of $v(a)$ as $a \to -\infty$; the boundedness of $\beta(x)$ from above and below].

The stated theorem allows us to investigate a single-server queueing system considered above under the condition of a heavy load ($\rho \to 1$ remaining less than 1).

Assume that the random variable ζ_m defined in Section 5.1 depends on a small parameter ε so that

$$\mathsf{M}\zeta_n = -\varepsilon, \tag{4}$$

$$\mathsf{D}\zeta_n = \sigma^2\varepsilon, \quad \sigma^2 > 0, \tag{5}$$

$$\mathsf{M}\{\zeta_n^2[E(\zeta_n - \tau) + E(\tau - \zeta_n)]\} = o(\varepsilon). \tag{6}$$

Clearly under these conditions $\rho \to 1$.

As it was shown previously, the stationary distribution of waiting time of an arbitrary customer equals the distribution of the maximum of a sum of independent random variables ζ_n under the condition that $\zeta_0 = 0$. Hence, we have here the case when the lower barrier is at the point $-\infty$ and the upper at point x. However, one can assume that $a = -\infty$ and $b = 0$; we then have $1 - F(x) = v(-x)$. Substituting formulas (3) and (4) into Eq. (2) yields $\alpha(x) = -1$, $\beta(x) = \sigma^2$,

$$\frac{\sigma^2}{2}\frac{d^2v}{dx^2} - \frac{dv}{dx} = 0, \tag{7}$$

$$\left.\begin{aligned} v(0) &= 1, \\ v(-\infty) &= 0. \end{aligned}\right\} \tag{8}$$

Equation (7) whose general solution is of the form

$$v(x) = Ce^{2x/\sigma^2} + C_1$$

with the boundary conditions (8) has a unique twice continuously differentiable solution

$$v(x) = e^{2x/\sigma^2},$$

whence

$$F(x) = 1 - v(-x) = 1 - e^{-2x/\sigma^2}. \tag{9}$$

We thus see that under conditions of a heavy load the distribution of waiting time approaches exponential.

We note yet another elementary probabilistic approach to the solution of this problem. In Section 5.1 it was shown that the limiting distribution of w_n as $n \to \infty$ coincides with the distribution of the variable

$$w = \sup_{n \geqslant 0} x_n,$$

where $x_0 = 0$, $x_n = x_{n-1} + \zeta_n$, $n \geqslant 1$. Denote by p the probability of the event $A_1 = \{\max_{n \geqslant 1}\{x_n\} > 0\}$, by ω_1 the first positive element of the sequence $\{x_1, x_2, \ldots\}$, by v_1 the index of this element in the given sequence. Thus, when the event A occurs, we have $\omega_1 = x_{v_1} > 0$. One can write

$$w = \omega_1 + \sup_{n \geqslant 1} x_n^{(1)},$$

where $x_n^{(1)} = x_{n+\nu_1} - x_{\nu_1} = x_{n+\nu_1} - \omega_1$. Analogously to the previous event, one can define the event $A_2 = \{\max_{n\geq 1}|x_n^{(1)}| > 0\}$; when this event occurs, we have $w = \omega_1 + \omega_2 + \sup_{n\geq 1} x^{(2)}$ and so on. We thus obtain

$$w = \sum_{k=1}^{\sigma} \omega_k,$$

where σ is the number of "records" in the sequence $\{x_n, n \geq 0\}$. Note that

$$\mathsf{P}\{\sigma = j\} = (1-p)^j, \quad j \geq 0,$$

and under the condition $\{\sigma = j\}$ the random variables ω_k which are the "record achievements" possess a common distribution $G(x)$.

For the random variable w we have obtained the same representation as for the interval between the events of a rarefied renewal stream (Chapter 2).

A direct corollary of the result in Chapter 2 concerning the distribution of this variable is the following:

Theorem. *Let p and $G(x)$ depend on the parameter $\varepsilon > 0$, $p \to 0$ as $\varepsilon \to 0$, $\tau = \int x\, dG(x)$ and for some $\varepsilon_0 > 0$*

$$\sup_{0<\varepsilon<\varepsilon_0} \frac{1}{\tau}\int_T^\infty \bar{G}(x)\,dx \xrightarrow[T\to\infty]{} 0. \tag{10}$$

Then

$$\mathsf{P}\{pw > \tau x\} \xrightarrow[\varepsilon\to 0]{} e^{-x}, \quad x \geq 0, \tag{11}$$

uniformly in $x \geq 0$.

5.6.2. *Utilization of the Invariance Principle*

Let $\{\xi_N(t)\}$ be a sequence of random processes weakly convergent to the random process $\xi(t)$. [The weak convergence signifies the convergence

$$\mathsf{P}\{\xi_N(t_i) < x_i,\ 1 \leq i \leq k\} \xrightarrow[N\to\infty]{} \mathsf{P}\{\xi(t_i) < x_i,\ 1 \leq i \leq k\}$$

for any k; $t_1, \ldots, t_k$; $x_1, \ldots, x_k$ such that $P\{\xi(t_i) = x_i\} = 0$.]

Let there be a functional $f(x(\cdot))$ defined for the functions $x(t)$, $0 \leq t \leq T$. Consider the random variables $X_N = f(\xi_N(\cdot))$, $X = f(\xi(\cdot))$. Under certain additional conditions, the distribution of X_N weakly converges to the distribution of X as $N \to \infty$.

Let $x = x(t)$, $y = y(t)$ be functions on the interval $0 \leq t \leq T$, $\rho(x, y) = \sup_{0\leq t\leq T}|x(t) - y(t)|$, and the functional is continuous in the given uniform metric:

$$f(x) \xrightarrow[\rho(x,y)\to 0]{} f(y).$$

[In particular functionals of the form $\sup_{0\leq t\leq T} x(t)$, $\inf_{0\leq t\leq T} x(t)$ and $\int_0^T \varphi(x(t))\,dt$ for uniformly continuous function $\varphi(z)$ satisfy this condition.]

The Donsker-Prokhorov invariance principle ensures the weak convergence of X_N to X in particular in the case when $\xi(t)$ is a standard Wiener process, $\xi_N(t)$ is a random broken function with the nodes at the points k/N, where $\zeta_{N_k} = \xi_N(k/N) - \xi_N((k-1)/N)$ are independent identically distributed random variables; $\mathsf{M}\zeta_{N_k} = 0$, $\mathsf{D}\zeta_{N_k} = 1/N$,

$$\mathsf{M}\zeta_{N_k}^2 I(|\zeta_{N_k}| > \varepsilon\sqrt{N}) \underset{N\to\infty}{\longrightarrow} 0$$

for any $\varepsilon > 0$ where $I(A)$ is the indicator of the event A. (The last condition is the well-known Linderberg condition.)

Let the distribution $K(x)$ of random variables ζ_k in the system $GI|G|1$ depend on the parameter N and the previously stated conditions be fulfilled in the notation $\zeta_k = \zeta_{N_k}$. Since in this case $w_{N,T} = \max_{0\leqslant t\leqslant T} \xi_N(t)$, the distribution of $w_{N,T}$ as $N \to \infty$ converges to the distribution of the variable $\max_{0\leqslant t\leqslant T} \xi(t)$. [Recall that $K(x)$ also depends on N.] If the condition $\zeta_{N_k} = 0$ is replaced by the condition $M\zeta_{N_k} = a/N$, the distribution of $w_{N,T}$ converges to the distribution of the variable $\max_{0\leqslant t\leqslant T}\{\xi(t) + at\}$.) Thus it is possible to investigate the system with critical load ($\rho = 1$) as well as systems with undercritical ($\rho < 1$) and overcritical ($\rho > 1$) loads. In the numerous papers of various authors, limit distributions for many stationary and nonstationary characteristics of systems under conditions that are close to critical ones are obtained.

5.6.3. *Borovkov's Theorem*

In Borovkov's monograph (1980) a general theory of limiting behavior of queueing processes was developed under very general conditions on the queueing system. Borovkov characterizes the behavior of a queueing system by a three-dimensional random process $S(t) = (e(t), r(t), s(t))$, where $e(t)$ is the number of customers arriving at the system before time t, $r(t)$ is the number of failures and $s(t)$ is the number of serviced customers. These functions are arbitrary; it is only assumed that the variable $q(t) = e(t) - r(t) - s(t)$ is always nonnegative. This variable is called the queue size. Quite general conditions are imposed on the local behavior (in an appropriate time scale) of the process $S(t)$. These conditions pertain to the behavior in the region where $q(t)$ is larger than the critical value. If, however, $q(t)$ is less than this value, a great variety of behavior patterns depending on the structure of the system and specific distributions arise. The limiting results are, however, invariant to these structures. Borovkov obtained limit theorems for multiserver systems with waiting and with losses in particular when the number of servers increases indefinitely and the intensity of the incoming stream tends to infinity. The controlling sequences (the intervals between the arrival of customers and service times) are generally stationary in the narrow sense random sequences satisfying conditions of a very general nature (for example, the condition of strong mixing). The limiting processes are of a very complex nature; they are reduced to diffusion processes only in some particular cases.

5.6.4. *Asymptotic Invariance*

Kuznetsov (1981) considered the case when the given invariance condition is fulfilled up to quantities $O(\varepsilon)$. Based on an analytic–statistical method, an algorithm for computing corrections to stationary probabilities of states up to quantities $o(\varepsilon)$ has been developed.

In Kuznetsov's papers (1982) and (1984) a method for computing stationary probabilities of states of the system $GI|G|m|0$ for which the incoming stream of customers in a certain sense is close to the simplest is proposed. The proximity is in the following sense.

Let the distribution function $F(x)$, which defines a recursive stream of customers, depend on a small parameter $\varepsilon > 0$ and ξ be a random variable with the distribution function $F(x)$. Under certain conditions ξ can be represented in the form

$$\xi = g(\xi_0, \eta_1, \ldots, \eta_m),$$

where $\xi_0, \eta_1, \ldots, \eta_m$ are independent random variables where ξ_0 possesses an exponential distribution and $\eta_1, \ldots, \eta_m$ are suitably chosen random variables. The proximity condition of the recurrent stream of customers to a Poisson stream is that

$$\mathsf{P}\{g(\xi_0, \eta_1, \ldots, \eta_m) \neq \xi_0\} \xrightarrow[\varepsilon \to 0]{} 0.$$

Under this condition Kuznetsov devised an analytic–statistical method for calculating deviations of stationary probabilities of the states of a $GI|G|m|0$ system from the corresponding probabilities of an $M|G|m|0$ system.

5.7. Underloaded Queueing Systems

5.7.1. *Introductory Remarks*

In Section 5.6 we considered the theory of heavily loaded queueing systems. This theory is applicable in practical problems characterized by accumulation of large queues. On the other hand, in many situations accomulation of queues is inadmissible, since it may cause material losses, breakdowns, and other undesirable effects. Problems related to such situations are of special interest in reliability theory, when it is required to construct highly reliable systems. In estimating the efficiency of such systems it is essential to investigate the case when the intensity of the incoming stream is considerably smaller than the critical value; in this case stationary conditions are not available. Furthermore, not only is a numerical value of efficiency for a given set of parameters (such as the intensity of the incoming stream) required, but also the changes in the efficiency as a function of parameters when these parameters vary. This is necessary in order to select characteristics of the system in an optimal manner. In this section we present one approach to this problem.

5.7.2. *Statement of the Problem*

Let a system $M|G|m|r$ with intensity λ of the incoming stream and the distribution function $B(x)$ of service time, $\tau = \int_0^\infty x\,dB(x) < \infty$ be given. Denote by t_k the time of the kth loss of a customer. It is required to investigate the limiting distribution of the random variables $t_1, t_2, \ldots$ as $\lambda \to 0$. This problem has many useful interpretations: in reliability theory, in the theory of calculating memories of computers, etc. The stated problem reduces directly to another one by means of the following lemma.

Lemma. *Let $x(T)$ be the total time in $(0, T)$ during which all the waiting locations are occupied. Then if for some function $\varphi(\lambda)$*

$$\lambda x(T/\varphi(\lambda)) \xrightarrow[\lambda\to 0]{p} T \tag{1}$$

for any fixed $T > 0$ ($\xrightarrow{p}$ denotes convergence in probability), the random variables $\varphi(\lambda)t_1$, $\varphi(\lambda)(t_2 - t_1), \ldots$ are asymptotically independent and asymptotically exponential with parameter 1.

Proof. Let $N(z)$ be the number of lost customers up to the time when the variable $\lambda x(T)$ attains the value z. Then independently of the process of the system's operation, $N(z)$ is the number of events of the simplest stream with parameter 1 in the interval of length z. Since the function $x(t)$ is monotone, it follows from (1) that on any finite interval with probability, arbitrarily close to 1,

$$T - \varepsilon \leqslant \lambda x(T/\varphi(\lambda)) \leqslant T + \varepsilon. \tag{2}$$

Let z_k be the time of the kth loss of a customer in variable z. In view of (2) if z_k is located in the bounded interval

$$|\varphi(\lambda)t_k - z_k| \leqslant \varepsilon,$$

i.e., the stream $\{t_k\}$ is a result of an infinitely small displacement of events of the simplest stream with parameter 1. Whence the assertion of the lemma follows. □

5.7.3. *Investigation of the Process $\lambda x(t)$*

Let $y(t)$ be the indicator of the event $\{\nu(t) = m + r\}$, where $\nu(t)$ is the number of customers in the system at time t. Then

$$x(t) = \int_0^t y(u)\,du. \tag{3}$$

If $y(t) = 1$, then a chain of transitions $0 \to 1 \to \cdots \to m + r$ with no intermediate state being 0 preceded the current state $m + r$ of the process $\nu(t)$. The chain can be monotone: $0 \to 1 \to 2 \to \cdots \to m + r - 1 \to m + r$, or it may not be monotone. In the first case we set $y_0(t) = 1$, $y_1(t) = 0$, in the second $y_0(t) = 0$, $y_1(t) = 1$.

Let ζ be a busy interval of the system. Then $\mathsf{P}\{\zeta > t\} \leqslant \mathsf{P}\{\zeta^* > t\}$, where ζ^* is the corresponding interval in the $M|G|1$ system. Indeed in the first case the service durations overlap, in the second case they are arranged sequentially in time; during the extra time other customers may arrive who will only increase ζ^*. In accordance with Chapter 4 $\mathsf{M}\zeta^* = \tau/(1 - \lambda\tau)$. If $B^*(x) = \mathsf{P}\{\zeta^* < x\}$, $p_0(t) = \mathsf{P}\{\nu(t) = 0\}$, then

$$1 - p_0(t) \leqslant \lambda \int_0^t \bar{B}^*(t - x)\, dx \leqslant \lambda \mathsf{M}\zeta^* = \lambda\tau/(1 - \lambda\tau).$$

Whence

$$p_0(t) \geqslant 1 - \lambda\tau/(1 - \lambda\tau). \tag{4}$$

Denote by $A(t, \Delta)$ the probability of the event $\{y_0(u) = 1, t \leqslant u \leqslant t + \Delta\}$. The transitions $0 \to 1 \cdots \to m + r$, which yield the current state, may occur at times $t_1 < t_2 < \cdots < t_{m+r} < t$; between the times $t_1, t_2, \ldots, t_{m+r}, t$ there were no other customers; at time t_1 the system should be free. The differential of probability of such an event equals $p_0(t_1)\lambda^{m+r}e^{-\lambda(t_{m+r}-t_1)}\, dt_1 \cdots dt_{m+r}$. Furthermore, servicings which commenced at times $t_1, \ldots, t_m$ should continue until the time $t + \Delta$. Thus,

$$\lambda^{-(m+r)}A(t, \Delta) = \int \cdots \int p_0(t_1)e^{-\lambda(t_{m+r}-t_1)}\bar{B}(t + \Delta - t_1) \cdots \bar{B}(t + \Delta - t_m)\, dt_1 \cdots dt_{m+r}, \tag{5}$$

where the domain of integration was defined previously. Two useful bounds follows immediately from formula (5). Replacing the exponent and $p_0(t)$ by 1 we arrive at the upper bound. On the other hand, denoting $\bar{B}_\lambda(x) = e^{-\lambda x}\bar{B}(x)$ we have the inequality $e^{-\lambda(t_{m+r}-t_1)}\bar{B}(t + \Delta - t_1) \cdots \bar{B}(t + \Delta - t_m) \geqslant \bar{B}_\lambda(t + \Delta - t_1) \cdots \bar{B}_\lambda(t + \Delta - t_m)$. For $p_0(t)$ we use the bound (4). In both cases we arrive at the integral of a function symmetric in $t_1, \ldots, t_m$ and $t_{m+1}, \ldots, t_{m+r}$.

Thus one can take the integral over the region $\{0 < t_{m+1} < t; 0 < t_k < t_{m+1}, 1 \leqslant k \leqslant m;\ t_{m+1} < t_k < t,\ m + 2 \leqslant k \leqslant m + r\}$ dividing the expression by $m!(r - 1)!$. We thus obtain

$$\begin{aligned} \frac{1 - 2\lambda\tau}{1 - \lambda\tau} \int_0^\infty x^{r-1}\, dx \left(\int_0^{t-x} \bar{B}_\lambda(x + y + \Delta)\, dy \right)^m dx \\ \leqslant m!(r - 1)!\,\lambda^{-(m+r)}A(t, \Delta) \\ \leqslant \int_0^\infty x^{r-1} \left(\int_0^{t-x} \bar{B}(x + y + \Delta)\, dy \right)^m dx, \quad r \geqslant 1. \end{aligned} \tag{6}$$

For $r = 0$ we have

$$\frac{1 - 2\lambda\tau}{1 - \lambda\tau} \left(\int_0^{t-x} \bar{B}_\lambda(y + \Delta)\, dy \right)^m \leqslant m!\,\lambda^{-m}A(t, \Delta) \leqslant \left(\int_0^{t-x} \bar{B}_\lambda(y + \Delta)\, dy \right)^m. \tag{7}$$

Assume that

$$\bar{B}(x) \leqslant cx^{-\beta}, \quad x > 0, \tag{8}$$

where $\beta > 1 + r/m$. Then it is easy to verify that

$$\lim_{\lambda \to 0, t \to \infty} (m!(r-1)!\,\lambda^{-(m+r)} A(t, \Delta)) = \int_0^\infty x^{r-1} \bar{\bar{B}}^m(x + \Delta)\, dx, \quad r \geqslant 1, \tag{9}$$

$$\lim_{\lambda \to 0, t \to \infty} (m!\,\lambda^{-m} A(t, \Delta)) = \bar{\bar{B}}^m(\Delta), \quad r = 0, \tag{10}$$

where $\bar{\bar{B}}(x) = \int_x^\infty \bar{B}(t)\,dt$. *Let* $x_i(t) = \int_0^\infty y_i(u)\,du$, $i = 0, 1$. Since $\mathsf{M}y_0(t) = A(t, 0)$, it follows that

$$\mathsf{M}x_0(t) \sim \frac{t}{m!(r-1)!} \lambda^{m+r} \int_0^\infty x^{r-1} \bar{\bar{B}}^m(x)\, dx, \quad r \geqslant 1, \tag{11}$$

$$\mathsf{M}x_0(t) \sim \frac{t}{m!} \tau^m, \qquad r = 0, \tag{12}$$

where in both cases $\lambda \to 0$, $t \to \infty$. We bound $\mathsf{M}y_0(t)y_0(t + \Delta)$. If $y_0(t) = y_0(t + \Delta) = 1$, then either $y_0(u) = 1$ for $t \leqslant u \leqslant t + \Delta$ or in the interval $(t, t + \Delta)$ the busy interval commences. Whence

$$\mathsf{M}y_0(t)y_0(t + \Delta) \leqslant A(t, \Delta) + A^*(t, 0)A^*(t + \Delta, 0), \tag{13}$$

where in view of (6) and (7) $A^*(t, 0)$ is the upper bound of $A(t, 0)$.

Consequently,

$$\mathsf{M}x_0^2(t) \leqslant \left(\int_0^t A^*(u, 0)\, du \right)^2 + 2tJ, \tag{14}$$

where

$$J = \frac{1}{m!(r-1)!} \lambda^{m+r} \int_0^\infty x^{r-1} \int_0^t \bar{\bar{B}}^m(x + \Delta)\, d\Delta\, dx, \quad r \geqslant 1; \tag{15}$$

$$J = \frac{1}{m!} \lambda^m \int_0^t \bar{\bar{B}}(\Delta)\, d\Delta, \quad r = 0. \tag{16}$$

The first summand on the right-hand side of (14) in accordance with the above is equivalent to $(\mathsf{M}x(t))^2$; the expression $\lambda^{-(m+r)}J$ is bounded under the bound (8) where $\beta > 1 + (r + 1)/m$. Thus under the last condition

$$(\mathsf{D}x_0(t))/(\mathsf{M}x_0(t))^2 = o(1), \tag{17}$$

provided $t\lambda^{m+r} \to \infty$, $\lambda \to 0$. Hence, for $r = 0$, say,

$$\lambda x_0(t) \Big/ \left(\frac{t}{m!} \lambda^{m+1} \tau^m \right) \xrightarrow{P} 0 \quad (\lambda \to 0, t \to \infty). \tag{18}$$

From (18) we obtain

$$\lambda x_0(T/\varphi(\lambda)) \to T \tag{19}$$

for $\varphi(\lambda) = \lambda^{m+1}\tau^m/m!$. Analogously (19) is fulfilled for $r \geqslant 1$,

$$\varphi(\lambda) = \frac{\lambda^{m+r+1}}{m!(r-1)!} \int_0^\infty x^{r-1} \bar{\bar{B}}^m(x)\,dx.$$

Consider now the random function $x_1(t)$. With probability 1, $x_1(t)$ does not exceed the total length of the busy intervals that started in $(0, t)$ during each one of which at least $m + r + 1$ customers were served. Indeed, if the trajectory $0 \to 1 \to \cdots \to m + r$ is not monotone it contains at least $m + r + 1$ overjumps. Thus,

$$\mathsf{M}x_1(t) \leqslant \lambda t \mathsf{M}\sigma, \tag{20}$$

where σ is the length of the busy interval if at least $m + r + 1$ customers were served during this interval and is zero otherwise.

We have noted the relation between the busy intervals in $M|G|m|r$ and $M|G|1$ systems. Under this correspondence, as it is easy to see, in the last case at least, as many customers are served during a busy interval as in the first case. Whence $\mathsf{M}\sigma \leqslant \mathsf{M}\sigma^*$, where σ^* is a quantity analogous to the σ for the system $M|G|1$.

We shall consider the probability structure of the random variable σ^*. We arrange in succession the service times $\eta_1, \ldots, \eta_{m+r}$ of the first $m + r$ customers. Let $s = \eta_1 + \cdots + \eta_{m+r}$. If we envision the inverse order of servicing as presented in Section 4.3, then the busy intervals $\zeta_1^*, \ldots, \zeta_k^*$ will be attached to the interval of length s as long as at least $k + m + r - 1$ customers ($k \geqslant 1$) arrive in the interval of length s. For a fixed value of s the mathematical expectation of σ^* is

$$\begin{aligned} &e^{-\lambda s} \sum_{i=m+r}^{\infty} \frac{(\lambda s)^i}{i!}(s + (i - m - r + 1)\tau/(1 - \lambda\tau)) \\ &\leqslant \frac{(\lambda s)^{m+r}}{(m+r)!}(s + (m + r)\tau/(1 - \lambda\tau)). \end{aligned} \tag{21}$$

If $\int_0^\infty x^{m+r+1}\,dB(x) < \infty$, then $\mathsf{M}s^{m+r+1} < \infty$, and in that case averaging (21) with respect to the distribution of s we obtain the bound

$$\mathsf{M}\sigma^* = O(\lambda^{m+r}). \tag{22}$$

It now follows from (20) that

$$\mathsf{M}x_1(t) = O(\lambda^{m+r+1}t), \qquad \mathsf{M}x_1(T/\varphi(\lambda)) \xrightarrow[\lambda \to 0]{p} 0. \tag{23}$$

From (19) and (23) we obtain

$$\lambda x(T/\varphi(\lambda)) \to 1. \tag{24}$$

By the lemma proved in subsection 2 it follows that the stream of losses is simplest in the limit; the random variables $\varphi(\lambda)t_1$, $\varphi(\lambda)(t_2 - t_1)$, $\ldots$ have asymptotically an exponential distribution with parameter 1.

It remains to note that condition (8) follows from the existence of the $(m + r + 1)$th moment of the distribution $B(x)$. Indeed

$$x^{m+r+1}\bar{B}(x) \leqslant \int_x^\infty y^{m+r+1}\,dB(y) \xrightarrow[x \to \infty]{} 0.$$

5.8. Little's Theory and its Corollaries

5.8.1. *General Statements*

Let an arbitrary queueing system be given. Denote by v_n the duration of the nth customer in the system, by $\nu(t)$ the number of customers in the system at time t, by $N(t)$ the number of customers who arrived in the time interval $(0, t)$.

The total time spent by the customers in the system during the interval $(0, T)$ is equal to $\int_0^T \nu(t)\,dt$. This time differs from $v_1 + \cdots + v_{N(T)}$ by no more than the total duration R of the customers who are in the system at time T. Under sufficiently general conditions R is a variable with a bounded mathematical expectation; hence, for large T

$$\frac{1}{T}\int_0^T v(t)\,dt \approx \frac{N(T)}{T}\frac{1}{N(T)}\sum_{k=1}^{N(T)} v_k.$$

Thus, if L and V are the mean stationary number of customers in the system and the mean duration of a customer in the system, respectively, and $\lambda = \lim_{T\to\infty} N(T)/T$, then

$$L = \lambda V. \tag{1}$$

Analogously the total waiting time of the customers in the interval $(0, T)$ is equal to $\int_0^T \nu_1(t)\,dt$, where $\nu_1(t)$ is the number of customers in the queue (that are not being served) at time t. Whence

$$L_0 = \lambda W, \tag{2}$$

where L_0 is the mean stationary value of $\nu_1(t)$, W is the mean waiting time of a customer. The assertions of formulas (1) and (2) are called the *Little theorem*.

We make the following assumptions:

1. Let $N(T)$ be the number of events of the incoming stream in the interval $(0, T)$ for $T > 0$ and in the interval $(T, 0)$ for $T < 0$. Then

$$\mathsf{P}\left\{\frac{N(T)}{|T|} \xrightarrow[|T|\to\infty]{} \lambda\right\} = 1.$$

2. The process $\nu(t)$ and the sequence $\{v_k\}$ are stationary and ergodic.

5.8.2. *Little's Theorem*

Theorem. *Under the conditions* 1 *and* 2 *the equality* (1) *is fulfilled.*

PROOF. We order the times t_k of arrival of customers as follows: $\cdots \leqslant t_{-1} \leqslant t_0 \leqslant 0 < t_1 \leqslant t_2 \leqslant \cdots$. Denote by $\Delta(a, b, c)$ the total duration time in the interval (b, c) of the customers who arrived in the interval (a, b). Evidently this function is nonincreasing in a and is nondecreasing in c and we shall prove below that Δ is finite. One can write

$$\int_0^T v(t)\,dt = \sum_{k=1}^{N(T)} v_k + \Delta(-\infty, 0, T) - \Delta(0, T, \infty), \quad T > 0.$$

Whence

$$\frac{1}{T}\int_0^T v(t)\,dt \leqslant \frac{N(T)}{T}\frac{1}{N(T)}\sum_{k=1}^{N(T)} v_k + \Delta(-\infty, 0, \infty)/T.$$

Thus, if $\Delta(-\infty, 0, \infty)$ is a nonsingular random variable, then $L \leqslant \lambda V$. Analogously

$$\int_{-T}^0 v(t)\,dt = \sum_{k=0}^{N(-T)-1} v_{-k} + \Delta(-\infty, -T, 0) - \Delta(-T, 0, \infty),$$

whence

$$\frac{1}{T}\int_{-T}^0 v(t)\,dt \geqslant \frac{N(-T)}{T}\frac{1}{N(-T)}\sum_{k=0}^{N(-T)-1} v_{-k} - \Delta(-\infty, 0, \infty)/T,$$

and hence $L \geqslant \lambda V$. Thus, it remains to prove that $\Delta(-\infty, 0, \infty)$ is a nonsingular random variable. By definition $\Delta(-\infty, 0, \infty) \leqslant \sum_{k=0}^{\infty}(v_{-k} + t_{-k})^+$. In view of condition 1, $t_{-k} < -ck$, $k \geqslant k_1(\omega)$, where ω is an elementary event. The ergodicity of $\{v_k\}$ implies that

$$\frac{1}{n}\sum_{k=0}^{n-1} v_{-k} \to V$$

with probability 1 and in particular $(1/n)v_{-n} \to 0$ with probability 1. Hence, only a finite number of summands in the sum $\sum(v_{-k} + t_{-k})^+$ differ from zero. Whence $\Delta(-\infty, 0, \infty)$ is a nonsingular random variable.

Remark. Formula (2) is established analogously.

Corollary. *In a $GI|G|m$ system under steady-state conditions, the mean waiting time (duration time in the system) does not depend on the rule of selection from the queue. In particular, for random equiprobable selection and inverse selection (when a server becomes free, the customer who was the last one to arrive enters the server) the mean waiting time is the same as in the case of first come first served discipline.*

We present an example of a queueing system where this property is violated (clearly also the mean length of the queue is changed). Consider the system $M|G|1$ in which, when a customer arrives at the service, the current service is interrupted and the server starts servicing the newly arrived customer. When a service is completed, the customer who arrived last is selected for the new service. This queueing discipline is called the *discipline of inverse service with interruption.*

As usual we denote: λ as the parameter of the incoming stream, $B(x)$ as the distribution of service time, ζ (possibly with the index) as the duration of a busy period.

For a customer whose service time η equals x, the duration time v in the system consists of the service time and the random number γ_x of busy intervals (with customers who arrived after the given one). The variable γ_x has a Poisson distribution with parameter λx. Whence

$$\mathsf{M}\{v|\eta = x\} = x + \mathsf{M} \sum_{k=1}^{\gamma_x} \zeta_k = x + \lambda x \mathsf{M}\zeta = x + \lambda x\tau/(1 - \rho)$$

$$= x/(1 - \rho), \quad \mathsf{M}v = \tau/(1 - \rho).$$

If we envision the waiting time w as the time when the customer awaits resumption of service (and not its beginning as it is in the case of "the usual" queueing discipline), then $v = w + \eta$. Whence

$$\mathsf{M}w = \rho\tau/(1 - \rho) = \lambda\tau^2/(1 - \rho).$$

It is of interest to note that if the service is on the first-come–first-served basis, then $\mathsf{M}w = \lambda(\tau^2 + \sigma^2)/[2(1 - \rho)]$, where σ^2 is the variance of the service time, hence this discipline is advantageous if $\sigma > \tau$. For the exponential distribution we have $\sigma = \tau$ and both disciplines yield the same value of $\mathsf{M}w$. This property is evident also from probabilistic considerations: in this case switching a server from one customer to another does not affect the behavior of the queue length.

5.8.3. *Notes*

A vast amount of literature is denoted to the solution of Eq. (12) in Section 5.1.2. and similar equations. Combinatorial methods were developed in this area in close connection with the theory of functions of complex variables and harmonic analysis (Wiener-Hopf and Krein methods). The greatest contributions to the probability-analytic aspect of the problem are due to Pollaczek, Spitzer, Borovkov, Zolotarev, Korolyuk, Rogozin, and Gusak. For the current state of the problem see Borovkov (1972).

Rossberg and Siegel (1974), (1975) and Klimova (1968) generalized the analytic theory of the system $GI|G|1$. The development of Wiener-Hopf's method and its applications to queueing theory is the subject of the papers by Prabhu (1974) and Malyshev (1976) (in addition to the works cited in the text). Numerous results on bounded random walks, which have interesting applications in queueing theory, are due to Rogozin (1964), Zolotarev, Borovkov, Presman, Korolyuk, Gusak and Ezhov.

We mention the papers by Whitt (1972) and Presman (1965) in connection with the study of $GI|G|m$ systems.

For ergodic theorems see Prokhorov and Rozanov (1973), Loynes (1962), Shurenkov (1981), Kovalenko et al. (1980). Afanas'yeva and Martynov's paper

(1969) is an example of the application of ergodic theorems to systems with constraints.

See König, Schmidt, and Stoyan (1976) and Franken, König, Arndt and Schmidt (1984) for the relations between stationary distributions and distributions of an imbedded queueing process.

Azlarov and Khusainov (1974) discuss heavily loaded systems.

The first rigorous proof of Erlang's formula for the $M|G|m|0$ system was given by Sevast'yanov (1957). Fortet (1956) generalized Erlang's formulas under the assumption that the distribution of service time is continuous.

The following papers deal with invariance theorems in queueing theory: Basharin and Kokotyshkin (1971), Chaiken and Ignall (1972), Chandy et al. (1977), Guseĭnov (1970), (1974), Kovalenko (1971), Zhuk (1968), and Takács (1975).

The result in Section 5.4.6 is due to Bogdantsev et al. (1984) using the method of semi-Markov processes with continuous phase space. Note that the result follows from the book by Franken et al. (1984).

Stoyan (1978) studies the invariance probelm for a network of queues. The network consists of N service centers. Incoming customers form a simplest stream and arrive at one of the centers with given probabilities. After the service is completed at the ith center the customer either arrives with probability p_{ij} at the jth service center or with probability $p_{i,N+1}$ leaves the system. The service time at the ith center has the distribution function $B_i(x)$, which for some i is exponential and for the others of a general form.

Assume that for the service centers with a nonexponential $B_i(x)$ one of the following three cases is valid:

1. The number of servers is infinite.
2. The number of servers is 1.
3. The number of servers is arbitrary and a customer with the smallest remaining service time possesses an absolute priority (Schrage's discipline; cf. Schrage and Miller (1966), Schrage (1968)).

The stationary state of the system is invariant to the form of $B_i(x)$ for "nonexponential" service centers for fixed $\int_0^\infty x\,dB_i(x)$.

Pellaumail (1979) investigated a network of queues consisting of several centers with an exponential service time with parameter depending on the state of the given center in the network "as a whole". Transition probabilities of customers between centers may also depend on the state of the network so that the whole network is described by an irreducible Markov chain. For the networks of this kind, a condition for statistical equilibrium is stated.

Jansen and König (1980) described a network that encompasses a wide class of open, closed, and mixed networks of queues studied in the literature. Stationary probabilities of states are the subject of investigation. Conditions are established under which these probabilities are of multiplicative form and a close connection between these properties and the properties of the model under consideration at individual nodes—for example, the simple nature of the outgoing streams of customers—is emphasized.

In Ivnitskiĭ's paper (1982) for networks of queues with a given number of customers of different classes necessary and sufficient conditions for independence of stationary probabilities of the service time distribution of customers when the mean values are fixed are established. Sufficient conditions for the existence of an ergodic distribution of the process that describes the variation in the states of the system are presented.

Köeningsberg (1981) presents a survey of the basic papers devoted to invariance principles in queueing theory that is a general basis for numerous models possessing properties of "product decomposition" of the stationary distribution of queues in a network consisting of n nodes

$$P_G(k_1, \ldots, k_n) = \prod_{i=1}^{n} P_i(k_i).$$

I.N. Kovalenko's theorem and its generalization due to Guseĭnov (1970) is discussed in detail. A number of corollaries of this theorem, some of which were independently obtained by other authors are presented. In particular, the fact that, if the stationary distribution $P_0(k_1, \ldots, k_n)$ depends on the distribution of service times $B_i(t)$, $i = 1, \ldots, n$, only through their mean values, then the product decomposition is valid, is noted. Some new applications of the invariance principle are discussed.

Basharin and Tolmachev's (1983) survey reviews the basic results in the theory of networks of queues and its applications to analysis of productivity of information-computational systems obtained before 1982. Special attention is devoted to a possible complete description of models of networks for which closed expressions of stationary distributions are available. Characteristic properties of these networks are studied and efficient computational algorithms for calculation of characteristics are presented. A number of approximation methods utilized for analytic modeling of real-world information–computational systems—such as networks of data transmission, computer networks, and their components—are described. We also note the papers by Basharin and Kokotushkin (1971) and Rybko (1982). In connection with queueing networks, see also Kelly (1976) and Walrand and Varaiya (1981).

Borovkov (1980) based on the innovation method developed by him, unified the proofs of theorems of stability and ergodicity of sequences. In his model $\{\xi(n)\}$ is an arbitrary stationary in the narrow sense metrically transitive sequence whose components are the intervals between the arrivals and service times of customers. Under the assumption of the existence of innovation times (clearing of the system) and rather mild additional conditions, the convergence of finite-dimensional distributions $\xi^*(n)$ to the distributions $\xi(n)$ implies the analogous property for characteristics of the service process (the vector of waiting times in a multiserver system with waiting, indicator of whether the servers are busy at the times of arrival of customers into a system with losses). Further results are given in Franken et al. (1984). The papers of Prokhorov

(1963), Borovkov (1972), Prokhorov and Viskov (1964) and Viskov (1964) are the first papers devoted to the application of general principles of queueing theory to the investigation of critical modes of queueing systems. In earlier papers by Kingman (1966) and Rise (1962) limit theorems were derived by means of investigating analytic expressions for characteristics of queueing systems. The previously mentioned works of Soviet scientists have shown the possibility of using general theorems on the convergence of distributions of functionals on random processes.

Asymptotic invariance is considered in the afterword to the book by Franken et al. (1984) written by Kovalenko and Kuznetsov.

Grishchenko (1983) generalized the basic theorem is Section 5.6.

Numerous authors have studied underloaded systems. We mention only some monographic literature: chapters written by Solov'yev in the book by Barzilovich et al (1983), Kovalenko (1965d), (1971), Korolyuk and Turbin (1976), (1978), Anisimov (1976), Sil'vestrov (1971). See also Stoyan (1973) and Gnedenko (1961).

See Stidham (1974) and the book by Franken et al. (1984) concerning general conditions for validity of Little's theorem.

6
Statistical Simulation of Systems

6.1. Principles of the Monte Carlo Method

6.1.1. *Foundation of the Methodology*

The Monte Carlo method is one of the best known computational methods. It involves utilization of random trials in solving mainly computational problems. To apply this method, a "source of randomness" is required, i.e., a device that produces realizations of random numbers. Such devices are called *random number generators* (RNG). RNG generate a sequence $\{\omega_n\}$, which is viewed as a sequence of independent random variables with a given distribution. Usually uniform random numbers, $\{\omega_n\}$, uniformly distributed on the interval $(0, 1)$, are used. We shall adopt this convention in what follows.

Random numbers are generated by a physical generator, i.e., a discrete transform of noises from a radio-engineering device; however, more often, in place of random numbers, *pseudorandom numbers* are used. This is a sequence $\{\omega_n\}$ produced by a computer program that simulates a random sequence in the frequency sense. Since $\{\omega_n\}$ are machine generated numbers, their distribution is actually discrete; however, for practical purposes substitution of a continuous distribution by a discrete one does not result in appreciable errors.

The method in a nutshell is as follows. A characteristic to be computed (a number, function, etc.) is represented in the form of a probability characteristic of a random process. Thus, if it is required to calculate the value of a numerical parameter a, one obtains a random process $\xi(t)$ such that $a = f(\xi(\cdot))$. (The dot in place of the argument indicates that f may be a functional of the trajectory of a process and not just a function of an instantaneous value.) Next we construct a realization of the random process $\xi(t)$ depending on a collection of independent random variables ω_i: $\xi(t) = \xi(t; \omega_1, \omega_2, \ldots)$. Evidently for physical implementation it is necessary that the number of operating variables ω_i be finite; however, this number may be different for different realizations. Reading a value of ω_i from GRN, we obtain a realization of $\xi(t)$ and with it a random variable $\eta = f(\xi(\cdot\,; \omega_1, \omega_2, \ldots))$. The experiment may be carried out N times, where N is a number that is either fixed a priori or is selected in the course of the experiment. In each realization new random numbers ω_i are utilized. We thus obtain N realizations $\eta_1, \ldots, \eta_N$. By the law of large numbers $\bar{\eta}_N = (1/N)(\eta_1 + \cdots + \eta_N)$ converges to the estimated quantity a in probability as $N \to \infty$; for a sufficiently large number of realizations N, the mean $\bar{\eta}_N$ will serve as an estimate of the quantity a with arbitrarily high precision: $\mathsf{P}\{|\bar{\eta}_N - a| > \delta\} < \varepsilon$. This estimator is an *unbiased* one; if $\sigma^2 = \mathsf{D}f(\xi(\cdot)) < \infty$, then it is also *asymptotically normal.*

Thus the error of estimating a quantity by means of the Monte Carlo method (in the case when a finite variance exists) is of order $1/\sqrt{N}$.

6.1.2. *Weighted Simulation*

In many cases this precision is not satisfactory, especially in calculations associated with rare events. In these cases *weighted simulation* is used. Let $a = \int f(x)\mathsf{P}(dx)$, where $\mathsf{P}(dx)$ is a probability measure. Then $a = \mathsf{M}f(\xi)$, where ξ is a random variable with the distribution $\mathsf{P}(dx)$. If now $q(x) \geqslant 0$ is a function such that $\int q(x)\mathsf{P}(dx) = 1$, then $Q(dx) = q(x)\mathsf{P}(dx)$ is the probability distribution of the random variable η and thus

$$a = \int \frac{f(x)}{q(x)} Q(dx) = \mathsf{M}\{f(\eta)/q(\eta)\}. \tag{1}$$

It is assumed that the integral $\int |f|\mathsf{P}(dx)$ over the set of values of x such that $q(x) = 0$ equals 0.

We tend to select the *weight function* $q(x)$ in such a manner that the variance of $f(\eta)/q(\eta)$ will be as small as possible. Let it be known, for example, that $|f(x) - f_0(x)| \leqslant \Delta(x)$, where $a_0 = \int f_0(x)\mathsf{P}(dx)$ can be calculated without simulation. Then it remains to calculate $a - a_0 = a_1 = \int g(x)\mathsf{P}(dx)$, where $g(x) = f(x) - f_0(x)$, $|g(x)| \leqslant \Delta(x)$. We have

$$\mathsf{D}\{g(\eta)/q(\eta)\} \leqslant \mathsf{M}\{g^2(\eta)/q^2(\eta)\} \leqslant \mathsf{M}\{\Delta^2(\eta)/q^2(\eta)\} = \int \{\Delta^2(x)/q(x)\}\,\mathsf{P}(dx).$$

The solution of minimizing the variational problem results in the formula

$$q(x) = \Delta(x) \bigg/ \int \Delta(y)\mathsf{P}(dy), \tag{2}$$

provided the integral in the denominator on the right-hand side of (2) converges. For $q(x)$ of this form we have

$$\mathsf{D}\{g(\eta)/q(\eta)\} \leqslant \left(\int \Delta(x)\mathsf{P}(dx)\right)^2. \tag{3}$$

6.2. Simulation of Some Classes of Random Processes

6.2.1. *Preliminary Remarks*

In order to be able to express a random process $\xi(t)$ in terms of a sequence $\{\omega_n\}$, it is required to define this process constructively in the form of a computable function of an element of the sample space. Numerous classes of random processes considered in theory and applied in practice by their very definition are constructively posed; these processes serve as a basis of mathematical models of queueing systems. This makes the process of constructing realizations an intuitive one. Examples are presented below. The method of statistical simulation usually requires not a realization of a process, but a value of a functional f on such a realization. We shall not dwell on this problem nor will we tackle the problem of a stopping rule for realizations based on time or a number of cycles.

6.2.2. *Simulation of Random Trials and Variables*

Let E be a trial with possible outcomes A_i, $\mathsf{P}\{A_i\} = p_i$. It is required to implement a trial E from the realization of a random variable ω uniform on the interval $(0, 1)$. We subdivide the interval $(0, 1)$ into arbitrary measurable sets Δ_i of the Lebesgue measure p_i. If ω falls into Δ_i, we say that the outcome A_i occurred. When operating with a uniform distribution on the set of m-digit number generated by computer, for the Lebesgue measure we must take the fraction of numbers belonging to the given set. Most often we use the subdivision of the interval $(0, 1)$ into the intervals of length p_i. A random trial with two equiprobable outcomes is implemented by means of a particular bit of a binary representation of a random number, for example, by the first bit after the decimal point. In the case of k equiprobable outcomes, one can choose different values of a vector from the first r bits where $2^{r-1} < k \leqslant 2^r$ and in the case of the appearance of a noncoded r-gram to repeat the trial (for examples, using the next r bits).

A universal method of obtaining a realization of the random variable ξ with the distribution function $F(x)$ by means of ω is as follows. Let $\psi(\omega) = \min\{x: F(x) \geqslant \omega\}$. [If the function $y = F(x)$ is continuous and strictly increasing at any point where it is greater than 0 or less than 1, then $\psi(x)$ is the inverse to $F(x)$.]

Then $\xi = \psi(\omega)$ is a random variable with the distribution function $F(x)$. Indeed the events $\{\omega < F(x)\}$ and $\{\psi(\omega) < x\}$ are equivalent (Fig. 7); it remains to note that $\mathsf{P}\{\omega < F(x)\} = F(x)$.

Let $\xi = (\xi_1, \ldots, \xi_n)$ be an n-dimensional vector. The distribution of ξ can be defined by the distribution functions

$$F_{\xi_1}(x),\ F_{\xi_2|\xi_1}(x|x_1), \ldots,\ F_{\xi_n|\xi_1,\ldots,\xi_{n-1}}(x|x_1, \ldots, x_{n-1}),$$

where $F_{\xi_1}(x)$ is the distribution function of ξ_1, $F_{\xi_k|\xi_1,\ldots,\xi_{k-1}}(x|x_1, \ldots, x_{k-1})$ is the conditional distribution function of ξ_k given that $\xi_i = x_i$, $i < k$.

The random variable ξ_1 is realized as in the one-dimensional case by means of the random variable ω_1. When $\xi_1, \ldots, \xi_{k-1}$ are realized, then ξ_k is determined in the same manner with $F(x) = F_{\xi_k|\xi_1,\ldots,\xi_{k-1}}(x|\xi_1, \ldots, \xi_{k-1})$.

Example. Let the random vector (ξ_1, ξ_2) have the density $p(x, y)$ equal to 2 for $x > 0$, $y > 0$, $x + y < 1$ and 0 otherwise. Then

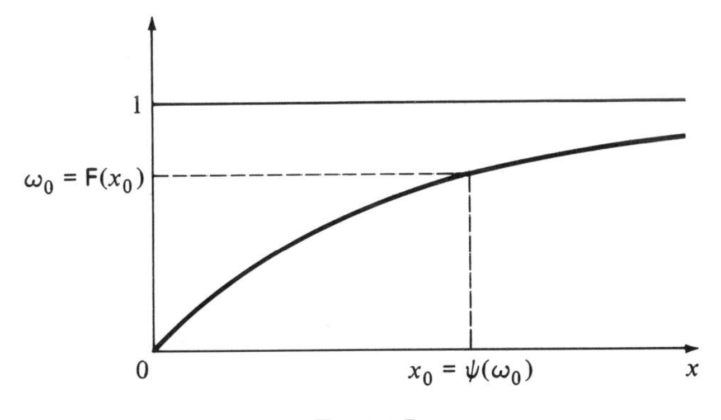

FIGURE 7

$$F_{\xi_1}(x) = 2x - x^2, \qquad 0 \leqslant x \leqslant 1;$$

$$F_{\xi_2|\xi_1}(x|y) = x/(1-y), \qquad 0 \leqslant x \leqslant 1-y.$$

Whence

$$\omega_1 = 2\xi_1 - \xi_1^2, \qquad \xi_1 = 1 - \sqrt{1-\omega_1};$$

$$\omega_2 = \xi_2/(1-\xi_1), \qquad \xi_2 = (1-\xi_1)\omega_2.$$

Besides this method numerous other methods are used. For example, let ξ be a random variable with density $p(x)$ where $\int_a^b p(x)\,dx = 1, p(x) \leqslant c$. Consider a sequence of pairs $(\omega_1, \omega_2), (\omega_3, \omega_4), \ldots$. Let $(\omega_{2n-1}, \omega_{2n})$ be the first pair for which $p(a + (b-a)\omega_{2n-1}) \geqslant c\omega_{2n}$. Then set $\xi = a + (b-a)\omega_{2n-1}$. The average number of pairs of random variables in calculations per one realization of ξ equals $(b-a)c$.

There exist special methods for generation of the most commonly used random variables. Thus, a generation of random variables ξ with the distribution close to normal is obtained by summing a number of uniform random variables (here the central limit theorem is utilized). Given an exponential variable γ with density $\frac{1}{2}e^{-x/2}$, $x \geqslant 0$ and variable ω independent of it, we obtain two independent variables ξ_1 and ξ_2 having the standard normal distribution by using the formula

$$\xi_1 = \sqrt{\gamma}\cos 2\pi\omega, \qquad \xi_2 = \sqrt{\gamma}\sin 2\pi\omega.$$

The Erlang random variable is generated as a sum of independent exponential variables. If $F(x) = \sum_i p_i F_i(x)$ where $p_i > 0$. $F_i(x)$ is a distribution function, then one can either first by means of a random number ω_1 realize a trial and at the ith outcome of this trial proceed to the realization of the variable with distribution $F_i(x)$ by means of ω_2 or to verify the condition

$$\{p_1 + \cdots + p_{i-1} \leqslant \omega_1 < p_1 + \cdots + p_i\};$$

when this condition is satisfied we proceed as in the first case utilizing the same ω_1: $\omega_2 = (1/p_i)(\omega_1 - p_1 - \cdots - p_{i-1})$.

6.2.3. *Simulation of a Homogeneous Markov Chain*

Let $P = \|p_{ij}\|$ be a transition matrix of a Markov chain $\{\nu_n, n \geqslant 0\}$: $p_{ij} = \mathsf{P}\{\nu_n = j | \nu_{n-1} = i\}$; $\{p_i^{(0)}\}$ be the initial distribution. In accordance with subsection 1, for each i a subprogram for simulating a random variable whose distribution is given by the ith row of the matrix P is constructed. We shall call this subprogram A_i. In addition a subprogram $A^{(0)}$ for simulating a random variable with distribution $\{p_i^{(0)}\}$ is constructed. At the zero step the subprogram $A^{(0)}$ is evoked and upon its implementation we obtain ν_0. We then evoke A_i with $i = \nu_0$. The result of execution of this subprogram is ν_1. Next we evoke A_i with $i = \nu_1$ and obtain ν_2 and so on. The binary coding of states of a Markov chain is convenient for programming the conditional transitions described above.

This simulation algorithm is universal; however, for simulating Markov chains that arise in practice, it is necessary to use specific features of the problem to save storage in a computer since a Markov chain may possess very many states; in such cases it is impossible to write down not only the elements of the matrix P but even their states. We shall describe two practical methods.

Often a Markov chain represents a bounded random walk in m-dimensional space and then the transition from $\nu_n = (\nu_{n1}, \ldots, \nu_{nm})$ to ν_{n+1} is performed in two stages: first we determine $\mu_n = \nu_n + \eta_n$, where η_n is a random vector with a known distribution independent of the preceding one. After that we check the condition of positivity of all the μ_n components. If this condition is fulfilled, then $\nu_{n+1} = \mu_n$; otherwise η_{n+1} is determined from μ_n in a more complicated manner (possibly from realization of new random variables). We see that the very same algorithms allow us to realize the transitions for numerous different initial states. This process can be continued utilizing the homogeneity of transitions over components $\mu_{nj}, j \in R$, if these components turn out to be positive.

A second method is based on the fact that components of the vector ν_n usually reflect the states of different objects appearing in a queueing system (such as sources of customers, servers, logical devices, etc.) Each transition from ν_n to ν_{n+1} is associated with a change in the state of a limited number of objects (often one or two). Thus, it is sufficient to assign an algorithm of a random choice of an object that induces a change in the state and then the algorithm of the corresponding local transformations. Here the process is often simplified by homogeneity of the objects and relations among them: in such a case in the transformation algorithm only the address part will be variable.

6.2.4. *Simulation of a Markov Process with a Finite Set of States*

Such a process is often defined by an initial distribution $\{p_i^{(0)}\}$, the matrix of transition intensities $\lambda(t) = \|\lambda_{ij}(t)\|$, where $\lambda_{ij}(t)$ for $j \neq i$ is an instantaneous intensity of the transition from i into j at time t, $\lambda_{ii}(t) = -\sum_{j\neq i}\lambda_{ij}(t) = -\lambda_i(t)$. The common feature of all these methods is the realization of the initial random state and a stepwise construction of trajectories of the process, i.e., its extension to the intervals $[0, t_1)$, $[t_1, t_2)$, $[t_2, t_3)$, and so on. The difference is in the method of realizations of the instants t_n: these may be either deterministic or random.

Method Δt. In this case $\Delta t = t_n - t_{n-1}$ is a constant quantity Δ; $t_n = n\Delta$, $n \geqslant 1$. The constant Δ is chosen in such a manner that the possibility of two or more transitions during time Δ can be neglected. Then in place of the trajectory $\nu(t)$ a Markov chain $\{\nu_n\}$ is simulated where, ν_n approximates $\nu(n\Delta)$. Transition probabilities are defined by the approximate formula

$$\mathsf{P}\{\nu_{n+1} = j \mid \nu_n = i\} = \int_{n\Delta}^{(n+1)\Delta} \lambda_{ij}(t)\,dt, \quad j \neq i;$$

$$\mathsf{P}\{\nu_{n+1} = i \mid \nu_n = i\} = 1 - \int_{n\Delta}^{(n+1)\Delta} \lambda_i(t)\,dt,$$

where in the case of continuous functions $\lambda_{ij}(t)$ the integral is most often replaced by the value of the function at the point $n\Delta + \Delta/2$. In the case when $\nu_{n+1} \neq \nu_n$ the transition time is approximately represented by a random variable uniformly distributed in the interval $(n\Delta, (n+1)\Delta)$ (provided this is essential for investigation of the required characteristic of the process).

This method is simple to implement, since during a short time the transitions are usually quite simple. However, the method is not always convenient: if the value of Δ is too large, then the nature of the process is distorted; if it is too small, the computational realization is dragged on due to the slow advancement along the trajectory.

When simulating a queueing system with $\nu(t) = (\nu_1(t), \ldots, \nu_m(t))$, where $\nu_i(t)$ is the state of the ith object at time t, one can utilize one of the two variations of the Δt method, which are well illustrated by the following example. Let $\nu_i(t) = 1$ if the ith server is busy, $\nu_i(t) = 0$ if it is free. The service time of the ith server is exponentially distributed with parameter μ_i. One can perform a random trial for each i separately. If the outcome is success (with probability $\mu_i\Delta$), then the ith server is disengaged. After that we analyze the situation: which servers were disengaged during time Δ and form a new state of a Markov chain. Alternatively, a single trial is performed with the probability of success being $\mu\Delta$, where $\mu = \nu_1(t)\mu_1 + \cdots + \nu_m(t)\mu_m$. The success of this trial means that during time Δ yet another operation was completed. The serial number of a disengaged server is chosen in accordance with a distribution with probabilities that are proportional to $\nu_i(t)\mu_i$. In the case of a compound system consisting of objects of the same type, the second method is often preferable to the first one.

The method of nodal moments. Any time of change of a state of the process $\nu(t)$ will be called a nodal moment. A sequence of nodal moments will be denoted by t_n, $n \geqslant 0$; $t_0 = 0$. Set $\nu_n = \nu(t_n)$ assuming that the process $\nu(t)$ is right continuous with probability 1. Then $\{(v_n, t_n), n \geqslant 0\}$ is a homogeneous Markov chain. Let $v_n = i$, $t_n = t$. The density function of the random variable $t_{n+1} - t$ is equal to $\lambda_i(t + x)\exp\{-\int_0^x \lambda_i(t + u)\,du\}$. If this variable takes on the value x, then the next state equals j with probability $\lambda_{ij}(t + x)/\lambda_i(t + x)(j \neq i)$.

This method is especially convenient in the case of a homogeneous Markov process: the internodal intervals are distributed in accordance with the exponential distribution with a parameter depending on the current state.

The drawback of this method is associated with the realization of the random variable $t_{n+1} - t_n$, whose distribution in general depends on time; this can be overcome in the following manner. Let $\lambda_i(t) \leqslant a_i$, $a_i > 0$ for all t. If $(v_n, t_n) = (i, t)$, we simulate an exponential random variable ξ with parameter a_i and set $v(u) = i$, $t \leqslant u < t + \xi$. After that we carry out a random trial with probability of the outcome j equal to $\lambda_{ij}(t + \xi)/a_i$ for $j \neq i$, $1 - \lambda_i(t + \xi)/a_i$ for $j = i$. The value of $v(t + \xi)$ is determined depending on the outcome and the procedure is repeated cyclically.

If $a_i \leqslant a$, one can proceed in a simpler manner. We simulate a simple stream with intensity a [i.e., by throwing randomly aT points into the interval $(0, T)$ where T is sufficiently large] and set ξ to be equal to the time until the next event of the stream occurs. Thus the process $v(t)$ may change its state only at the times of occurrences of events of a simple stream.

If the mean value $\lambda_i(t)$ in the given interval is substantially smaller than a, the previously described procedure becomes inefficient due to slow advancement along the trajectory. Various modifications of this method are available. Thus, for example, if $\lambda_i(t) \leqslant f(t)$, where $f(t)$ is an integrable function, then performing the time replacement $\tau = \int_0^t f(u)\,du$ we arrive at the preceding case with $a = 1$.

We indicate yet another useful approach. A random variable with the density

$$p(t) = \lambda(t)\exp\left\{-\int_0^t \lambda(u)\,du\right\}, \quad t > 0,$$

where $\lambda(t)$ is an integrable function, $\int_0^\infty \lambda(t)\,dt = \infty$, is realized as the value ξ of variable t for which $\Lambda(t) = \int_0^t \lambda(u)\,du$ attains the value γ, where γ is an exponential random variable with parameter 1. Utilization of this method avoids the necessity of application of random transforms in a small interval Δ. Instead, an integral (with a variable upper limit) of the function $\lambda_i(t)$ determined for the given state is calculated.

This method is efficient, in particular, when constructing realizations of processes with fast or slow transitions. Let the states of a given Markov process $v(t)$—"microstates"—be combined into classes—"macrostates"—so that within a given class transitions between states have a relatively high intensity while the intensity of transitions between macrostates is relatively low. Under certain conditions one could average the characteristics of transitions within a class and consider the consolidated process on the set of macrostates. However, in many cases it may be necessary to simulate the process "in detail."

We construct a realization of a random process $v_0(t)$ with the same intensities, as in the initial process, of transitions between the states of one class and prohibiting transitions between macrostates. Along with this, denoting by $\lambda_i^{(1)}(t)$ the intensity of transition from the state i outside the current macrostate, we determine the random process $x(t) = \int_{a(t)}^{t} \lambda_{v_0}^{(1)}(u)\,du$, where $a(t)$ is the time starting from which the process stays in a given macrostate. At the instant of the entry of the process into a given macrostate a random variable γ is realized, which determines the time t of the next change of the macrostate: $x(t) = \gamma$.

Thus, realization of $v(t)$ is reduced to realization of the behavior of the process inside a macrostate and a random transition during the interchange of macrostates.

6.2.5. *Simulation of a Semi-Markov Process with a Finite Set of States*

The very definition of SMP is well adapted to direct simulation. The successive states of an embedded Markov chain and the durations of stay are realized in succession. Often a different method is used which is as follows. During the entry into a state i a set of independent random variables $\eta_{i1}, \dots, \eta_{iN}$ is realized; the duration of the stay in the current state is equal to the minimum of these variables and the next state is determined by the serial number of the minimal variable. In the case when all η_{ij} are identically distributed, $\mathsf{P}\{\eta_{ij} \geqslant x\} = \bar{B}_i(x)$, the first method is preferable since $\mathsf{P}\{\min \eta_{ij} \geqslant x\} = \bar{B}_i^N(x)$. In the general case the following method can be used. If η_j are independent random variables with a continuous distribution, $F_j(x) = \mathsf{P}\{\eta_j < x\}$, $\lambda_j(x) = F_j'(x)/\bar{F}_j(x)$, then the $\min\{\eta_j\}$ is simulated as the value of ξ such that

$$\int_0^{\xi} \sum_j \lambda_j(x)\,dx = \gamma,$$

where γ is an exponentially distributed random variable with parameter 1. This method is expedient in the case of a large number of possible states of a process when for a given current state i the intensity of transition into the next state is

$$\lambda_i(x) = \sum_j e_{ij}\lambda_j^*(x), \qquad e_{ij} = 1, 0,$$

and owing to a large number of states it is either unreasonable or impossible to construct beforehand an algorithm for realization of the duration of stay in an arbitrary state.

6.2.6. *Simulation of a Piecewise-Linear Markov Process*

In accordance with the definition in Chapter 3 a piecewise-linear Markov process alternates a continuous variation of additional components in a linear manner with random jumps that occur spontaneously and at the times when any of these components vanish. When simulating by the method of nodal moments it is often expedient to apply this device without spontaneous jumps, which allows us to deal with different random durations in the same manner. The simulation algorithm is reduced to a realization of an instantaneous transform at the instant when the zero value is attained and a time shift: if after the instantaneous transformation, the additional components of the vector of states equal $\xi_1, \ldots, \xi_m$ and the rates of their decrease form $\alpha_1, \ldots, \alpha_m$, then the duration until the next jump is

$$\tau = \min\{\xi_j/\alpha_j\},$$

where $\xi_j/\alpha_j = \infty$ for $\alpha_j = 0$. At the instant preceding this jump, ξ_j is replaced by $\xi_j - \alpha_j\tau$.

6.3. Statistical Problems Associated with Simulation

As in any statistical experiment, in the course of simulation of the system problems of processing statistical data arise—to obtain the most accurate estimates of parameters and the most reliable statistical inference concerning the object under investigation. At the same time simulation differs from an observation of an actual object, first because it is possible to select an arbitrary incoming stream, an arbitrary variation of parameters of the system up to a change of its structure. The most typical problems are the following:

1. Choice of the length of realization of a process and the number of repetitions of realizations when estimating the service characteristics.
2. Estimation of the time of the entry of the system into a stationary state when estimating stationary characteristics.
3. Choosing a statistics that will serve as an estimator of the parameter. For example, in the case of an $M|G|m|0$ system, the probability of loss of a customer is equal to the fraction of time during which all the servers are busy. From here two competing statistics for estimating the loss emerge: the relative number of lost customers among the first n and the relative time during which all the servers are busy in the interval of a given length.

Deeper problems whose solution requires special knowledge of the systems under investigation (engineering, economic, biological, etc.) involve verification of the adequacy of a mathematical model, tying up the model to the results of a full-scale experiment, joint dynamic planning of a full-scale and statistical experiments, statistical estimation of parameters of a model (e.g., λ and τ) based on a specific statistic.

In each realization produced by means of a simulation algorithm, the independent variables ω_n are utilized as the source of randomness, while at the output dependent variables, for example, waiting times w_n, are obtained. This presents some difficulties in the course of processing the results. In those rare cases when the correlation function $R(n)$ of the outgoing sequence is known or at least the upper bound on $|R(n)|$ is available, one can estimate the variance of, say, the mean statistical value $\bar{w}_n = (1/n)\left(\text{or } \frac{1}{n}\sum_{i=1}^{n} w_i\right)(w_1 + \cdots + w_n)$.

The same state of affairs exists when one observes a continuous process, for example, the total time lost in waiting by the customers that arrived in a given interval. In such a case an estimate of the correlation function $R(t)$ of a process with continuous time is required.

In the literature on this problem (cf. the papers by Crane and Iglehart (1974), (1975) previously cited) a theory of precision of statistical estimates has been developed based on the method of regenerative processes. If the process that describes the service is regenerative, then the functionals used for estimation of the characteristics of the quality of service—such as the number of lost customers, the total busy period of all the customers, the total waiting time, etc.—are represented as a sum of the corresponding variables over regeneration cycles (which are usually the busy intervals), i.e., in the form of a sum of independent random variables.

Let u_n be an indicator of the quality of servicing the nth customer (i.e., an indicator of the loss of this customer, its waiting time or its duration in the system, the number of interruptions of its service), $a = \mathsf{M}u_n$ in the stationary mode. Then it follows from ergodic considerations that

$$a = \mathsf{M}U/\mathsf{M}N, \tag{1}$$

where N is the number of customers serviced in one busy interval, U is the sum of u_n over these customers. Similarly, if $f(x)$ is the function of the state of the system and $\xi(t)$ is a random process that defines the state at time t, then in the stationary mode we have

$$\mathsf{M}f(\xi(t)) = \mathsf{M}\int_{t_0}^{t_1} f(\xi(t))\,dt/\mathsf{M}(t_1 - t_0), \tag{2}$$

where (t_0, t_1) is a regeneration cycle. In both cases the estimated characteristic is in the form $a = \mathsf{M}U/\mathsf{M}V$, where U and V are random variables that are defined by the trajectory of the process on the regeneration cycle. Assume that $\sigma_u^2 = \mathsf{D}U$ and $\sigma^2 = \mathsf{D}V$ are finite and consider the estimator

$$a_n = (U_1 + \cdots + U_n)/(V_1 + \cdots + V_n),$$

where U_i, V_i are the values of U and V on n regeneration cycles. For large n we have the following approximate formulas.

Denote by m_U, m_V the mathematical expectations of U and V; $U_i^0 = U_i - m_U$, $V_i^0 = V_i - m_V$, $1 \leqslant i \leqslant n$. Then

$$\begin{aligned} a_n &= \left(nm_U + \sum U_i^0\right)\Big/\left(nm_V + \sum V_i^0\right) \\ &= \frac{m_U}{m_V}\left(1 + \frac{1}{nm_U}\sum U_i^0\right)\Big/\left(1 + \frac{1}{nm_V}\sum V_i^0\right) \\ &\approx \frac{m_U}{m_V}\left(1 + \frac{1}{n}\sum_{i=1}^{n}\left(\frac{1}{m_U}U_i^0 - \frac{1}{m_V}V_i^0\right)\right). \end{aligned} \tag{3}$$

Hence the relative error Δ_n of the estimator a_n of the parameter a is represented in the form $\Delta_n = (1/\sqrt{n})\zeta_n$, where ζ_n is an asymptotically normal random variable;

$$\mathsf{M}\zeta_n \xrightarrow[n\to\infty]{} 0;$$

$$\mathsf{D}\zeta_n \xrightarrow[n\to\infty]{} \mathsf{D}\left(\frac{1}{m_U}U - \frac{1}{m_V}V\right) = \sigma_\zeta^2 = \left(\frac{\sigma_U}{m_U}\right)^2 + \left(\frac{\sigma_V}{m_V}\right)^2 - 2\frac{\operatorname{cov}(U,V)}{m_U m_V}. \quad (4)$$

If the value of σ_ζ^2 were known, it would be possible, based on the statistical mean a_n, to construct an approximate confidence interval $(a_n(1-\Delta), a_n(1+\Delta))$ that covers the true value of the parameter a with the probability approximately equal to $1 - 2\varepsilon$ by setting

$$\Delta = \frac{1}{\sqrt{n}}\sigma_\zeta z_\varepsilon, \quad (5)$$

where z_ε is the root of the equation

$$\frac{1}{\sqrt{2\pi}}\int_{z_\varepsilon}^{\infty} e^{-x^2/2}\,dx = \varepsilon.$$

When σ_ζ^2 is unknown, the result, which is asymptotically equivalent to that described above, is valid; here the value of σ_ζ^2 in formula (5) should be replaced by its statistical estimator. The latter is of the same form as the right-hand side of (4), where m_U and m_V are replaced by their means $\bar{U}_n$ and $\bar{V}_n$ based on n realizations, σ_U^2, σ_V^2 and $\operatorname{cov}(U,V)$ are replaced by

$$\frac{1}{n-1}\sum_{i=1}^{n}(U_i - \bar{U}_n)^2,$$

$$\frac{1}{n-1}\sum_{i=1}^{n}(V_i - \bar{V}_n)^2,$$

$$\frac{1}{n-1}\sum_{i=1}^{n}(U_i - \bar{U}_n)(V_i - \bar{V}_n)$$

respectively. For a given confidence coefficient $1 - 2\varepsilon$ and an admissible relative error Δ_0, the number of trials is calculated as follows. Based on the first N trials (e.g., $N = 100$) we calculate the estimator $\hat{\sigma}_\zeta^2$ of the variance σ_ζ^2 as described previously. Then the total number n of trials (including the preliminary ones) is determined by the relation

$$n \approx \hat{\sigma}_\zeta^2 z_\varepsilon^2/\Delta^2. \quad (6)$$

A more accurate method uses a refined expression for the variance and with it also the number of realizations after every N realizations. A sequential stopping rule was developed by Lavenberg and Sauer (1977).

6.4. Simulation of Queueing Systems

6.4.1. *General Principles of Simulation of Systems*

A model of a system is usually constructed purposefully: the problem of investigating specific characteristics of the system is posed beforehand with the aim to check how well the system fits its purpose, to estimate various characteristics of efficiency, to study its behavior in a given environment, to discover possible accidents, to optimize the structure of the system and its parameters, and so on. Thus, when constructing a model of a system, we have in mind specific characteristics to be estimated. Weighing the possible consequences of errors of estimation of these characteristics, we thus determine the precision with which the model reproduces the actual process and the required duration of the experiment (the number of realizations).

For example, let the mean virtual waiting time in an $M|G|m$ system with periodically varying instantaneous intensity $\lambda(t)$ of the incoming stream be investigated. If the period $\lambda(t)$ is comparable to the mean service time, then it is permissible to replace $\lambda(t)$ by its mean value over the period $\bar{\lambda}$. Another characteristic—the distribution of the queue length at the time of a customer's arrival—is more crucial to the variations in intensity: customers have a tendency to accumulate in intervals of higher intensity.

In practice neither too simple nor too complex a model is required: one should take into account the basic factors and discard the unessential ones (or to average their effect in a reasonable manner).

When studying a complex real-world system, one usually constructs a collection of models that would reveal various aspects of the object under investigation. The results of realizations of different models are compared among themselves, and often simple models allow us to discover gross errors in the complex ones.

Experts recommend simulation only in those cases when an analytic investigation is impossible.

A combined utilization of simulation and queueing theory yields sound results. Even if one decides to study a system using "pure" simulation, the knowledge of the theory will help us in constructing an appropriate process, in choosing the basic functional and in interpreting results. On the other hand, a formal application of queueing theory results (i.e. those dealing with the simple nature of the rarefied stream) when they are not justified often yields a complete misrepresentation of the results.

6.4.2. *Block Principle of Simulation*

When constructing a model of a complex system, it is hardly possible to encompass the entire process of its operation as a whole and to mirror it in a model. An applicaton of the block principle of simulation is a natural way to overcome this difficulty: the system is provisionally subdivided into blocks (units) each one of which admits construction of a corresponding model. The simulation process would have thus been completed if the units were to operate independently. In a typical case, when the units are interconnected, this connection is schematized by the corresponding interaction among the units. Thus, operation of each unit involves both internal as well as external factors.

Buslenko (1961), (1969), building upon the wide experience of practical simulation, developed the notion of a complex system consisting of aggregates. Each aggregate is characterized by its inner state—a time function $z(t)$ that evolves as a random process. An aggregate is capable of sending signals $y \in Y$ and also receiving signals $x \in X$. Each signal is transmitted instantaneously; in the intervals between the arrival of signals the operation of an aggregate [i.e., the behavior of the state $z(t)$ and the outgoing signals] is autonomous (independent of other aggregates). The fact that the transmission of signals is discrete simplifies the construction of the model: a "continuous" signal may generate complex feedback.

Two operators G and H control the model of an aggregate's operation. The first determines a transformation associated with the arrival of an incoming signal, the second an autonomous operation of a given aggregate. It is convenient to represent the signals coming out of aggregates as consisting of an address and content parts.

6.4.3. *Piecewise-Linear Aggregates*

Let an operator H realize a piecewise-linear Markov process $z(t) = (v(t), \bar{\xi}(t))$, where in general the dimensionality of the vector of additional components $\bar{\xi}(t)$ depends on the state of $v(t)$. The outgoing signal is sent at the instant when an additional component vanishes and is a function of the values of the remaining components at this instant. As long as no input signals arrive and all the additional components remain positive, these components vary in a linear fashion.

Let the system consist of s aggregates. The state of such a system will be $z(t) = (z_1(t), \ldots, z_s(t))$, where $z_i(t)$ is the state of the ith aggregate. Assume that the signals arrive only from one aggregate to another but not from outside the system. Then the whole system will be equivalent to an aggregate with a wider space of states. A trajectory of the process $z(t)$ is constructed by means of a transition from one nodal moment t_n to the next one t_{n+1}. The basic component of the process $z(t)$ remains unchanged in the interval (t_n, t_{n+1}) and the additional ones vary in a linear fashion. Transition from $z(t_n - 0)$ to $z(t_n + 0)$ is implemented as follows. At time t_n an additional component of an aggregate vanishes. Its state at time $t_n + 0$ is determined by means of the operator H as well as the outgoing signal if it comes up. Following an address(es) of this signal, we transform the states of aggregates that received this signal. This process may be continued.* If the model is correctly constructed, there will be no infinite accumulation of signals at a single instant of time, and we shall arrive at the state at time $t_n + 0$.

We construct provisional trajectories $z_i^*(t)$ of processes $z_i(t)$—each one up to time t_i^* when an outgoing signal emerges in the corresponding aggregate.† The next nodal time will be $t_{n+1} = \min[t_1^*, \ldots, t_s^*]$.

Note that at nodal times usually the states of only some aggregates are changed. Here the previously constructed trajectory $z_i^*(t)$ is "cancelled," i.e., the actual trajectory does not coincide with the provisional one; at the same time, for the aggregates that did not receive a signal, the provisional trajectory is "confirmed" becoming an actual one at least until the next nodal time.

6.4.4. *A Typical Element of the Model*

Software implementation is one of the most important and most difficult stages of simulation. This process is substantially facilitated by the availability of a collection of rather simple elements—elementary aggregates that serve as building blocks for construction of a complex system in accordance with definite rules. Simulation software must include standard subroutines of elements and a control part that implements the operation of the system as a whole. The elements can be chosen at different levels of aggregations. Thus, in Lifshits and Mal'ts' (1978) book the principle of choosing typical queueing systems and their combined elements—a queue, priority schemes, etc.—serves as the starting point.

We shall describe an approach based on the initial level of aggregation that has some degree of universality.

Consider an element (an elementary aggregate) possessing the following properties.

The state of the aggregate is $z(t) = (v(t), \xi(t))$, where $v(t)$ is an element of a finite set Z_0, $\xi(t)$ takes on nonnegative numerical values. To each $i \in Z_0$ there corresponds a number α_i: if $v(t) = i$, $\xi(t) > 0$, then $\xi'(t) = -\alpha_i$. The incoming and outgoing signals of an aggregate are in the form (i, x), where i is an element of a finite set X_0 in the case of input and of Y_0 in the case of output and x is a number.

Sometimes it is more convenient to consider two types of states and signals: (i, x) and simply i. Clearly this does not result in a greater generality of the scheme.

Let $z(t-0) = (i, 0)$, $\xi(t - \varepsilon) > 0$ for sufficiently small $\varepsilon > 0$. Then $v(t+0) = j$ with probability p_{ij}. If $v(t-0) = i$, $v(t+0) = j$, then $\xi(t+0) = \varphi_{ij}(\omega)$, where ω is a uniform $(0, 1)$ random variable. If, moreover, $\xi(t, 0) = z'$, the aggregate

* The time t_n is split into several times; this can be expressed by the notation (t_n, j), $0 \leqslant j \leqslant m$, where $(t_n, 0) = t_n - 0$ $(t_n, m) = t_n + 0$.

† In practice the given quantity is bounded by a preassigned number Δ_i.

sends a signal $\psi_{ij}(z', \omega')$, where ω' is a uniform $(0, 1)$ random variable independent of the preceding one.

Now let $z(t - 0) = (i, z)$ and let, at time t, the signal (k, x) arrive at the input of a given aggregate. Then $v(t + 0) = j$ with probability $p_{ij}^{(k)}$. For fixed i, j, k, and z' we have $\xi(t + 0) = \varphi_{ij}^{(k)}(z', x, \omega)$ where ω is as above. At the same time an outgoing signal $\psi_{ij}^{(k)}(z', x, \omega')$ is sent where $z' = \xi(t + 0)$, ω' is a uniform $(0, 1)$ random variable independent of the preceding one.

Note that the signals may be empty. Note also that ψ_{ij}, $\psi_{ij}^{(k)}$ are functions with values (y_0, y_1), where y_1 is a number, y_0 is an element of a finite set that in particular determines the address(es) of the transmission of the signal.

Summarizing we conclude that an elementary aggregate performs a probabilistic–automatic transformation and is also capable of delaying a signal for a time determined by its state. The identification of an elementary aggregate involves the definition of functions φ and ψ with indices. If these functions always belong to a finite set, then one can include instead a code of the corresponding procedure into the description. If the functions φ_{ij}, $\varphi_{ij}^{(k)}$ and the second components of the functions ψ_{ij}, $\psi_{ij}^{(k)}$ are linear, then identification is simplified: it is sufficient to assign coefficients that determine these functions. In many problems replacement of ω and ω' by their functions from a given finite set leads to a linear case. Then the identification of an elementary aggregate involves transition probabilities, coefficients of linear functions, and programs of procedures for obtaining random numbers with given distributions. This approach is taken in the computer package of applied programs AMOS (Aggregated Simulation of Systems) developed by Krivutsa (1980).

6.4.5. *Interpretation of Elements of Queueing Systems*

A server of a system (including a priority system without interruption) with a constant rate of service can be interpreted in two ways depending on whether the service time is given by an external signal or is realized in the server. In the first case the element receives a signal (i, x) (where i is the type of customer, x is the service time) and after x units of time the element emits the signal $(j(i), x)$. It is assumed that the signal determines the server at which the customer arrives; the signal may also serve as a record for accumulation of statistics. In the second case, the input signal is of the form i or equivalently $(i, 0)$. By means of the signal the random variable $\eta^{(i)} = \varphi^{(i)}(\omega)$ is realized; the rest is the same as above with x replaced by $\eta^{(i)}$.

The source of customers is interpreted by an element of the preceding type with feedback: the outgoing signal goes—apart from the address at which the customer arrives—to the input of the element. We thus obtain a source that generates customers of different types at the transition times of a semi-Markov process. Evidently a computer program implementation of the source of customers is simpler: the feedback is superfluous.

A server with a variable rate and possible interruptions in service is interpreted by an element possessing the same properties as previously mentioned with the following modification. The state (i, z) of an element stipulates the type of service i and the remaining work z required to complete this service. The variable z decreases in time with the rate $\alpha_i \geqslant 0$. The input signal (k, x) signifies that a new customer arrives. Denote by (i, z) the state of an element at time $t - 0$ when t is the instant of arrival of the signal. We then have two customers—the "old" and the "new." The "fate" of these customers depends on the pair (i, k). If the service is not interrupted, the state of the element remains unchanged, and the signal (k, x) is directed to an appropriate address necessarily acquiring characteristic i. Otherwise the new customer remains, the state of the element will be (k, x) and, at the same instant, the signal (i, z) is transmitted to the address corresponding to the other server, waiting location or accumulator of lost customers.

A waiting location is a memory cell containing information of the form (i, x), where i is a qualitative and x is a quantitative characteristic of a customer. The outside information contained in the element is transmitted by means of the signal to the address indicated depending on i; after that it may be replaced by new information contained in the signal or retained in the previous form.

A customer is an object that is characterized in general by the vector (i, x), where i is the type of a customer and x is the amount of work required for his or her servicing. Thus, the state of a server is the state of a customer to be found in it. Customers available at a given instant may be described by elements, but it is almost always simpler to encode a customer in terms of the state of the element (server, waiting location) in which he or she is located.

An "impatient" customer with limited waiting time is characterized by the vector (i, x_1, x_2), where x_1 is the required remaining waiting time (or the amount of work), x_2 is the remaining admissible waiting time. To describe such a customer it is sufficient to take two elements that are characterized also by additional variables x_1 and x_2, respectively. The "fate" of a customer is determined by the first vanishing variable. Schematization of other types of constraints is completely analogous.

6.5. Calculation of Corrections to Characteristics of Systems

6.5.1. *Introductory Remarks*

In calculations associated with service systems one is often required to analyze the characteristics of systems that differ only slightly from the systems for which the corresponding characteristics are known (for example, those for which analytic formulas or numerical estimates are available). It is thus natural to pose the problem of calculation of corrections rather than simulating the whole system anew. We shall consider some typical examples.

EXAMPLE 1. A standby system consists of m elements and a single renewal channel. The renewal (repair) time has an arbitrary distribution; trouble-free service time has a density $p(x)$, which is arbitrary for $0 < x < x_0$ and equals $ce^{-\lambda x}$ for $x > x_0$. In this case

$$p(x) = \lambda e^{-\lambda x} + [(c - \lambda)e^{-\lambda x} + (p(x) - ce^{-\lambda x})E(x_0 - x)], \tag{1}$$

where $E(x) = 0$ for $x \leqslant 0$, $E(x) = 1$ for $x > 0$. Formula (1) indicates that the actual distribution is a perturbation of an exponential distribution with parameter λ. The unperturbed system is a system with a Markov incoming stream: given k faulty elements, the probability of yet another faulty element during time dt equals $(m - k)\lambda\,dt$ independently of the foregoing: for this systems an analytic theory of stationary characteristics and of mean duration in a set of states is available.

EXAMPLE 2. The system $M|G|m|0$ with an unreliable server may be viewed as a perturbation of the well-known $M|G|m|0$ system.

Numerous examples of this kind are available. In all of them there exists a definite unperturbed mode of a service system which is being perturbed from time to time.

6.5.2. *Statement of the Problem*

Consider a homogeneous Markov chain $\{\xi_n\}$ on a measurable space $(X, \mathfrak{G})$, $0 \in X$ is a marked state. The transition function $\{\xi_n\}$ will be denoted by

$$P(A|y) = P\{\xi_n \in A | \xi_{n-1} = y\}, \quad A \in \mathfrak{G},$$

assuming that it is measurable in y. Suppose that the sequence $\{\xi_n\}$ returns to the state 0 with probability 1, the mean return time T is finite and the aperiodicity condition is fulfilled. Denote by $\pi(A)$ the ergodic distribution of $\{\xi_n\}$. This distribution must be estimated.

Consider also a Markov chain $\{\xi_{0n}\}$ with the transition function $P_0(A|y)$ and a known ergodic distribution $\pi_0(A)$.

Let $P_1(A|y) = P(A|y) - P_0(A|y)$. Set $P_1(A|y) = \varepsilon(y)(P^+(A|y) - P^-(A|y))$, where $\varepsilon(y) \geqslant 0$, $P^\pm(A|y)$ are probability measures for any $y \in X$ such that $\varepsilon(y) > 0$. The problem is to estimate

$$a = \int f(x)\pi(dx) - \int f(x)\pi_0(dx), \tag{2}$$

where $f(x)$ is a given function; it is assumed that both integrals on the right-hand side of (2) are finite. We shall assume without loss of generality that $f(0) = 0$, $P^\pm(\{0\}) = 0$.

Denote

$$\bar{P}^{(n)}(A|y) = \mathsf{P}\{\xi_n \in A; \xi_k \neq 0, 0 < k < n | \xi_0 = 0\},$$

and let $\bar{P}_0^{(n)}(A|y)$ be the analogous characteristic of $\{\xi_{0n}\}$ in place of $\{\xi_n\}$. Then for $0 \notin A$

$$\pi(A) = \pi(0) \sum_{n=1}^{\infty} \bar{P}^{(n)}(A|0). \tag{3}$$

The expression for $\bar{P}^{(n)}(A|0)$ is an integral (over a certain set) of an expression that can symbolically be written as

$$P^n = P(dx_1|0)P(dx_2|x_1)\cdots P(dx_n|x_{n-1}).$$

We have

$$P^n = (P_0 + P_1)^n = P_0^n + \sum_{k=0}^{n-1} P_0^{n-k-1}P_1 P^k.$$

Whence

$$\pi(A) = \pi(0)\sum_{n=1}^{\infty} \bar{P}_0^{(n)}(A|0) + \pi(0)\int_{z\neq 0}^{\infty} \sum_{n=1}^{\infty}\Bigg[P_1(dz|0)\bar{P}^{(n-1)}(A|z) + \sum_{k=0}^{n-2}\int_{y\neq 0} \bar{P}_0^{(n-k-1)}(dy|0)P_1(dz|y)\bar{P}^{(k)}(A|z)\Bigg]. \tag{4}$$

Along with this we have the identity

$$\pi_0(B) = \pi_0(0)\left[\chi(B) + \sum_{n=1}^{\infty} \bar{P}_0^{(n)}(B|0)\right] \tag{5}$$

where $\chi(B) = 1$ for $0 \in B$, $\chi(B) = 0$ otherwise. Substituting this identity into (4) we obtain

$$\pi(A) = \frac{\pi(0)}{\pi_0(0)}\left(\pi_0(A) + \int \pi_0(dy)P_1(dz|y)\sum_{k=0}^{\infty} \bar{P}^{(k)}(A|z)\right). \tag{6}$$

Whence

$$\int f(x)\pi(dx) = \frac{\pi(0)}{\pi_0(0)}\left(\int f(x)\pi_0(dx) + \int \pi_0(dy)P_1(dz|y)\sum_{k=0}^{\infty}\int f(x)\bar{P}^{(k)}(dx|z)\right). \tag{7}$$

Consider an auxiliary Markov chain $\{\xi_1(z), \xi_2(z), \dots\}$ with a transition function $P(A|y)$ and the initial state $\xi_1(z) = z$. Denote by $v(z)$ the smallest (serial) number of the step for which $\xi_n(z) = 0$ and set

$$g(z) = \mathsf{M}\sum_{k=1}^{v(z)} f(\xi_k(z)). \tag{8}$$

It is easy to verify that $g(z)$ equals the last sum on the r.h.s. of (7).

Consider the following random variables: ξ_0 with distribution $c(y)\pi_0(dy)$, where $c(y) \geqslant 0$, $\int c(y)\pi_0(dy) = 1$, $\int_{c(y)=0} \pi_0(dy) = 0$; θ with distribution $P\{\theta = 1\} = P\{\theta = -1\} = \frac{1}{2}$, ξ_1 with distribution $P^{\pm}(dz|y)$ for $\xi_0 = y$, $\theta = \pm 1$. It follows from (7) and (8) that

$$\int \pi_0(dy)P_1(dz|y)\sum_{k=0}^{\infty}\int f(x)\bar{P}^{(k)}(dx|z) = 2\mathsf{M}\left\{\frac{\theta\varepsilon(\xi_0)}{c(\xi_0)}\sum_{k=1}^{v} f(\xi_k)\right\}, \tag{9}$$

where $\{\xi_k\}$ is a Markov chain with the transition function

$$\mathsf{P}\{\xi_n \in A | \xi_{n-1} = y\} = P(A|y), \quad n \geqslant 2,$$

and the distributions of ξ_0 and ξ_1, $v = \min\{n: \xi_n = 0\}$ defined previously. Thus

$$a = \frac{\pi(0) - \pi_0(0)}{\pi_0(0)} \int f(x)\pi_0(dx) + 2\frac{\pi(0)}{\pi_0(0)} \mathsf{M} \frac{\theta\varepsilon(\xi_0)}{c(\xi_0)} \sum_{k=1}^{v} f(\xi_k). \tag{10}$$

In this formula we set $f(x) = 1$ for $x \neq 0$ and $f(0) = 0$. We thus obtain the equality

$$\pi_0(0) - \pi(0) = \frac{\pi(0) - \pi_0(0)}{\pi_0(0)}(1 - \pi_0(0)) + 2\left(1 + \frac{\pi(0) - \pi_0(0)}{\pi_0(0)}\right) \mathsf{M}\left\{\frac{\theta\varepsilon(\xi_0)v}{c(\xi_0)}\right\}.$$

Whence

$$\pi(0) - \pi_0(0) = -\frac{2\pi_0(0)\mathsf{M}\{\theta\varepsilon(\xi_0)v/c(\xi_0)\}}{1 + 2\mathsf{M}\{\theta\varepsilon(\xi_0)v/c(\xi_0)\}}. \tag{11}$$

Thus, a correction a to the integral $\int f(x)\pi_0(dx)$ when π_0 replaces π is obtained in the form of a mathematical expectation of a function of random variables. This allows us to evaluate the correction by means of the Monte Carlo method. If f and $\mathsf{M}v$ are bounded and $\varepsilon(x) \leqslant \varepsilon$, then the correction is of order ε as $\varepsilon \to 0$. In such a case

$$a \sim -2f_0\mathsf{M}\{\theta\varepsilon(\xi_0)v/c(\xi_0)\} + 2\mathsf{M}\left\{\theta\varepsilon(\xi_0) \sum_{k=1}^{v} f(\xi_k)/c(\xi_0)\right\}, \tag{12}$$

where $f_0 = \int f(x)\pi_0(dx)$.

Let $\delta = \int \varepsilon(x)\pi_0(dx)$. We thus obtain from (12)

$$a \sim 2\delta\mathsf{M}\left\{\theta \sum_{k=1}^{v} (f(\xi_k) - f_0)\right\} \tag{13}$$

for $c(x) = \varepsilon(x)/\delta$.

6.5.3. *Remark*

A somewhat different method for calculating corrections may be suggested. The random variable ξ_0 is realized as above. Instead of $\xi_1, \ldots, \xi_v$ two sequences $\xi_1^+, \ldots, \xi_{v+}^+$ and $\xi_1^-, \ldots, \xi_{v-}^-$ of a Markov chain with the transition function $P(A|y)$ and the initial distributions

$$\mathsf{P}\{\xi_1^\pm \in A | \xi_0 = y\} = P^\pm(A|y) \tag{14}$$

are realized. Denote by $v^\pm$ the time of the first entry of $\xi_k^\pm$ into the zero state. In this case we obtain in place of (13)

$$a \sim \delta\mathsf{M}\left\{\sum_{k=1}^{v+} (f(\xi_k^+) - f_0) - \sum_{k=1}^{v-} (f(\xi_k^-) - f_0)\right\}. \tag{15}$$

In many practical cases the sequences $\{\xi_n^+\}$ and $\{\xi_n^-\}$ are synchronized, i.e.,

$\xi_n^+ = \xi_n^-$ with a positive probability. If such an event occurs then in the formula (15) the summation can be taken only from 1 to $n - 1$. It is of interest to note that the Markov chains $\{\xi_n^+\}$ and $\{\xi_n^-\}$ may be arbitrarily connected as long as condiion (14) is fulfilled. This fact is used to reduce the mean synchronization time as much as possible by a suitable chice of a general probability space.

6.5.4. *Notes*

See Ermakov and Mikhaĭlov (1976), Hammersley and Handscomb (1964), Pollyak (1971), Buslenko and Shreider (1961), Crane and Iglehart (1974a, b), (1975), and Sobol' (1973) concerning theoretical foundations of the Monte Carlo method.

The works by Cooper (1981), Pugachev (1973), Krein and Lemuan (1982), and Sharakshane et al. (1977) deal with statistical problems associated with simulation.

The effect of nonstationarity of a stream for standby systems is discussed in Kovalenko (1965d).

The notion of a piecewise-linear aggregate is due to Kovalenko (1975).

See Shannon (1978) concerning methodological problems of simulation.

The books by Glushkov et al. (1975), Fishman (1973), Genchev (1975), and Lambrecht (1976) are devoted to computer programming of simulation systems. We also note the important monographs by Pirogov (1977), Chetverikov et al. (1978), and Klimov (1964).

The principle of automated simulation was realized by Yarovitskiĭ and is also described in the book of Bakayev et al. (1978). Buslenko (1961), Lutkov (1978), and Krivutsa (1980) deal with Buslenko's aggregate approach and its developments.

We note that formula (11) in Section 6.5.2. is a corollary to formula (7) in Kartashov's paper (1981). (In Kartashov's paper problems of statistical simulation were not discussed.)

Sil'vestrov (1971) developed the synchronization theory of random sequences preserving their marginal distributions.

Results presented in Section 6.5 can be generalized to algorithms for statistical simulation of continuous time Markov processes.

Bibliography

Afanas'eva, L.G.

Existence of a limit distribution in queueing systems with bounded sojourn time, *Teoriya Veroyatnosteĭ i ee Primeneniya,* **X** (3), 1965, pp. 570–578, (English translation: *Theory of Probab. and Applic.* **10**, 1965, pp. 515–522).

Afanas'eva, L.G. and Martynov. A.V.

On ergodic properties of queueing systems with a limited sojourn time, *Teoriya Veroyatnosteĭ i ee Primeneniya,* **14** (1), 1969, pp. 102–112 (English translation: *Theory of Probab. and Applic.* **14**, 1969, pp. 105–114).

Anisimov, V.V.

Limit Theorems for Random Processes and their Application to Discrete Summation Schemes, Vishcha Shkola, Kiyev, 1976 (in Russian).

Annayev, T. and Manilov, N. Ya.

On a method of investigating a queueing system with an infinite number of servers, *Vistnik Kiivs'kogo Un-tu, Ser. Mat.-mech.* **17**, 1975, pp. 127–134 (in Ukrainian).

Arentov, V.A.

[1980a] Evaluation of reliability estimators of certain systems using the method of successive consolidation of states, *Electron. Modelirovaniye*, No. 1, 1980, pp. 86–89 (in Russian).

[1980b] Some estimators of the distribution function of the time of attaining a remote state by a semi–Markov process, *Mat. metody issled. oper. i teorii nadezhnosti.* Institute of Cybernetics AN UkrSSR, 1980, pp. 8–15 (in Russian).

[1983] Estimation of reliability of systems taking spare parts into account, *Electr. Modelirovaniye*, No. 6, 1983, pp. 70–72 (in Russian).

[1984] Estimation of reliability of a multichannel system with a variable operational mode, *Kiberntika*, No 3, 1984, pp. 108–110 (in Russian).

Artamonov, G.T. and Brekhov, O.M.

Analytic Probabilistic Models of Computer Operation. Energiya, Moscow, 1979 (in Russian).

Athreya, K.B., McDonald, D., and Ney, P.E.

Construction of a class of stationary processes with application in reliability, *Zastos. Matem.*, **16**, 1979, p. 397.

Azlarov, T.A. and Khusainov, Ya. M.

Limit theorems for queueing systems with absolute priority under heavy load, *Izv. AN UzSSR, Ser. Fiz.-mat. nauk*, No. 6, 1974, pp. 15–21 (in Russian).

Bakayev, A.A., Kostina, N.I., and Yarovitskií, N.V.
Simulation Models in Economics, Naukova Dumka, Kiev, 1978 (in Russian).

Barlow, R.E. and Proschan, F.
Statistical Theory of Reliability and Life Testing: Probability Models, To Begin With, Silver Spring, MD, 1981.

Barrer, D.V.
[1957a] Queuing with impatient customers and indifferent clerks, *Oper. Res.* **5** (5), 1957, pp. 644–649.
[1957b] Queuing with impatient customers and ordered service, *Oper. Res.* **5** (5), 1957, pp. 650–656.

Barzilovich, Ye. Yu., Belyayev, Yu. K., Koshtanov, V.A., et al.
Problems of Mathematical Reliability Theory, Radio i Svyaz, Moscow, 1983 (in Russian).

Basharin, G.P.
[1958a] Distribution of the number of busy lines in the second cascade switchboard in a telephone system with refusals, *DAN SSSR*, **121** (2), 1958, pp. 280–283 (in Russian).
[1958b] A probabilistic investigation of a two stage trunkhunting telephone system with refused calls, *DAN SSSR*, **121** (1), 1958, pp. 101–104 (in Russian).
[1960] The congestion time limiting distribution for a fully available group of trunks, *Teoriya Veroyatnosteĭ i ee Primeneniya*, **5** (2), 1960, pp. 246–252 (English translation: *Theory of Probab. and Appl.* **5**, 1960, pp. 223–228).
[1961] Analytical determination and computational methods of the probability of losses in switching circuits, *Problemy peredachi informatsii*, Issue 9, 1961, pp. 5–47 (in Russian).
[1962] Tables of probabilities and mean square deviations of losses for a fully available group of trunks, *Izdatel'stvo AN SSSR*, Moscow, 1962 (in Russian).
[1964] On complex queueing systems with several finite queues and impatient customers, *Kibernetiku—na sluzhbu kommunizmu*, Vol **2**, Moskva-Leningrad, *Energiya*, 1964, pp. 274–302 (in Russian).
[1965] A server with a finite queue and customers of several types, *Teoriya Veroyatnosteĭ i ee Primeneniya*, **X** (2), 1965, pp. 282–296 (English translation: *Theory of Probab. and Appl.* **10**, 1965, pp. 261–274).
[1967a] On servicing two streams with relative priority on a fully available system with a limited number of waiting locations, *Tekhn. Kibernet.* No. 2, 1967, pp. 72–86 (English translation: *Eng. Cybern.*).
[1967b] On servicing two streams with absolute priority, *Tekhn. Kibern.* No. 5, 1967, pp. 106–116 (English translation: *Eng. Cybern.*).

Basharin, G.P. and Gromov, A.I.
A matrix approach to the determination of the stationary distribution for some nonstandard queueing systems, *Avtomatika i telemekh.* No. 1, 1978, pp. 29–39 (in Russian).

Basharin, G.P. and Kokotushkin, V.A.
Conditions of strong statistical equilibrium for complex queueing systems, *Probl. peredachi inform.* **7** (3), 1971, pp. 67–75 (in Russian).

Basharin, G.P. and Tolmachev, A.L.
Theory of queueing networks and its application to an analysis of

information-computational systems. *Itogi Nauki i Tekhniki, Theory of Probab. and Math. Statist.*, **21**, VINITI, Moscow, 1983, pp. 3–119.

Basharin, G.P., Kharkovich, A.D. and Shneps, N.A.
Queueing System in Telephone Communication, Nauka, Moscow, 1966 (in Russian).

Beichelt, F. and Franken, P.
Zuverlässigkeit und Instandhaltung, Veb. Verlag Technik, Berlin, 1983.

Belyayev, Yu. K.
[1963] Limit theorems for dissipative flows, *Teoriya Veroyatnosteĭ i ee Primeneniya*, **VIII** (2), 1963, pp. 175–184 (English translation: *Theory of Probab. and Appl.* **8**, 1963, pp. 165–173).
[1962] Markov line processes and their application to problems of reliability, *Trudy VI Vsesoyuznogo soveshchaniya po teorii veroyatnosteĭ i matematicheskoĭ statistike, Vil'nyus*, 1960, pp. 302–323. Gos. izdat. politich. i nauchnoi literatury LitSSR, 1962.
[1964] Efficiency in the presence of two types of failure, *Kibernetika—na sluzhbu kommunizmu*, **2**, Energiya, Moscow-Leningrad, 1964.
[1969] New results and generalizations of problems of the crossing type. Addendum to the Russian translation of H. Cramér and M. Leadbetter's book: *Stationary Stochastic Processes*, Mir, Moscow, 1969, pp. 341–388.

Beneš, V.E.
[1963] *General Stochastic Processes in the Theory of Queues*, Addison-Wesley, Reading, MA, 1963.
[1960] General stochastic processes in traffic systems with one server, *Bell. System. Techn. J.*, **34** (1), 1960, pp. 127–160.

Bharucha-Reid, A.T.
Elements of Markov Processes and their Applications, McGraw-Hill, New York, 1960.

Bocharov, P.P.
[1968] On an unreliable server with several types of customers and completion of the customer's service interrupted by a breakdown, *Probl. peredachi inform.* **4** (2), 1968, pp. 53–61.
[1980] On Poisson multiphase system of finite capacity with blocking and reservicing, *Teor. teletraf. i seti s upravlyaemymi elementami*, Nauka, Moscow, 1980, pp. 130–137.

Bocharev, P.P. and Lysenko, V.T.
On a single-server system with relative priority and a limited number of waiting locations, *Veroyatn. zadachi v struct.-slozhn. sistimakh kommutatsii*, Nauka, Moscow, 1969, pp. 60–66.

Bogdantsev, Ye. N., Tsaturyan, G. Zh., and Sukiasyan, A.A.
Model of analysis of a system with a variable loading on subsystems, *Kibernetika*, No. 6, 1984, pp. 92–95.

Bokuchava, I.T., Donadze, N.K. and Geldiashvili, N.I.
On an outgoing stream in a system with losses, *Soobshch. AN GruzSSR*, No 53, 1969, pp. 37–40 (in Russian).

Borovkov, A.A.
[1972] *Stochastic Processes in Queueing Theory*, Nauka, Moscow, 1972 (English transl. *Applications of Mathematics*, Springer-Verlag, New York, 1976, Vol. 4).

[1980] *Asymptotic Methods of Queueing Theory*, Nauka, Moscow, 1980 (in Russian).

Borozdin, O.P. and Ezhov, I.I.
On a class of limiting functions for strongly regenerating random processes. *DAN UkrSSR*, No. 9, 1976, pp. 771–773.

Breiman, L.
The Poisson tendency in traffic distribution, *Ann. of Math. Statist.*, **34** (1), 1963, pp. 308–311.

Brémaud, P.
[1978] On the output theorem of queueing theory via filtering, *J. Appl. Prob.*, **15**, No. 2, 1978, pp. 397–406.
[1981] *Point Processes and Queues: Martingale Dynamics*, Springer-Verlag, Heidelberg, 1981.

Brockwell, P.J.
Stationary distribution for dams with additive input and content-dependent release rate, *Adv. App. Prob.*, **9**, No. 3, 1977, pp. 645–663.

Brockwell, P.J. and Resnick, S.I.
Storage processes with general release rule and additive inputs, *Adv. Appl. Prob.*, **14**, No. 2, 1982, pp. 392–433.

Brodi, S.M.
[1959] On integro-differential equations for systems with τ-waiting, *Dop. AN URSR*, **6**, 1959 (in Ukrainian).
[1960] A queueing problem, *Trudy V Vsesoyuznogo soveshchaniya po teorii veroyatnosteĭ i matematicheskoĭ statistike*, Erevan, Izdatel'stvo AN ArmSSR, 1960, pp. 142–147 (in Russian).
[1962] A single-server system with τ-waiting and Erlang incoming stream, *Dop. AN URSR*, 1962, pp. 1425–1428 (in Ukrainian).

Brodi, S.M. and Pogosyan, I.A.
Imbedded Stochastic Processes in Queueing Theory, Naukova Dumka, Kiyev, 1973 (in Russian).

Buslenko, N.P.
[1961] Solution of problems of queueing theory by simulation on digital computers, *Problemy peredachi informatsii*, Issue 9, 1961, pp. 48–69.
[1964] *Mathematical Simulation of Production Processes*, Nauka, Moscow, 1964.
[1968] *Modelling of Complex Systems*, Nauka, Moscow, 1968.

Buslenko, N.P. and Shreider, Yu.A.
The Method of Statistical Trials, Fizmatgiz, Moscow, 1961.

Chaiken, J.M. and Ignall, E.
An extension of Erlang's formulas which distinguishes individual servers, *J. Appl. Prob.* **9** (1), 1972, pp. 192–197.

Chandy, K.M., Howard, J.H., and Towesley, D.F.
Product form and local balance in queueing networks, *Journ. ACM* **24**, 1977, pp. 250–263.

Chetverikov, V.N., Bakanovich, Ye.A., and Men'kov, A.V.
Computation Technology for Statistical Simulation, Sov. Radio, Moscow, 1978 (in Russian).

Çinlar, E.
[1967] Time dependence of queues with semi-Markovian services, *J. Appl. Prob.* **4** (2), 1967, pp. 350–364.

[1972] Superposition of point processes, In: *Stochastic Point Processes: Statist. Anal. Theory and Appl.*, Wiley-Interscience, New York, 1972, pp. 549–606.

[1975] *Introduction to Stochastic Processes*, Prentice-Hall, Englewood Cliffs, New Jersey, 1975.

Çinlar, E. and Pinsky, M.

[1971] A stochastic integral in storage theory, *Z. Wahrscheinlichkeitstheorie und verw. Geb.* **17** (2), 1971, pp. 227–240.

[1972] On dams with additive inputs and a general release rule, *J. Appl. Prob.* **9** (2), 1972, pp. 422–429.

Cohen, J.W.

[1968] Single server queue with uniformly bounded virtual waiting time, *J. Appl. Prob.* **5** (1), 1968, pp. 93–122.

[1969] *The Single Server Queue*, North Holland, Amsterdam, 1969.

[1976] *On Regenerative Processes in Queueing Theory*, Lecture Notes in Econ. Math. Systems **121**, Springer-Verlag, New York, 1976.

Cooper, R.B.

Introduction to Queueing Theory, 2nd ed., North Holland, New York, 1981.

Cox, D.R.

[1955] The analysis of non-Markovian stochastic processes by the inclusion of supplementary variables, *Proc. Camb. Phil. Soc.* **51** (3), 1955, pp. 433–441.

[1962] *Renewal Theory*, Methuen, London, 1962.

Cox, D.R. and Lewis, P.A.W.

The Statistical Analysis of a Series of Events, Wiley, New York, 1966.

Crane, M.A. and Iglehart, D.L.

[1974a] Simulating stable stochastic systems I: General multiserver queues, *Journ. ACM* **21** (1), 1974, pp. 103–113.

[1974b] Simulating stable stochastic systems II: Markov chains, *Journ. ACM* **21** (1), 1974, pp. 114–123.

[1975] Simulating stable stochastic systems III: Regenerative processes and discrete-event simulations, *Oper. Res.* **23** (1), 1975, pp. 33–45.

Daley, D.J.

Queueing output processes, *Adv. Appl. Probab.* **8** (12), 1976, pp. 395–415.

Ditkin, V.A. and Prudnikov, A.P.

Integral Transformations and Operational Calculus, Nauka, Moscow, 1974. (in Russian).

Dobrushin, R.L.

The Poisson law for the distribution of particles in space, *Ukr. Mat. Zhurnal*, **8**, 1965, pp. 127–134.

Einstein, A. and Smoluchowski, M.

Papers on the Theory of Brownian Motion (Russian translation: ONTI, 1936).

Englund, G.

Remainder term estimate for the asymptotic normality of the number of renewals, *Scand. J. Statist.*, **7** (1), 1980, pp. 197–202.

Epstein, B. and Hosford, T.

Reliability of some two-unit redundant systems, *Proc. 6th Nat. Symposium on Reliability and Quality Control*, 1960, pp. 466–476.

Ermakov, S.M. an Mikhailov, G.A.

A Course in Statistical Modeling, Nauka, Moscow, 1976. (in Russian).

Ezhov, I.I.
On the distribution of the maximum for a class of random processes with step trajectories, *Trudy In-ta Prikl. Matem., Tbilisi Univ., Tbilisi*, **2**, 1969, pp. 207–214.

Ezhov, I.I. and Skorokhod, A.V.
Markov processes homogeneous in the second component, *Teoriya Veroyatnosteĭ i ee Primeneniya*, I. **14** (1), 1969, pp. 4–14; II. **14** (4), 1969, pp. 679–692 (English translation: *Theory of Probab. and Applic.*, pp. 1–13, 652–667).

Faddeyev, D.K. and Faddeyeva, V.N.
Computational Methods of Linear Algebra, Fizmatgiz, Moscow, 1963.

Feller, W.
[1940] On the integro-differential equations of purely discontinuous Markov processes, *Trans. Am. Math. Soc.*, **48**, 1940, pp. 488–515.
[1957] On boundary conditions for the Kolmogorov differential equations, *Annals of Math.*, **65**, 1957, pp. 527–570.
[1950] *An Introduction to Probability Theory and its Applications*, 1st ed., Wiley, New York, 1950 (3rd ed., 1968).

Fishman, G.S.
Concepts and Methods in Discrete Event Digital Simulation, Wiley, New York, 1973.

Fortet, R.M.
Random distributions with an application to telephone engineering.—*Proc. Third Berkeley Sympos. Math. Stat. and Prob.*, University of California Press, **2** 1956, pp. 81–88.

Foster, F.G
On Stochastic Matrices Associated with Certain Queueing Processes. *Ann. Math. Statist.*, **24** 1953, pp. 355–360.

Franken, P., König, D., Arndt, U., and Shmidt, F.
Queues and Point Processes, Naukova Dumka, Kiyev, 1984 (Russian translation).

Franken, P. and Kerstan, J.
Bedeinungssystems mit unendlich vielen Bedienungs-apparaten, *Operationsforsch. und Math. Statist.*, 1968, pp. 67–76.

Franken, P. and Streller, A.
Reliability analysis of complex repairable systems by means of marked point processes, *J. Appl. Prob.*, **17** (1), 1980, pp. 154–167.

Fry, T.C.
Probability and its Engineering Uses, 2nd ed., New York, Van Nostrand, 1965.

Gantmakher, F.R.
The Theory of Matrices, Fizmatgiz, Moscow, 1953 (English translation Chelsea, New York, 1959).

Gel'fand, I.M. and Shilov, G.E.
Generalized Functions: Properties and Operations, 2nd ed., Fizmatgiz, Moscow, 1959. (English translation: Academic Press, New York-London, 1964).

Genchev, S.
Simulation Modelling using GPSS/360, VSSSO, Sofiya, 1975 (in Bulgarian).

Gergely, T. and Ezhov, I.I.
Asymptotic behavior of stochastic processes modelling an ordinary Poisson process, *Periodica Math. Hung.*, **6**, No. 2, 1975, pp. 203–211.

Gikhman, I.I. and Skorokhod, A.V.
Theory of Stochastic Processes, Nauka, Moscow, 1971, Vol. 1 (English translation, Springer-Verlag, New York).

Glushkov, V.M., Gusev, V.V., Mar'yanovich, T.P., and Sakhnyuk, M.A.
Software for Simulation of Continuously-Discrete Systems, Naukova Dumka, Kiyev, 1975 (in Russian).

Gnedenko, B.V.
[1961] *The Theory of Probability*, 3rd ed., Moskva, Fizmatgiz, 1961 (English translations, 4th ed., Chelsea, New York, 1967).
[1959] Some remarks on two papers of D.V. Barrer, *Buletinul Institutuli din Jasi*, **5** (1–2), 1959, pp. 111–117 (in Russian).
[1960] Über einige Aspekte der Entwicklung der Theorie der Warteschlangen (On some aspects of the development of queueing theory), *Math. Technik Wirtschaft.*, No. 3, 1960, pp. 162–166 (in German).
[1939] Limit theorems for sums of independent random variables, *Izv. AN SSSR, Math. ser.*, 1939, pp. 181–232 (in Russian).

Gnedenko, B.V., Belyaev, Yu. K. and Solov'yev, A.D.
Mathematical Methods in Reliability Theory, Nauka, Moscow, 1965 (in Russian).

Gnedenko, B.V. and König, D.
Handbuch der Bedienungstheories, Akademie-Verlag, Berlin, Vol. I. Grundlagen und Methoden, 1983; Vol. **II**. Formeln und andere Ergebnisse, 1984.

Gnedenko, B.V. and Zubkov, M.N.
Determination of the optimal number of mooring lines, *Morskoĭ Sbornik*, 1964, pp. 39–59 (in Russian).

Gnedenko, B.V., et al.
Priority Queueing Systems, Moscow University Press, Moscow, 1973 (in Russian).

Gnedenko, B.V. and Kovalenko, I.N.
Lectures on Queueing Theory, Kiyev, 1963 (in Russian).

Gnedenko, D.B.
On a generalization of Erlang's formula, *Zastos. Matem.* **12**, 1971, pp. 239–249 (in Russian).

Grigelionis, B.I.
[1962a] Accuracy of approximation of a superposition of renewal processes by a Poisson process, *Litovskiĭ Matematicheskiĭ Sbornik*, **II** (2), 1962, pp. 135–143.
[1962b] A refined multivariate limit theorem on convergence to a Poisson law, *Litovskiĭ Matematicheskiĭ Sbornik*, **II**, 2, 1962, pp. 143–148.
[1963] On the convergence of sums of random step processes to a Poisson process, *Teoriya Veroyatnosteĭ i ee Primenenia*, **VIII** (2), 1963, pp. 189–194 (English translation: *Theory of Probab. and Appl.* **8**, pp. 177–182).
[1964] Limit theorems for sums of renewal processes, *Kibernetika–na sluzhbu kommunizmy*, Energiya, Moskow-Leningrad, 1964, pp. 246–265.
[1975] Random point processes and martingales, *Litovsk. Matemat. Sbornik*, **15** (3), 1975, pp. 101–114.

Grishchenko, V.A.
Stream of lost customers in a multiserver queueing system with rare losses, *Ukr. Mat. Zhurn.* **35** (4), 1983, pp. 422–426.

Gross, D. and Harris, C.M.
Fundamentals of Queueing Theory, Wiley, New York, 1974.

Gusak, D.V.
Distribution of sojourn time of a homogeneous process with independent increments over an arbitrary level, *Teoriya Veroyatnosteĭ i ee Primeneniya*, **28** (3), 1983, pp. 478–488 (English translation: *Theory of Probab. and Applic.* **28**, pp. 503–514).

Guseĭnov, B.T.
[1970] Generalization of Kovalenko's theorems on the invariance of state probabilities of service systems with respect to service time distribution, *Lecture at the 6th Internat. Teletraffic Congress*, München, 1970, pp. 312/7–312/10.
[1974] On a multiserver queueing system with losses and mutual assistance, *Probl. peredachi inform.* **10** (4), 1974, pp. 78–84.

Haji, R. and Newell, G.F.
A relation between a stationary queue and waiting time distributions, *J. Appl. Prob.*, **8** (3), 1971, pp. 617–620.

Hammersley, J.M. and Handscomb, D.C.
Monte Carlo Methods, Methuen, London, 1964.

Harrison, J.M. and Resnick, S.I.
The stationary distribution and first exit probabilities of a storage process with general release rule, *Math. Oper. Res.*, **4** (1), 1976, pp. 347–358.

Ivchenko, G.I., Kashtanov, V.A., and Kovalenko, I.N.
Queueing Theory, Vysshaya Shkola, Moscow, 1982 (in Russian).

Ivnitskiĭ, V.A.
[1965] Asymptotic investigation of the stationary queue for generalized incoming streams, *Kibernetika*, No. 5, 1965, pp. 60–65.
[1966] A single-server queueing system with variable intensity of the incoming stream and service rate, *Litovskiĭ Matematicheskiĭ Sbornik*, **VI**, No. 1, 1966, pp. 122–128.
[1969] On reconstructing the characteristics of a single-server system with limitations on waiting time, *Tekhnich. Kibern.*, No. 3, 1969. pp. 60–65 (English translation: *Eng. Cybern.*).
[1977] On reconstruction of characteristics of a GI/M/1 system from observations on the outgoing stream, *Teoriya Veroyatnosteĭ i ee Primeneniya*, **22** (1), 1977, pp. 188–191 (English translation: *Theory of Probab. and Applic.* **22**, 1977, pp. 184–187).
[1982] On the invariance condition for stationary probabilities of queueing networks, *Teoriya Veroyatnosteĭ i ee Primeneniya*, **27** (1), 1982, pp. 188–192 (English translation: *Theory of Probab. and Applic.* **27**, 1982, pp. 196–201).

Jansen, U. and König, D.
Insensitivity and steady-state probabilities in product form for queuing networks, *Elektron. Informationsverarb. und Kybern.* **16** (3), 1980, pp. 385–397.

Jayswal, N.K.
Priority Queues, Academic Press, New York, 1968.

Kabanov, Yu. M., Liptser, R. Sh., and Shiryaev, A.N.

Martingale methods in the theory of point processes, *Trudy shkoly-seminara sluch. protsessov*, Druskininkaĭ, 1974; Part II, Vil'nyus, 1975, pp. 269–354.

Kalashnikov, V.V.

Qualitative Analysis of Behavior of a Complex System Using the Method of Trial Functions, Nauka, Moscow, 1978.

Karlin, S. and McGregor, J.

[1957a] The classification of birth and death processes, *Trans. Amer. Math. Soc.*, **86**, 1957, pp. 366–400.

[1957b] The differential equations of birth and death processes and the Stieltjes moment problem, *Trans. Amer. Math. Soc.*, **85**, 1957, pp. 489–546.

Kartashov, N.V.

Strongly stable Markov processes, *Prob. ustoĭch. stokhast. model.* VNIISI, Moscow, 1981, pp. 54–59.

Kelly, F.P.

Networks of queues, *Adv. Appl. Prob.* **8** (2), 1976, pp. 416–432.

Kendall, D.G.

[1953] Stochastic processes occurring in the theory of queues and their analysis by means of the imbedded Markov chains, *Ann. Math. Statist.*, **24**, 1953, pp. 338–354.

[1960] Geometric Ergodicity and the Theory of Queues, in: *Mathematical Methods and Social Sciences*, Stanford Univ. Press, 1960.

[1951] Some problems in the theory of queues, *J. Roy. Statist. Soc., Ser B*, **13** (2), 1951, pp. 151–185.

Khinchin, A. Ya.

[1963] *Papers on the Mathematical Theory of Queues*, Fizmatgiz, Moscow, 1963.

[1955] *Mathematical Methods in the Theory of Queueing*, Trudy Matematicheskogo Instituta im. V.A. Steklova, **49**, Azd. AN SSSR, 1955 (English translation: *Griffin's Statistical Monographs and Courses*, No. **7**, London, 1960).

[1956] On Poisson sequence of random events, *Teoriya Veroyatnosteĭ i ee Primeneniya*, **1** (3), 1956, pp. 320–327 (English translation: *Theory of Prob. and Appl.* **1**, pp. 291–297).

[1932] Mathematical theory of a stationary queue, *Matematicheskiĭ Sbornik*, **39** (4), 1932, pp. 73–84.

Kiefer, J. and Wolfowitz, J.

[1956] On the characteristics of the general queueing process with application to random walk, *Ann. Math. Statist.*, **27** (1), 1956, pp. 147–161.

[1955] On the theory of queues with many servers, *Trans. Amer. Math. Soc.*, **78**, 1955, pp. 1–18.

Kindler, Ye.

Simulation Languages, Energoatomizdat, Moscow, 1965 (translated from Czech).

Kingman, J.F.C.

[1966] *On the Algebra of Queues*, Methuen, London, 1966.

[1970] Inequalities in the theory of queues, *J. Roy. Statist. Soc., Ser. B.*, **32**, 1970, pp. 102–110.

[1962] On queues in heavy traffic, *J. Roy. Statist. Soc., Ser. B* **24**, 1962, pp. 381–392.

Kleinrock, L.
Queueing Systems, Wiley, New York, 1975, Vols. 1 and 2.
Klimov, G.P.
[1964] "Extremal problems in queueing theory", *Kibernetika-na sluzhbu kommunizmu*, **2**, Energiya, Moscow, 1964, pp. 310–325.
[1966] *Stochastic Queueing Systems*, Nauka, Moscow, 1966.
[1970] Some solved and unsolved problems in servicing by a consecutive chain of servers, *Tekhn. Kibern.* No. 6, 1970, pp. 88–92 (English translation: *Eng. Cybern.*).
Klimova, Ye. Z.
Investigation of a single server system with "warm-up," *Tekhn. Kibern.* No. 1, 1968, pp. 91–97 (English translation: *Eng. Cybern*).
Köeningsberg, E.
Invariance properties of queueing networks and their application to computer-communications systems, *Inform. Can. J. Oper. Res. and Inf. Process*, **19**, No. 3, 1981, pp. 185–204.
König, D., Matthes, K., and Navrotzky, K.
Verallgemeinerungen der Erlangschen und Engsetchen Formeln, Akademie-Verlag, Berlin, 1967.
König, D., Schmidt, V. and Stoyan, D.
On some relations between stationary distributions of queue length in $G|G|S$ systems, *Math. Oper. und Statist.* **7** (4), 1976, pp. 577–586.
König, D. and Stoyan, D.
Methoden der Bedienungstheorie, Akademie-Verlag, Berlin, 1976.
Korolyuk, V.S.
Holding time of a semiMarkov process in a fixed set of states, *Ukr. Math. Journ.* **3**, 1965, pp. 123–128.
Korolyuk, V.S. and Turbin, A.F.
[1976] *Semi-Markov Processes and their Applications*, Naukova Dumka, Kiyev, 1976.
[1978] *Mathematical Foundations of Phase Consolidation of Complex Systems*, Naukova Dumka, Kiyev, 1978.
Kovalenko, I.N.
[1960] Study of a many-server queueing system with limited holding time, *Ukr. Mat. Zhurn.* **XII**, No 3, 1960, pp. 471–476.
[1961a] Queueing systems with restrictions, *Teoriya Veroyatnosteĭ i ee Primeneniya*, **VI** (1), pp. 222–228 (English translation *Theory of Probab. and Appl.* **6**, 1961, pp. 204–208.)
[1961b] *On a Queueing System with Limited Waiting Time*, Kiev Institute of Math., AN UkrSSR, 1961.
[1962] Sur la condition pur que, en régime stationnaire la distribution soit indépendante des lois des durées de conversation (The condition under which the stationary distribution is independent of the law of the duration of the call) *Ann. des Télécommunications*, **17** (7–8), 1962, pp. 190–191.
[1963] The condition under which stationary distributions are independent of the form of the service time distribution, *Problemy peredachi informatsii*, Issue 11, 1963, pp. 147–151.
[1964a] Some analytical methods in queueing theory, *Kibernetika-na sluzhbu kommunizmu*, **2**, Energiya, Moscow, 1964, pp. 325–337.

[1964b] Some problems of reliability theory of complex systems, *ibid.*, pp. 194–205.

[1965a] On classes of complex systems, *Tekhn. Kibern.*, No. 6, 1964, pp. 3–9; No. 1, 1965, pp. 14–20; No. 3, 1965, pp. 3–11. (English translation: *Eng. Cybern.*)

[1965b] On a class of limiting distributions for thinning streams of homogeneous events, *Litovskiĭ Matematicheskiĭ Sbornik*, **V** (4) 1965, pp. 569–573.

[1965c] On a class of limiting distributions for a sequence of series of sums of independent renewal processes, *ibid.* V.

[1965d] On renewal of characteristics of the system by means of observations of the outgoing stream, *Dokl. Akad. Nauk SSSR*, **164** (5), 1965, pp. 979–981.

[1971] On a queueing system with service rate dependent on the number of customers in a system and periodical detachment of the channel, *Probl. peredachi inform.* **7** (2), 1971, pp. 106–111.

[1975] *Studies in the Analysis of Reliability of Complex Systems*, Naukova Dumka, Kiyev, 1975.

[1980] *Analysis of Rare Events in Estimation of Efficiency and Relability of Systems*, Sov. Radio, Moscow, 1980.

Kovalenko, I.N. and Kuznetsov, N. Yu.

Construction of an imbedded renewal process for multidimensional processes of queueing theory and its application to derivation of limit theorems, Preprint 80-12, *Institute of Cybernetics*, AN UkrSSR, Kiyev, 1980.

Kovalenko, I.N. and Yurkevich, O.M.

[1970] New results in the theory of queueing systems with limitations, *Teor. Veroyatn. i Mat. Statistika*, **2**, 1970, pp. 45–51.

[1972] Queueing system with simultaneously arriving "impatient clients," *Tekhn. Kibern.*, No. 1, 1972, pp. 52–56. (English translation: *Eng. Cybern.*)

Krein, M.G.

Integral equations on a half-line, *Uspehi Mat. Nauk*, **13** (5(83)), 1961, pp. 35–150. (Engl. translation: *Soviet Math. Surveys*).

Krein, M. and Lemuan, O.

Introduction to the Regenerative Method of Analysis of Models, Nauka, Moscow, 1982 (in Russian).

Krivutsa, V.G.

Package of Applied Programs AMOS (Aggregated Modeling of Complex Systems) Vol. **1**–**2**. Institute of Cybernetics AN UkrSSR, Kiyev, 1980.

Kuznetsov, N. Yu.

[1981] On construction of estimators of stationary probabilities of states of a system for which the conditions similar to the conditions of Kovalenko's theorem are fulfilled, *Prob. ustoĭchiv. stokhast. modeleĭ*, VNIISI, Moscow, 1981, pp. 70–81.

[1982] Quantitative estimates of stability of the Erlang system when the incoming stream of customers is perturbed, *Probl. ustoĭchiv. stokhast. modeleĭ*, VNIISI, Moscow, 1982, pp. 71–76.

[1984] Determination of stationary probabilities of states of a $GI|G|n|0$ system with incoming stream of customers similar to the Poisson, *Kibernetika*, No. 2, 1984, pp. 74–79.

Lamprecht, G.

Einführung in die Programmiersprache SIMULA, Vieweg-Verlag, Braunschweig, 1976.

Lavenberg, S.S. and Sauer, C.H.
Sequential stopping rules for the regenerative method of simulation, *IBM J. Res. Devel.*, **21**, No. 6, 1977, pp. 545–558.

Le Gall, P.
Les systemes avec ou sans attente et les proces stochastiques [Systems with or without waiting and stochastic processes], Dunod, Paris, 1952, Vol. 1.

Lindley, D.V.
The theory of queues with a single server, *Proc. Cambr. Phil. Soc.* **48** (2), 1952, pp. 277–289.

Lifshits, A.L. and Mal'ts, E.A.
Statistical Simulation of Queueing Systems, Sov. Radio, Moscow, 1978 (in Russian).

Loynes, R.M.
The stability of a queue with nonindependent inter-arrival and service times, *Proc. Cambr. Phil. Soc.*, **58** (3), 1962, pp. 494–520.

Lutkov, V.I.
Simulation Modelling and Production Management, MTsNTI, Moscow, 1978 (in Russian).

Malyshev, V.A.
Wiener–Hopf equations and their application in probability theory, *Itogi nauki i tekhniki*, VINITI, Moscow, **13**, 1976, pp. 5–35.

Malyshev, V.A. and Men'shikov, M.V.
Ergodicity, continuity and analyticity of countable Markov chains, *Tr. Moskovskogo Mat. Obshchestva*, **39**, 1979. pp. 3–48.

Marcienkiewicz, J.
Quelques théorèmes sur les fonctions indépendantes, *Studia Math.*, **7** (1), 1937, pp. 104–120.

Marshall, W.G. and Harris, C.M.
A modified Erlang approach to approximating $GI|G|1$ queues, *J. Appl. Prob.*, **13** (1), 1976, pp. 118–125.

Mar'yanovich, T.P.
[1960] Generalization of Erlang's formulas when the servers may fail and be renewed, *Ukr. Mat. Zhurn.* **XII**, No. 3, 1960 (in Russian).
[1961a] Reliability of redundant systems, *Dop. AN URSR*, No. 7, 1961 (in Ukrainian).
[1961b] Reliability of partially active redundant systems, *Dop. AN URSR*, No. 8, 1961 (in Ukrainian).
[1962a] A single-server queueing system with an unreliable server, *Ukr. Mat. Zhurn.* **XIV**, No. 4, 1962 pp. 417–422 (in Russian).
[1962b] Service with failure to the server, *Trudy $\underline{\text{VI}}$ Vsesoyuznogo soveshchaniya po teorii veroyatnosteĭ i matematicheskoĭ statistike*, Vil'nyus, 1962 (in Russian).
[1963] Some problems on the reliability of redundant systems, *Ukr. Mat. Zhurn.* **XV**, (2), 1963 pp. 213–219 (in Russian).

Matthes, K.
[1960] *Zur Theorie des Verlustverkehrs* (On the Theory of Traffic with Losses), Lecture at a Mathematical Symposium on the 150th Anniversary of Humboldt University in Berlin, 1960.
[1962] Zur Theorie der Bedienungsprozesse (On the theory of queueing processes). *Trans. Third Prague Confer. Inform. Theory, Statistical Decision Functions, Random Processes*, Prague 1964, pp. 513–528.

Miller, R.G. Jr.
Priority queues, *Ann. Math. Statist.*, **31** (1), 1960, pp. 86–103.

Mills, R.C.
Models of Stochastic Service Systems with Batched Arrivals, Ph.D. Dissertation, Columbia University, 1980.

Moran, P.A.P.
A theory of dams with continuous input and a general release rule, *J. Appl. Prob.*, **6** (1), 1969, pp. 88–98.

Morozov, Ye. V.
Investigation of service systems with limited waiting time, *Avtomatizir. sistemy plan. raschetov*, **9**, 1977, pp. 48–57.

Mova, V.V. and Ponomarenko, L.A.
Controlling a queueing system with a finite queue, *Prom. Kibernetika*, Kiyev, 1971, pp. 328–338 (in Russian).

Nagayev, S.V.
Ergodic theorems for Markov processes with discrete time, *Sibirsk. Mat. Zhurn.* **6** (2), 1965, pp. 413–432.

Nasirova, T.I.
On a generalization of the Erlang problem, *Trudy Vychislitel'nogo Tsentra AN AzerbSSR*, **2**, 1963, pp. 3–18.

Neuts, M.F.
[1968] Two queues in a series with a finite intermediate waiting room, *J. Appl. Prob.*, **5** (1), 1968, pp. 123–142.
[1975] Computational uses of the method of phases in the theory of queues, *Computers Math. Appl.*, No. 1, 1975, pp. 151–166.

Newell, G.F.
Applications of Queueing Theory, Chapman and Hall, London, 1971.

Nollau, V.
Semi-Markowsche Prozesse, Akademie-Verlag, Berlin, 1980.

Ososkov, G.A.
A limit theorem for flows of similar events, *Teoriya Veroyatnosteĭ i ee Primeneniya*, **1** (2) 1956, pp. 274–282 (English translation: *Theory of Probab. and Appl.* **1** (2) 1956, pp. 248–255).

Palm, C.
[1947] The distribution of repairment in servicing automatic machines, *Industrition, Norden*, **75**, 1947, pp. 75–80, 90–94, 119–123.
[1943] Intensitätsschwankungen im Fernsprechverkehr, *Ericsson Technics*, **44** (**1**), 1943, pp. 1–189.

Pellaumail, J.
Formule de produit et decomposition de reseaux de files d'attente, *Ann. Inst. H. Poincare*, **15** (3), 1979, pp. 261–286.

Pirogov, V.V.
Dialog Simulation Systems, Zinatiye, Riga, 1977.

Pogozhev, I.B.
[1964] Estimating deviations of a stream of failures in heavy-duty equipment from a Poisson stream. In: "*Kibernetiku—na sluzhbu kommunizmu*," Vol. **2**, Energiya. Moscow–Leningrad, 1964, pp. 228–245.

Pollaczek, F.
[1957] Problèmes stochastiques posés par le phénomène de formation d'une

queue d'attente à un guichet et par des phénomènes apparentés (Stochastic Problems Involved in the Formation of a Queue at a Cash-Desk and in Similar Phenomena). *Memorial des Sciences Mathématiques.* Gauthier-Villars, Paris, 1957.

[1961] Théorie analytique des problèms stochastiques relative a un groupe de lignes téléphoniques avec dispositif d'attente, Gauthier-Villars, Paris, 1961.

Pollyak, Yu. G.

Probabilistic Simulation on Computers, Sov. Radio, Moscow, 1971 (in Russian).

Polyaev, L.N.

Multichannel queueing system with input stream depending on the state of the system, finite queue and bounded waiting time, *Veroyatn. protsessy i ikh prilozheniye*, MIEM, Moscow, 1983, pp. 70–73.

Popov, N.N.

Conditions of geometric ergodicity for countable Markov Chains, *Dokl. AN SSSR*, **234**, No. 2, 1977, pp. 316–319.

Prabhu, N.U.

[1965] *Queues and Inventories*, Wiley, New York, 1965.

[1974] Wiener–Hopf techniques in queueing theory, *Mathem. Methods in Queueing Theory, Lect. Notes in Economics and Math. Systems*, Springer-Verlag, Heidelberg, 1974, pp. 81–90.

Presman, E.L.

On waiting time for a many-server queueing system, *Teoriya Veroyatnosteĭ i ee Primeneniya*, **10** (1), 1965, pp. 69–81 (English translation. *Theory of Probab. and Appl.* **10** (1), 1965, pp. 63–73).

Prokhorov, Yu. V.

Transient phenomen in queueing processes, *Litovskiĭ Matematicheskiĭ Sbornik* **III** (1), 1963, pp. 199–206.

Prokhorov, Yu. V. and Viskov, O.V.

The probability of loss of calls in heavy traffic, *Teoriya Veroyastnostei i ee Primeneniya*, **IX** (1), 1964, pp. 99–104 (English translation: *Theory of Probab. and Appl.* **9**, 1964, pp. 92–96).

Prokhorov, Yu. V. and Rozanov, Yu. A.

Probability Theory: Basic Notions: Limit Theorems: Random Processes, 2nd ed., Nauka, Moscow, 1973.

Pugachev, V.N.

Combinatorial Methods for Determination of Probabilistic Characteristics, Sov. Radio, Moscow, 1973.

Redheffer, R.M.

A note on the Poisson law, *Math. Magazine*, **26**, 1953, pp. 185–188.

Reich, E.

On an integrodifferential equation of Takács, I, *Ann. Math. Statist.*, **29**, 1958, pp. 563–570.

Rényi, A.

Poisson-folyamat egy jellemzëse (A characteristic of the Poisson stream), *Proc. Math. Inst., Hungarian Acad. Sci.* **1** (4), 1956, pp. 563–570.

Rise, O.

Single server systems, *The Bell Techn. Journ.* No. 1 1962, pp. 269–278.

Riordan, J.

Stochastic Service Systems, Wiley, New York, 1962.

Rogozin, B.A.
On the distribution of the first jump, *Teoriya Veroyatnosteĭ i ee Primeneniya*, **9** (3), 1964, pp. 498–515 (English translation: *Theory of Probab. and Appl.* **9**, 1964, pp. 450–465).

Rossberg, H.-J. and Siegel, G.
[1974] On Kingman's integral inequalities for approximation of the waiting time distribution in the queueing model $GI|G|1$ with and without warming-up time, *Zastos. Matem.* **14** (1), 1974, pp. 27–30.
[1975] New method of investigating the distribution of waiting time under stationary conditions for the queueing system $GI|G|1$ with "warm-up," *Zastos. Matem.* 1975, pp. 537–547.

Rozental', G.O. and Tolmachev, A.L.
On sequential service with blocking, *Metody teorii teletrafika*, Nauka, Moscow, 1975, pp. 55–59.

Rudlovchak, V.
Palm's problem for systems with a limited queue, *Vychisliteln. Metody i Programmirovaniye*, **9**, 1967, pp. 131–142.

Rybko, A.N.
A condition for the existence of a stationary mode for two types of communication networks with intercommunication. *Probl. peredachi inform.*, **18** (1), 1982, pp. 94–103.

Rykov, V.V.
[1966] Markov decision processes with a finite space of states and decisions, *Teoriya Veroyatnosteĭ i ee Primeneniya*, **11** (2), 1966, pp. 343–351. (English translation: *Theory of Probab. and Appl.* **11**, 1966, pp. 302–311.)
[1970] On controllable queueing systems, *Teor. mass. obsluzh., Trudy II Vsesoyuzn. shkoly-soveshch. po teorii mass. obsluzh.* Dilizhan, Moscow University Press, 1970, pp. 106–114.
[1975] Controllable queueing systems, *Itogi nauki i tekhniki*, VINITI, Moscow, **12**, 1975, pp. 43–153.

Saaty, T.
Elements of Queueing Theory with Applications, McGraw-Hill, New York, 1961.

Sarymsakov, T.A.
Principles of the Theory of Markov Processes, Gostekhizdat, Moscow, 1954.

Shneps, M.A.
Numerical Methods of Teletraffic Theory, Svyaz', Moscow, 1974.

Schrage, L.E.
A proof of the optimality of the shortest remaining time discipline, *J. Oper. Res.*, **16** (3), 1968, pp. 687–690.

Schrage, L.E. and Miller, L.W.
The queue $M|G|1$ with the shortest remaining processing time discipline, *J. Oper. Res.*, **14** (4), 1966, pp. 670–684.

Shwab, N.D.
Virtual waiting time for systems with τ-interruption, *Mat. modeli slozhnykh sistem*, Cybernetics Institute of the AN UkrSSR, Kiyev, 1973, pp. 206–211.

Schwetlick, H.
Direkte Verfahren zur Lösung grosser linearer Gleichungssysteme, *Wiss. Informationen der TH Karl-Marx-Stadt*, **9**, 1979, pp. 22–42.

Sevast'yanov, B.A.

[1957] An ergodic theorem for Markov processes and its application to telephone systems with refusals, *Teoriya Veroyatnosteĭ i ee Primeneniya*, **II** (1): 106–116, 1957 (Engl. translation: *Teor. Prob.* and *Applic* **2**: 104–112, 1957.)

[1959] *Erlang's Formulas in Telephone Traffic with an Arbitrary Law of the Distribution of the Length of Calls.*—Trudy III Vsesoyuznogo matematicheskogo s"ezda, **4**, pp. 68–70, Moscow, AN SSSR. 1959.

Shakhbazov, A.A. and Samandarov, E.G.

On servicing non-ordinary streams, *Kibernetiku-na sluzhbu kommunizmu* **2**, Energia, Moscow, 1964, pp. 338–353 (in Russian).

Shannon, R.E.

Simulation Systems: The Art and Science, Prentice-Hall, Englewood Cliffs, N.J., 1975.

Sharakshane, A.S., Zheleznov, I.G., and Ivnitskii, V.A.

Complex Systems, Vysshaya shkola, Moscow, 1977 (in Russian).

Shurenkov, V.M.

Ergodic Theorems and Related Problems of the Theory of Random Processes, Naukova Dumka, Kiev, 1981 (in Russian).

Sil'vestrov, D.S.

SemiMarkov Processes with a Discrete Set of States, Kiyev Univ. Press, 1971 (in Russian).

Simonova, S.N.

On the outgoing stream of single server queueing systems, *Ukr. Mat. Zhurn.*, **21** (3), 1969, pp. 501–510.

Smith, W.L.

[1958] Renewal theory and its ramifications, *J. Roy. Statist. Soc.*, Ser B, **20** (2), 1958, pp. 243–302.

[1972] The infinitely-many server queue with semi-Markovian arrivals and customer-dependent exponential service times, *J. Oper. Res.*, **20** (4), 1972, pp. 907–913.

Smith, N.M. and Yeo, G.F.

On a general storage problem and its approximating solution, *Adv. Appl. Prob.*, **13** (3), 1981, pp. 567–602.

Sobol', I.M.

Numerical Monte-Carlo Methods, Nauka, Moscow, 1973.

Solov'yev, A.D.

The asymptotic behavior of the time of the first occurrence of a rare event in the regenerative process, *Tekhn. Kibern.* No. 6, 1971, pp. 79–89 (English translation: *Eng. Cybern.*).

Stephan, F.F.

Two queues under preemptive priority, *J. Oper. Res.*, **6** (2), 1958, pp. 399–418.

Stidham, S., Jr.

A last word on $L = \lambda W$, *J. Oper. Res.* **22** (2), 1974, pp. 417–421.

Stoyan, D.

[1973] Queueing systems of type $M|G|1$ under light load, *Kibernetika*, No. 2. 1973, pp. 108–110.

[1977] Further stochastic order relations among $GI|G|1$ queues with a common traffic intensity. *Math. Operationsforsch. und Statist.* **8**, No. 4, 1977, pp. 541–548.

[1978] Queueing networks insensitivity and heuristic approximation, *Electron. Informationsverarb. und Kybern.* **14**, No. 1, 1978, pp. 135–143.

Syski, R.

Introduction to Congestion Theory in Telephone Systems, Oliver and Boyd, Edinburgh and London, 1960.

Takács, L.

[1967] *Combinatorial Methods in the Theory of Stochastic Processes*, J Wiley, New York, 1967.

[1969] On Erlang's formula, *Ann. Math. Statist.*, **40** (1), 1969, pp. 71–78.

[1974] A single server queue with limited virtual waiting time, *J. App. Prob.*, **11** (3), 1974, pp. 612–617.

[1975] Combinatorial and analytic methods in the theory of queues, *Adv. Appl. Prob.*, **7** (3), 1975, pp. 607–635.

Taylor, R.L.

Some steady-state solutions for a tandem queue, *New J. Statist. and Oper. Res.*, **8** (1), 1972, pp. 1–5.

Tien, T.M.

Verallgemeinerte stückweise lineare Markowsche Prozesse für die Bedienungs- und Zuverlässigkeitstheorie, *Math. Operationsforsch. und Statist. Ser. Optimization* **8** (3), 1977, pp. 419–431.

Tsitsiashvili, G. Sh.

Construction of Lyapunov functions for study of the stability of piecewise-linear Markov chains, *Tekh. Kibern.* no. 2, 1979, pp. 74–84 (English translation: *Eng. Cybern.*)

Tweedie, R.L.

Relations between ergodicity and mean drift for Markov chains, *Austral. J. Stat.* **17** (2), 1975, pp. 293–305.

Vêre-Jones, D.A.

[1962] Geometric ergodicity in denumerable Markov chains, *Quart. J. Math.* **13** (49), 1962, pp. 7–28.

[1964] A rate of convergence problem in the theory of queues, *Teoriya Veroyatnosteĭ i ee Primeneniya*, **9** (4), 1964, pp. 104–112 (English translation: *Theory of Prob. and Appl.* **9**, pp. 96–103).

Veretennikov, A. Yu.

On the ergodicity of queueing systems with infinitely many servers, *Mat. Zametki*, **22**, 1977, pp. 561–569 (Engl. translation: *Mathematical Notes*).

Viskov, O.V.

Asymptotic formulas in the theory of queues, *Teoriya Veroyatnosteĭ i ee Primeneniya* **9** (1), 1964, pp. 177–8 (Engl. translation: *Th. Prob. and Appl.* **9**).

Viskov, O.V. and Ismailov, A.I.

On the distribution of a queue length in mixed queueing systems, *Izv. AN UzSSR*, no. 3, 1970, pp. 68–9.

Walrand, J. and Varaiya, P.

Flows in queueing networks: a martingale approach, *Math. Oper. Res.* **16** (2), 1981, pp. 387–404.

Whitt, W.

Embedded renewal processes in the $GI|G|s$ queue, *J. Appl. Prob.* **9** (3), 1972, pp. 650–658.

Williams, T.M.
Nonpreemptive multiserver priority queues, *J. Oper. Res. Soc.* **31** (4), 1967, pp. 1105–1107.

Yaroshenko, V.N.
On a certain queueing problem, Dopovidi AN URSR, No. 2, 1962, pp. 153–156 (in Ukrainian).

Yarovitskiĭ, N.V.
[1961] On an outgoing stream of a one server queueing system with losses, *Dopovidi AN UkrSSR*, No. 10, 1961, pp. 1251–1254 (in Russian).
[1966] Probabilistic-automated simulation of discrete systems, *Kibernetika*, No. 5, 1966, pp. 35–43 (in Russian).

Yemel'yanov, S.V., Kalashnikov, V.V., Lutkov, V.I. and Nemchinov, B.V.
Methodological Problems of Constructing Simulation Systems: Survey, MTsNTI, Moscow, 1978 (in Russian).

Yurkevich, O.M.
[1970] Investigation of multiserver queueing systems with bounded waiting time, *Tekhn. Kibern.* No. 5, 1970, pp. 50–58. (English translation *Eng. Cybern.*)
[1971] On multiserver systems with random limitations on waiting time, *Tekhn. Kibern.*, No. 4, 1971, pp. 63–69 (English translation *Eng. Cybern.*)

Zakusilo, O.K.
[1972a] Rarefied semi-Markov processes, *Teor. Veroyat. i Matem. Statistika*, **6**, 1972, pp. 54–59 (in Russian).
[1972b] Necessary conditions for convergence of a rarefied semi-Markov processes, *ibid.* **7**, 1972, pp. 65–69 (in Russian).
[1973a] Conditions for convergence of a sum of a random number of random variables defined on a Markov chain, *ibid.*, **8**, 1973, pp. 59–64 (in Russian).
[1973b] On convergence of a sum of a random number of random variables defined on a process with a finite number of states, *Dopovidi URSR*, No. 5, 1973, pp. 390–392 (in Ukrainian).
[1973c] On the convergence of a sum of random number of random variables defined on an ergodic process, *Teor. Veroyat. i Matem. Statistika*, **9**, 1973, pp. 90–99 (in Russian).

Zakusilo, O.K. and Meleshchuk, I.V.
On the convergence of a superposition of independent streams to a simple one, *Visnik KDU, Ser. mat.-mekh.* **18**, 1976, pp. 122–125 (in Ukrainian).

Zavadskaya, L.A.
[1979] On application of an analytic-statistical method to the investigation of a highly reliable standby system, *Metody issled. operats. i teoriya nadezhn. v analize sistem*, Institute of Cybernetics, AN UkrSSR, 1979, pp. 15–25.
[1981] Estimation of reliability of a system with control and maintenance using analytic-statistical methods, *Kibernetika*, No. 2, 1981, pp. 56–59.
[1984] On an approach to accelerating simulation of standby problems, *Elektron. modelirovaniye*, **6** (3), 1984, pp. 57–60.

Zhuk, P.I.
On a transform of generalized Erlang formulas, *Kibernetika*, No. 4, 1968, pp. 146–147.

Zitek, F.
[1957] Remarks on a theorem of Korolyuk, *Czechosl. Math. J.*, **7**, 1957, pp. 318–319 (in Russian).

[1958] Contributions to the theory of orderly queues, *Czechosl. Math. J.*, **8**, 1958, pp. 448–459 (in Russian).

Zolotarev, V.M.

Metric distances in the space of random variables and their distributions, *Mat. Sbornik*, **101** (143), No. 3, 1976, pp. 416–454.

[1977] Quantitative estimates of the continuity property of queueing systems of the $G|G|\infty$ type, *Teoriya Veroyatnosteĭ i ee Primeneniya*, **22** (4), 1977, pp. 700–711 (English translation: *Theory of Probab. and Appl.* **22**, 1977, pp. 679–691).

Index

In view of the detailed table of contents, the items not explicitly mentioned therein are included in this Index. The reader is thus advised to consult the table of contents first.

Availability coefficient, 189, 241

Berk's theorem, 224
Blackwell's theorem, 116

"Clean up" time, 256
Controlling sequence, 257
Convergence in the Khinchin sense, 135
Cyclic service, 234

Degree of service, 213
Distance in variation, 254–255

Elementary renewal theorem, 116
Ergodic distribution, 166
Erlang distribution, 157
Erlang formulas, 38
Exit density, 54

Factorization components, 230
Factorization representation, 230
Feller theorem, 29–31
Finite stream, 82, 96
Franken theorem, 127

Generating function, 78

HyperErlang distribution, 57

Instantaneous value of a parameter, 76
Ivnitskiĭ theorem, 201

Kendall's method, 137
Key renewal theorem, 117
Khinchin formula, 173, 215

Laplace-Stieltjes transform, 31
Last-come–first-served discipline, 175
Leading function of a stream, 96, 121
Levy-Prokhorov distance, 258
Limited aftereffects, 109
Lindeberg conditions, 259, 262
Load on the system, 164, 192, 230
L-process, 145

Marked (graded) service, 248
Mar'yanovich system, 241
Moran model, 160

NBUE, 231
Nonhomogeneous-in-space random walks, 160
NWUE, 231

Ordering (regulating condition), 219
Overjump
  amount of, 118

Palm-type stream, 109
Parameter of a stream, 83
Poisson stream, 74
Pseudorandom numbers, 274
Pure birth process, 28

Queue size, 262

Random stream, 74
Record achievements, 261
Recurrent process, 223
Recurrent stream, 122
Regular point, 96
Regular PLMP, 157
Regular process, 82
Regular stream, 96
RNG, 274

Schrage's discipline, 271
Service with blocking, 256
Singular Poisson stream, 81
Smith's theorem, 117, 131
Sojourn phase, 160
Spitzer's identity, 231
Stable on the average in time
  Markov chains, 257
Step process, 120
Stochastic ordering, 231
Stream
  finite, 82, 96
  leading function of, 96
  Palm-type, 109
  parameter of, 83
  Poisson, 74
  recurrent, 122
  regular, 96
  singular Poisson, 81
  with limited aftereffects, 109

Underjump
  amount of, 118
Underload, 171

Warming-up, 189

MATHEMATICAL MODELING

Series editors:

William Lucas
Department of Mathematics
Claremont Graduate School
Claremont, CA 91711

Maynard Thompson
Department of Mathematics
Indiana University
Bloomington, IN 47405

Mathematical Modeling is a series of carefully selected books which present serious applications of mathematics for both the student and professional audience. The series aims to familiarize the user with new models and new methods and to demonstrate the art of constructing useful mathematical models of real-world phenomena.

We encourage preparation of manuscripts in LateX or AMS T_EX for delivery in camera-ready copy, which leads to rapid publication, or in electronic form for interfacing with laser printers or typesetters.

Proposals should be sent directly to the editors or to: Birkhäuser Boston, 675 Massachusetts Avenue, Suite 601, Cambridge, MA 02139.

MMO1 *Probability in Social Science,* Samuel Goldberg

MMO2 *Popularizing Mathematical Methods in China: Some Personal Experiences,* Hua Loo-Keng and Wang Yuan

MMO3 *Mathematical Modeling in Ecology,* Clark Jeffries

MMO4 *Newton to Aristotle: Toward a Theory of Models for Living Systems,* John Casti and Anders Karlqvist, eds.

MMO5 *Introduction to Queueing Theory, 2nd edition,* B.V. Gnedenko and I.N. Kovalenko (translated from Russian)